ENERGY HARVESTING

Solar, Wind, and Ocean Energy Conversion Systems

ENERGY HARVESTING

Solar, Wind, and Ocean Energy Conversion Systems

Alireza Khaligh
Omer C. Onar

Energy, Power Electronics, and Machines Series
Ali Emadi, Series Editor

CRC Press
Taylor & Francis Group
Boca Raton   London   New York

CRC Press is an imprint of the
Taylor & Francis Group, an **informa** business

CRC Press
Taylor & Francis Group
6000 Broken Sound Parkway NW, Suite 300
Boca Raton, FL 33487-2742

© 2010 by Taylor and Francis Group, LLC
CRC Press is an imprint of Taylor & Francis Group, an Informa business

No claim to original U.S. Government works

Printed in the United States of America on acid-free paper
10 9 8 7 6 5 4 3 2 1

International Standard Book Number: 978-1-4398-1508-3 (Hardback)

Library of Congress Cataloging-in-Publication Data

Khaligh, Alireza.
  Energy harvesting : solar, wind, and ocean energy conversion systems / Alireza Khaligh, Omer C. Onar.
    p. cm. -- (Energy, power electronics, and machines)
  Includes bibliographical references and index.
  ISBN 978-1-4398-1508-3 (hardcover : alk. paper)
  1. Renewable energy sources. 2. Solar energy. 3. Ocean energy resources. 4. Wind power. 5. Wind energy conversion systems 6. Energy
    conversion. I. Onar, Omer C. II. Title. III. Series.

TJ808.K515 2010
621.042--dc22   2009035958

Visit the Taylor & Francis Web site at
http://www.taylorandfrancis.com

and the CRC Press Web site at
http://www.crcpress.com

To our families,

Alireza Khaligh

and

Omer C. Onar

Contents

Preface xiii
Authors xv

1. **Solar Energy Harvesting** 1
   1.1   Introduction 1
       1.1.1   Structures of Photovoltaic Cells/Modules/Arrays 1
       1.1.2   Semiconductor Materials for PV Cells 3
       1.1.3   Active and Passive Solar Energy Systems 5
       1.1.4   Components of a Solar Energy System 5
   1.2   *I–V* Characteristics of PV Systems 7
   1.3   PV Models and Equivalent Circuits 9
       1.3.1   Single-Diode and Dual-Diode Models 9
       1.3.2   Single-Diode Model without Parallel Resistance 10
       1.3.3   Single-Diode Model without Resistances 12
       1.3.4   PV Model Performance under Nominal and Standard
              Conditions 13
       1.3.5   Effects of Irradiance and Temperature on PV Characteristics 15
   1.4   Sun Tracking Systems 15
   1.5   MPPT Techniques 21
       1.5.1   Incremental Conductance–Based MPPT Technique 21
       1.5.2   Perturb and Observe–Based MPPT 23
       1.5.3   MPPT Controller–Based on Linearized *I–V* Characteristics 24
       1.5.4   Fractional Open-Circuit Voltage-Based MPPT 26
       1.5.5   Fractional Short-Circuit Current-Based MPPT 27
       1.5.6   Fuzzy Logic Control–Based MPPT 28
       1.5.7   Neural Network–Based MPPT 30
       1.5.8   Ripple Correlation Control–Based MPPT 31
       1.5.9   Current Sweep–Based MPPT 32
       1.5.10 DC Link Capacitor Droop Control–Based MPPT 33
   1.6   Shading Effects on PV Cells 35
   1.7   Power Electronic Interfaces for PV Systems 40
       1.7.1   Power Electronic Interfaces for Grid-Connected PV Systems 40
            1.7.1.1 Topologies Based on Inverter Utilization 41
            1.7.1.2 Topologies Based on Module and Stage Configurations 42
       1.7.2   Power Electronic Interfaces for Stand-Alone PV Systems 61
            1.7.2.1 PV/Battery Connection: Type 1 61

		1.7.2.2	PV/Battery Connection: Type 2	63
		1.7.2.3	PV/Battery Connection: Type 3	64
		1.7.2.4	PV/Battery Connection: Type 4	65
		1.7.2.5	PV/Battery Connection: Type 5	65

1.8 Sizing the PV Panel and Battery Pack for Stand-Alone PV Applications 66

    1.8.1 Sun Hour 67

    1.8.2 Load Calculation 67

    1.8.3 Days of Autonomy 67

    1.8.4 Solar Radiation 68

    1.8.5 PV Array Sizing 68

        1.8.5.1 PV Module Selection 68

        1.8.5.2 System Losses 68

        1.8.5.3 Determination of the Number of Series-Connected PV Modules 68

        1.8.5.4 Determination of the Number of Parallel-Connected PV Modules 69

        1.8.5.5 Design Verification and a Sizing Example 69

1.9 Modern Solar Energy Applications 72

    1.9.1 Residential Applications 72

        1.9.1.1 Boost Converter 74

        1.9.1.2 MPPT Control of the PV Panel 77

        1.9.1.3 Bidirectional Inverter/Converter 78

        1.9.1.4 Power Conditioner 79

    1.9.2 Electric Vehicle Applications 81

        1.9.2.1 Solar Miner II 84

    1.9.3 Naval Applications 86

        1.9.3.1 Solar-Powered AUV 87

        1.9.3.2 Solar-Powered Boat, _Korona_ 88

    1.9.4 Space Applications................................... 89

        1.9.4.1 Solar Power Satellite Energy System 90

        1.9.4.2 _Pathfinder_, the Solar Rechargeable Aircraft 90

        1.9.4.3 Solar-Powered Aircraft 91

        1.9.4.4 Solar Airplane 91

        1.9.4.5 Solar-Powered Unmanned Aerial Vehicle 92

1.10 Summary 94

    References 94

2. Wind Energy Harvesting 101

2.1 Introduction 101

2.2 Winds 101

2.3 History of Wind Energy Harvesting 104

2.4 Fundamentals of Wind Energy Harvesting 105

    2.4.1 Wind Turbine Siting 105

    2.4.2 Wind Turbine Power 107

        2.4.2.1 Betz Law 107

        2.4.2.2 Power Curve 109

        2.4.2.3 Power Coefficient 110

2.5 Wind Turbine Systems 112
    2.5.1 Basic Parts of Wind Turbines 112
        2.5.1.1 The Tower 112
        2.5.1.2 Yaw Mechanism 113
        2.5.1.3 The Nacelle 113
        2.5.1.4 The Turbine 113
2.6 Wind Turbines 114
    2.6.1 Wind Turbines Based on Axis Position 114
    2.6.2 Wind Turbines Based on Power Capacity 114
2.7 Different Electrical Machines in Wind Turbines 116
    2.7.1 BLDC Machines 116
    2.7.2 Permanent Magnet Synchronous Machines 123
    2.7.3 Induction Machines 130
        2.7.3.1 Conventional Control Scheme 134
        2.7.3.2 Voltage and Frequency Control with Load Regulation 134
        2.7.3.3 Improved Voltage and Frequency Control with a VSI 135
        2.7.3.4 Advanced Voltage and Frequency Control Using VSI 136
        2.7.3.5 Back-to-Back Connected PWM VSI 136
        2.7.3.6 DFIG 137
        2.7.3.7 Voltage and Frequency Control Using Energy
               Storage Devices 143
2.8 Synchronous Generators 151
2.9 Wind Harvesting Research and Development 156
    2.9.1 Developments in Control Systems 156
    2.9.2 Developments in Machine Design 158
    2.9.3 Developments in Distribution and Grid-Connected Topologies 158
2.10 Summary 159
    References 159

3. **Tidal Energy Harvesting** 167
3.1 Introduction 167
    3.1.1 History of Tidal Energy Harvesting 169
    3.1.2 Physical Principles of Tidal Energy 170
3.2 Categories of Tidal Power and Corresponding Generation Technology . . . 173
    3.2.1 Potential Energy 173
    3.2.2 Tidal Barrages Approach 174
        3.2.2.1 Ebb Generation 174
        3.2.2.2 Flood Generation 174
        3.2.2.3 Two-Way Generation 174
    3.2.3 Tidal Lagoons Concept 176
    3.2.4 Tidal Turbines Used in Barrages 178
        3.2.4.1 Bulb Turbine 178
        3.2.4.2 Rim Turbine 179
        3.2.4.3 Tubular Turbine 179
3.3 Turbine and Generator's Control 180
    3.3.1 Modeling of Hydraulic Turbine Conduit Dynamics 181
    3.3.2 Hydro Turbine Controls 183
    3.3.3 Kinetic Energy 188

3.3.3.1 Energy Calculation for Tidal Current Energy
Harvesting Technique 189
3.3.3.2 Tidal Turbines Used in Tidal Current Approach 192
3.4 Tidal Energy Conversion Systems 199
3.4.1 Generators 199
3.4.1.1 Synchronous Generator 199
3.4.1.2 Asynchronous Generator 202
3.4.2 Gearbox 205
3.4.3 Optimal Running Principle of Water Turbine 205
3.4.4 Maximum Power Point Tracking 206
3.4.4.1 MPPT Method Based on Look-Up Table 207
3.4.4.2 MPPT Method Based on Current Speed Calculation 208
3.4.5 P&O-Based MPPT Method 208
3.5 Grid Connection Interfaces for Tidal Energy Harvesting Applications 211
3.5.1 Grid Connection Interfaces for Tidal Turbine Applications 211
3.5.1.1 Grid-Connected Systems 212
3.5.2 Grid Connection and Synchronization for Tidal Energy
Harvesting with Basin Constructions 216
3.6 Potential Resources 217
3.7 Environmental Impacts 218
3.7.1 Sediment 218
3.7.2 Fish 219
3.7.3 Salinity 219
3.8 Summary 219
References 219

4. Ocean Wave Energy Harvesting 223
4.1 Introduction to Ocean Wave Energy Harvesting 223
4.2 The Power of Ocean Waves 225
4.3 Wave Energy Harvesting Technologies 226
4.3.1 Offshore Energy Harvesting Topologies 227
4.3.1.1 Dynamics of Fixed Bodies in Water 227
4.3.1.2 Dynamics of Floating Bodies in Water 228
4.3.1.3 Air-Driven Turbines 229
4.3.1.4 Fixed Stator and Directly Driven PM Linear
Generator–Based Buoy Applications 230
4.3.1.5 Salter Cam Method 231
4.3.2 Nearshore Energy Harvesting Topologies 233
4.3.2.1 Nearshore Wave Energy Harvesting by the
Channel/Reservoir/Turbine Method 233
4.3.2.2 Air-Driven Turbines Based on the Nearshore Wave
Energy Harvesting Method 233
4.3.3 Wave Power Absorbers 234
4.3.4 Wave Power Turbine Types 236
4.3.4.1 The "Wells" Air Turbines 237
4.3.4.2 Self-Pitch-Controlled Blades Turbine for WEC 239
4.3.4.3 Kaplan Turbines for WEC 241
4.3.4.4 Other Types of Turbines Used for WEC 243

      4.3.5   Wave Power Generators 249
              4.3.5.1 A Wave-Activated Linear Generator Model 249
              4.3.5.2 Linear, Synchronous, Longitudinal-Flux PM Generators ... 255
              4.3.5.3 A Three-Phase Synchronous Generator for Ocean
                       Wave Applications 260
              4.3.5.4 Radial Flux PM Synchronous Generator for WEC 266
              4.3.5.5 Induction Machines for Ocean WEC 272
              4.3.5.6 Switched Reluctance Machines for Ocean WEC 274
              4.3.5.7 Ocean Energy Conversion Using
                       Piezoelectric/Electrostictive Materials 275
      4.3.6   Grid Connection Topologies for Different Generators Used in
              Wave Energy Harvesting Applications 282
              4.3.6.1 Grid Connection Interface for Linear and Synchronous
                       Generator Applications 282
              4.3.6.2 Grid Connection Interface for Induction Generator
                       Applications 283
              4.3.6.3 Grid Connection Interface for SRG Applications 286
              4.3.6.4 Grid Connection Interface for
                       Piezoelectric/Electrostrictive Power Generators 289
  4.4 Wave Energy Applications 289
      4.4.1   Oscillating Water Column 290
      4.4.2   Pelamis 291
      4.4.3   Wave Dragon 294
      4.4.4   AWS 294
      4.4.5   Wave Star Energy 296
      4.4.6   Magnetohydrodynamics Wave Energy Converter 298
  4.5 Wave Energy in Future 298
  4.6 Summary 299
       References 299

5. **Ocean Thermal Energy Harvesting** 305
  5.1 History 306
  5.2 Classification of OTECs 308
      5.2.1   Closed-Cycle OTEC Systems 308
              5.2.1.1 Structure and Principles of Closed-Cycle Systems 308
              5.2.1.2 Thermodynamic Principles of Closed-Cycle Systems 309
  5.3 Technical Obstacles of Closed-Cycle OTEC Systems 312
      5.3.1   Working Fluids and Its Potential Leakage 312
              5.3.1.1 Degradation of Heat Exchanger Performance by
                       Microbial Fouling 314
      5.3.2   Thermal Energy Conversion for OTEC Systems 314
      5.3.3   Open-Cycle OTEC Systems 315
              5.3.3.1 Structure and Principles of Open-Cycle Systems 315
              5.3.3.2 Technical Difficulties of Open-Cycle OTEC Systems 316
      5.3.4   Hybrid Cycle OTEC Systems 316
              5.3.4.1 Structure and Principles of Hybrid OTEC Systems 316
  5.4 Components of an OTEC System 317
      5.4.1   Heat Exchanger 318
      5.4.2   Evaporator 318

	5.4.3 Condenser	322
	5.4.4 Vacuum Flash Evaporator	325
5.5	Control of an OTEC Power Plant	326
5.6	Control of a Steam Turbine	329
5.7	Potential Resources	333
5.8	Multipurpose Utilization of OTEC Systems	337
	5.8.1 Desalination	337
	5.8.2 Aquaculture	337
	5.8.3 Air-Conditioning	338
	5.8.4 Mineral Extraction	338
5.9	Impact on Environment	338
5.10	Summary	339
	References	339
Index		343

Preface

Increasing demand for energy, decreasing conventional fossil-fuel energy sources, and environmental concerns are driving forces toward renewable energy sources. Energy sources such as oil, coal, and gas are being quickly depleted or have insufficient reserves for future demands. Moreover, they are not environmental friendly due to greenhouse gas emissions and other pollutants. Nuclear energy has a great establishment cost along with a number of safety concerns. On the other hand, hydroelectric power plants are inexpensive but have a limited life span and mostly cannot be utilized due to geo-political reasons and seasonal irregularity of available water. Therefore, the contribution from renewable energy sources is increasing. Since the world today is experiencing a great shortage of energy, it should be captured, stored, conditioned, and utilized by alternative techniques. Energy demand will always increase with the increase in technological developments while conventional sources will diminish and environmental concerns will gain increased attention.

Energy harvesting, also called energy scavenging, is a concept by which energy is captured, stored, and utilized using various sources by employing interfaces, storage devices, and other units. Unlike the conventional electric power generation systems, in the renewable energy harvesting concept, fossil fuels are not used and the generation units can be decentralized. Therefore, the transmission and distribution losses can be significantly reduced. There are many sources for harvesting energy. Solar, wind, ocean, hydro, electromagnetic, electrostatic, thermal, vibration, and human body motion are renewable sources of energy. Economic, environmental, and geopolitical constraints on global conventional energy resources started forcing the nation to accelerate energy harvesting from renewable sources. Therefore, advanced technical methods should be developed to increase the efficiency of devices in harvesting energy from various environmentally friendly resources and converting them into electrical energy. These developments have sparked interest in many communities such as science, engineering, and education to develop more energy harvesting applications and new curriculums for renewable energy and energy harvesting topics. This book describes various energy harvesting technologies such as solar, wind, ocean wave, ocean tidal, and ocean thermal energy harvesting along with many different topologies and many types of power electronic interfaces for the utilization and/or grid connection of energy harvesting applications. In addition, some simulation models are developed throughout the book in order to build an insight to system analysis and modeling. In the book, the concepts and theoretical background are built for energy harvesting applications.

Chapter 1 of the book focuses on solar energy harvesting since solar energy is one of the most important renewable energy sources that has gained increased attention in recent years. Solar energy is plentiful; it has the greatest availability among all the other energy sources. The chapter deals with $I-V$ characteristics of photovoltaic (PV) systems, PV models and equivalent circuits, sun tracking systems, maximum power point tracking systems, shading effects, power electronic interfaces for grid connected and stand-alone PV systems, sizing criteria for applications, and modern solar energy applications such as residential, vehicular, naval, and space applications. Wind energy harvesting techniques are analyzed in Chapter 2. Wind power is a clean way of energy conversion; it is renewable, widely distributed, and plentiful. In addition, it contributes toward reducing greenhouse gas emissions, since it can be used as an alternative to fossil-fuel-based power generation. Different types of wind turbines and electrical machines are reviewed throughout this chapter along with various power electronic interfaces. In Chapter 3, various features of the ocean tidal energy harvesting are explained that have great potential, however, are not widely utilized yet. Different energy generation technologies, their optimal operation principles, and possible utilization techniques are described throughout this chapter. Chapter 4 of the book deals with ocean wave energy harvesting in which the kinetic and potential energy contained in the natural oscillations of ocean waves are converted into electric power. Nearshore and offshore approaches along with required absorber, turbine, and generator types are discussed. Moreover, power electronic interfaces for grid connection scenarios are explained. In the final section of the chapter, commercialized ocean wave energy conversion applications are presented. In Chapter 5, ocean thermal energy conversion, an energy-generating technology taking advantage of the temperature difference between the ocean's shallow warm water and cold deeper water, is investigated. The chapter consists of closed, open, and hybrid-cycle ocean thermal energy conversion systems as well as their required components. In addition, potential resources and multipurpose ocean thermal energy conversion systems are presented in this chapter.

This book is recommended as a reference text for courses such as Renewable Energies, Alternative Energy Resources, Energy Harvesting, and many other similar courses. This book is also an in-depth source for engineers, researchers, and managers who are working in energy harvesting, renewable energies, electrical power engineering, power electronics, and related industries.

We would like to gratefully acknowledge Yao Da and Haojie Luan for their contributions to Chapters 3, 4, and 5. We would also like to acknowledge the efforts and assistance of the staff of CRC Press–Taylor & Francis Group.

Alireza Khaligh
Omer C. Onar

Authors

Alireza Khaligh (IEEE S'04–M'06) received his BS (1999) and MS (2001) degrees in electrical engineering (with highest distinction) from Sharif University of Technology, Tehran, Iran. He also received his PhD degree (2006) in electrical engineering from the Illinois Institute of Technology (IIT), Chicago, Illinois. He was a postdoctoral research associate in the Electrical and Computer Engineering Department of the University of Illinois at Urbana-Champaign (UIUC).

Dr. Khaligh is currently an assistant professor and director of the Energy Harvesting and Renewable Energies Laboratory (EHREL) at the Electric Power and Power Electronics Center (EPPEC) of the Electrical and Computer Engineering Department at IIT. Dr. Khaligh is the recipient of the Exceptional Talents Fellowship from Sharif University of Technology. He is also the recipient of the Distinguished Undergraduate Student Award in Sharif University of Technology, presented jointly by the Minister of Science, Research and Technology and the president of Sharif University. He is also a recipient of the NSF Fellowship in Energy Challenge and Nanotechnology.

Dr. Khaligh is a member of the Vehicle Power and Propulsion Committee, IEEE Vehicular Technology Society (VTS), IEEE Power Electronics Society (PELS), Industrial Electronics Society (IES), and Society of Automotive Engineers (SAE). He is the principal author/coauthor of more than 50 papers, books, and invention disclosures. His major research interests include modeling, analysis, design, and control of power electronic converters, energy scavenging/harvesting from environmental sources, electric and hybrid electric vehicles, and design of energy-efficient power supplies for battery-powered portable applications. Dr. Khaligh is an associate editor of the *IEEE Transactions on Vehicular Technology*. He is also a guest editor for the special section of the *IEEE Transactions on Vehicular Technology on Vehicular Energy Storage Systems*.

Omer C. Onar (IEEE S'05) received his BSc and MSc degrees in electrical engineering from Yildiz Technical University, Turkey, in 2004 and 2006, respectively. He was a research scholar in the Electrical and Computer Engineering Department at the University of South Alabama from August 2005 to August 2006 and is involved in U.S. Department of Energy projects based on power management for fuel cell applications. Currently, he is a doctoral research assistant at the EHREL at the EPPEC of the Electrical and Computer Engineering Department at IIT. He is author/coauthor of more than 35 publications including journals, conference proceedings, and text books. His research interests include power electronics, energy harvesting/scavenging, renewable energies, and electric and hybrid electric vehicles. He is a member of the IEEE Power & Energy Society, Vehicular Technology Society, and Power Electronics Society. He was the recipient of the IEEE Transportation Electronics Fellowship Award for 2008–2009.

1

Solar Energy Harvesting

1.1 Introduction

Solar energy is one of the most important renewable energy sources that has been gaining increased attention in recent years. Solar energy is plentiful; it has the greatest availability compared to other energy sources. The amount of energy supplied to the earth in one day by the sun is sufficient to power the total energy needs of the earth for one year [1]. Solar energy is clean and free of emissions, since it does not produce pollutants or by-products harmful to nature. The conversion of solar energy into electrical energy has many application fields. Residential, vehicular, space and aircraft, and naval applications are the main fields of solar energy.

Sunlight has been used as an energy source by ancient civilizations to ignite fires and burn enemy warships using "burning mirrors." Till the eighteenth century, solar power was used for heating and lighting purposes. During the 1800s, Europeans started to build solar-heated greenhouses and conservatories. In the late 1800s, French scientists powered a steam engine using the heat from a solar collector. This solar-powered steam engine was used for a printing press in Paris in 1882 [2]. A highly efficient solar-powered hot air engine was developed by John Ericsson, a Swedish-American inventor. These solar-driven engines were used for ships [3]. The first solar boiler was invented by Dr. Charles Greely, who is considered the father of modern solar energy [4]. The first working solar cells were invented in 1883 by Charles Fritts [5]. Selenium was used to build these prototypes, achieving efficiencies of about 1%. Silicon solar cells were developed in 1954 by researchers Calvin Fuller, Daryl Chapin, and Gerald Pearson. This accomplishment was achieved by following the fundamental work of Russel Ohl in the 1940s [6]. This breakthrough marked a fundamental change in the generation of power. The efficiency of solar cells increased from 6% up to 10% after the subsequent development of solar cells during the 1950s [7]; however, due to the high costs of solar cells ($300 per watt) commercial applications were limited to novelty items [6].

1.1.1 Structures of Photovoltaic Cells/Modules/Arrays

A photovoltaic (PV) cell converts sunlight into electricity, which is the physical process known as photoelectric effect. Light, which shines on a PV cell, may be reflected, absorbed, or passed through; however, only absorbed light generates electricity. The energy of absorbed light is transferred to electrons in the atoms of the PV cell. With their newfound energy, these electrons escape from their normal positions in the atoms of semiconductor PV material and become part of the electrical flow, or current, in an electrical circuit. A special electrical property of the PV cell, called "built-in electric field," provides the force or voltage required to drive the current through an external "load" such as a light bulb [8]. To induce the built-in electric field within a PV cell, two layers of different semiconductor materials are placed in contact with each other. One layer is an "n-type" semiconductor with an abundance of electrons, which have a negative electrical charge. The other layer is a "p-type" semiconductor with an abundance of "holes," which have a positive electrical charge.

Although both materials are electrically neutral, n-type silicon has excess electrons and p-type silicon has excess holes. Sandwiching these together creates a p/n junction at their interface, thereby creating an electric field. Figure 1.1 shows the p–n junction of a PV cell.

When n-type and p-type silicon come into contact, excess electrons move from the n-type side to the p-type side. The result is the buildup of positive charge along the n-type side of the interface and of negative charge along the p-type side.

The two semiconductors behave like a battery, creating an electric field at the surface where they meet, called the p/n junction. This is a result of the flow of electrons and holes. The electrical field forces the electrons to move from the semiconductor toward the negative surface to carry current. At the same time, the holes move in the opposite direction, toward the positive surface, where they wait for incoming electrons [8].

Additional structures and components are required to convert the direct current (DC) to alternate current (AC). Some systems also store some electricity, usually in batteries, for future use. All these items are referred to as the "balance of system" (BOS) components [9].

Combining modules with the BOS components creates an entire PV system. This system is usually all that is needed to meet a particular energy demand, such as powering a water pump, or the appliances and lights in a home. In Figure 1.2, a single cell, a module consisting of cells, and an array consisting of modules are presented.

A PV or solar cell is the basic building block of a PV (or solar electric) system. An individual PV cell is usually quite small, typically producing about 1 or 2 W of power [9]. To boost the

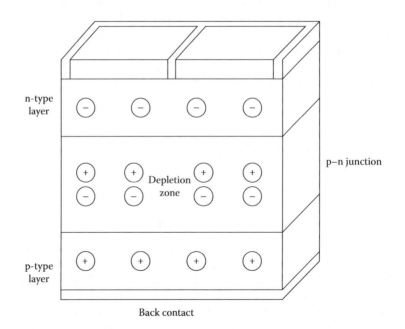

FIGURE 1.1   p–n junction of the PV cell.

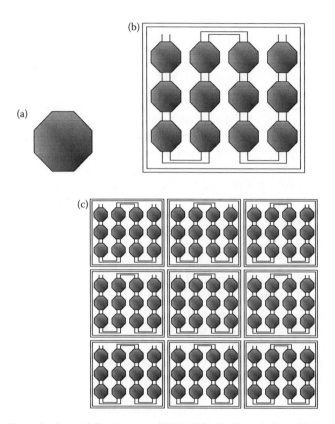

FIGURE 1.2 (See color insert following page 80.) (a) PV cell, (b) module, and (c) array.

power output of PV cells, they have to be connected together to form larger units called modules. The modules, in turn, can be connected to form larger units called arrays, which can be interconnected to produce more power. By connecting the cells or modules in series, the output voltage can be increased. On the other hand, the output current can reach higher values by connecting the cells or modules in parallel.

Based on the sunlight collection method, PV systems can be classified into two general categories: flat-plate systems and concentrator systems [9]. Flat panel PV systems directly capture the sunlight or they use the diffused sunlight from the environment. They can be fixed or combined with sun tracking systems. On the other hand, concentrator systems collect a large amount of sunlight and concentrate and focus the sunlight to target PV panels using lenses and reflectors. These systems reduce the size and required number of cells while the power output is increased. Moreover, by concentrating the solar light, the efficiency of the PV cell is increased.

1.1.2 Semiconductor Materials for PV Cells

PV devices can be made from various types of semiconductor materials, deposited or arranged in various structures [10].

The three main types of materials used for solar cells are silicon, polycrystalline thin films, and single-crystalline thin film [11]. The first type is silicon, which can be used in various forms, including single crystalline, multicrystalline, and amorphous. The second type is polycrystalline thin films, with a specific discussion of copper indium diselenide (CIS), cadmium telluride (CdTe), and thin-film silicon. Finally, the third type of material is single-crystalline thin film, focusing especially on cells made from gallium arsenide (GaAs) [10,12].

- *Silicon (Si)*—including single-crystalline Si, multicrystalline Si, and amorphous Si. Silicon, used to make some of the earliest PV devices, is still the most popular material for solar cells [13]. Outranked only by oxygen, silicon is also the second-most abundant element in the earth's crust [13]. However, to be useful as a semiconductor material in solar cells, silicon must be refined to a purity of 99.9999% [12,13].

  In single-crystal silicon, the molecular structure, which is the arrangement of atoms in the material, is uniform, because the entire structure is grown from the same crystal. This uniformity is ideal for transferring electrons efficiently through the material. To make an effective PV cell, silicon has to be "doped" with other elements to make it n-type or p-type [14].

  Semicrystalline silicon, in contrast, consists of several smaller crystals or grains, which introduce boundaries. These boundaries impede the flow of electrons and encourage them to recombine with holes to reduce the power output of the solar cell. Semicrystalline silicon is much less expensive to produce than single-crystalline silicon. Hence researchers are working on other methods to minimize the effects of grain boundaries [14].

- *Polycrystalline thin films*—including CIS, CdTe, and thin-film silicon. Another scientific discovery of the computer semiconductor industry, which vitally impacted the PV industry, is thin-film technology. The "thin-film" term comes from the method used to deposit the film, not from the thinness of the film: thin-film cells are deposited in very thin, consecutive layers of atoms, molecules, or ions. Thin-film cells have many advantages over their "thick-film" counterparts. For example, they use much less material—the cell's active area is usually only 1–10 μm thick, whereas thick films are typically 100–300 μm thick. Thin-film cells can usually be manufactured in a large-area process, which can be an automated, continuous production process. Finally, they can be deposited on flexible substrate materials [15].

- *Single-crystalline thin films*—including high-efficiency material such as GaAs. GaAs is a compound semiconductor: a mixture of two elements, gallium and arsenic. Gallium is a by-product of the smelting of other metals, notably aluminum and zinc, and it is rarer than gold. Arsenic is not rare, but it is poisonous [15]. GaAs was developed for use in solar cells at about the same time that it was developed for light-emitting diodes, lasers, and other electronic devices that use light.

  GaAs is especially suitable for use in multijunction and high-efficiency solar cells for the following reasons:

  - The GaAs band gap is 1.43 V, nearly ideal for single-junction solar cells.

  - GaAs has a very high absorptivity so that a cell requires only a few micrometers thickness to absorb sunlight (crystalline silicon requires a 100 μm or thicker layer) [15].

  - Unlike silicon cells, GaAs cells are relatively insensitive to heat. Cell temperatures can often be quite high, especially in concentrator applications.

  - Alloys made from GaAs and aluminum, phosphorus, antimony, or indium have characteristics that are complementary to those of GaAs, allowing great flexibility in cell design.

  – GaAs is highly resistant to radiation damage. This, along with its high
     efficiency, makes GaAs desirable for space applications.

One of the most important advantages of GaAs and its alloys as PV cell materials
is that it is amenable to a wide range of designs. A cell with a GaAs base can
have several layers of slightly different compositions; this allows a cell designer to
precisely control the generation and collection of electrons and holes [15].

This degree of control allows cell designers to push efficiencies closer and
closer to theoretical levels. For example, one of the most common GaAs cell
structures has a very thin window layer made of aluminum GaAs. This thin
layer allows electrons and holes to be created close to the electric field at the
junction [15].

1.1.3 Active and Passive Solar Energy Systems

Solar energy systems are typically classified into two systems: passive and active systems [16,17].

Passive solar energy systems do not involve panel systems or other moving mechanisms to produce energy. Passive systems utilize nonmechanical techniques to control the
amount of captured sunlight and distribute this energy into useful forms such as heating, lighting, cooling, and ventilation. These techniques include selecting materials with
favorable thermal properties to absorb and retain energy, designing spaces that naturally
circulate air to transfer energy, and referencing the position of a building to the sun to
enhance energy capture. In some cases passive solar devices can be composed of moving
parts, with the distinction that this movement is automatic and directly powered by the
sun. These systems can be used for heating and lighting purposes. This means that solar
power is used for powering the movement of the sun tracking system as well as heating
and lighting.

Active solar energy systems typically involve electrical and mechanical components
such as tracking mechanisms, pumps, and fans to capture sunlight and process it into
usable forms such as heating, lighting, and electricity. The panels are oriented to maximize
exposure to the sun. Depending on the system, the panels will then convert sunlight into
electricity, which is then transformed from DC electricity to AC electricity and stored in
batteries or fed into the local utility grid. Active systems are more expensive and complex.

1.1.4 Components of a Solar Energy System

Figure 1.3 demonstrates the block diagram of a solar energy system. In this system, sunlight
is captured by the PV array. The sun tracking system takes the photodiode or photosensor
signals and determines the sun tracking motor positions. Therefore, daily and seasonal solar
position changes are followed in order to face the sun directly and capture the most available
sunlight. The output of the PV panel is connected to a DC/DC converter to operate at the
desired current or voltage to match the maximum available power from the PV module.
This MPPT DC/DC converter is followed by a DC/AC inverter for grid connection or to
supply power to the AC loads. A battery pack can be connected to the DC bus of the system
to provide extra power that might not be available from the PV module during night and
cloudy periods. The battery pack can also store energy when the PV module generates more
power than that demanded.

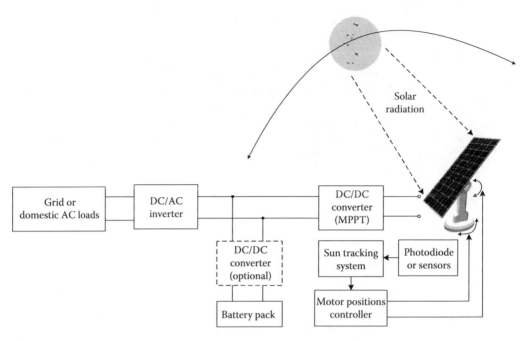

FIGURE 1.3   Solar energy system.

A solar energy system consists of several components, as shown in Figure 1.3. Several PV modules are connected in series or in parallel, depending on the voltage and energy requirements of the PV array. The direction of the PV arrays toward the sun might be controlled by the position motors allowing the motion of the panels horizontally and/or vertically. The sun tracking system takes the data from sun sensors or photodiodes and processes the data to determine the motor positions for the best illumination position of the PV array directed to the sun. This position detection and adjustment is similar to a mechanical power point tracking technique [18]. On the other hand, due to the voltage–current and current–power characteristics of the PV array, the electrical maximum power point (MPP) may vary based on operating current and voltage. Therefore, an electrical maximum power point tracking (MPPT) system is needed to operate the PV array at the MPP. This can be done through a DC/DC converter.

Due to the possible mismatches of the power generated by the PV array and the power requirements of the load, a battery pack might be used in order to compensate for these mismatches. This battery pack can be used as the energy buffer to store energy when the available power from the PV array is more than the required power. Contrarily, the battery pack can be discharged in order to satisfy the sustained load demands during periods when the solar energy is less than the required load power. The battery can be connected in various topologies. Each topology might have advantages and disadvantages in terms of number of required parts, control complexity, flexibility, battery size, and cost. Finally, the DC bus power should be converted to AC power for grid connection or for satisfying the power requirements of domestic AC loads in stand-alone applications. A grid connection is also useful to draw/inject power from/to the utility network to take advantage of the excess power or to recharge the batteries using grid power during the peak-off periods of the utility network.

1.2 *I–V* Characteristics of PV Systems

Current–voltage (*I–V*) curves are obtained by exposing the cell to a constant level of light, while maintaining a constant cell temperature, varying the resistance of the load, and measuring the produced current. The *I–V* curve typically passes through two points:

- *Short-circuit current* (I_{SC}): I_{SC} is the current produced when the positive and negative terminals of the cell are short-circuited, and the voltage between the terminals is zero, which corresponds to zero load resistance.
- *Open-circuit voltage* (V_{OC}): V_{OC} is the voltage across the positive and negative terminals under open-circuit conditions, when the current is zero, which corresponds to infinite load resistance.

The cell may be operated over a wide range of voltages and currents. By varying the load resistance from zero (a short circuit) to infinity (an open circuit), the MPP of the cell can be determined. On the *I–V* curve, the maximum power point (P_m) occurs when the product of current and voltage is maximum. No power is produced at the short-circuit current with no voltage, or at the open-circuit voltage with no current. Therefore, MPP is somewhere between these two points. Maximum power is generated at about the "knee" of the curve. This point represents the maximum efficiency of the solar device in converting sunlight into electricity [19].

A PV system consists of many cells connected in series and parallel to provide the desired output terminal voltage and current. This PV system exhibits a nonlinear *I–V* characteristic [20]. There are various models available to model the *I–V* characteristics of PV systems. Masoum et al. introduced a model for silicon solar PV panels: the PV cell equivalent model, which represents the dynamic nonlinear *I–V* characteristics of the PV system described in Equation 1.1. The output voltage characteristic of the PV system can be expressed as

$$V_{\text{PV}} = \frac{N_s \alpha k T}{q} \ln \left[\frac{I_{\text{sc}} - I_{\text{PV}} + N_p I_0}{N_p I_0} \right] - \frac{N_s}{N_p} R_s I_{\text{PV}}. \tag{1.1}$$

The parameters used in the PV output voltage equation are as follows:

α ideality or completion factor

I_0 PV cell reverse saturation current (A)

I_{PV} PV cell output current (A)

I_{SC} short-circuit cell current (representing insulation level) (A)

k Boltzmann's constant (J/°K) (1.380×10^{-23})

M_V voltage factor

N_p number of parallel strings

N_s number of series cells per string

q electron in charge (C) (-1.602×10^{-19})

R_s series resistance of PV cell (Ω)

T PV cell temperature (°K)

V_{MP} PV cell voltage corresponding to maximum power (V)

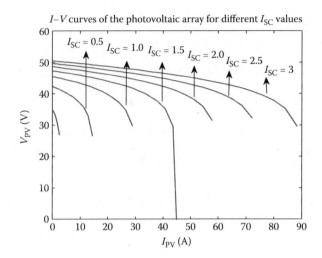

FIGURE 1.4   Current–voltage curves of the PV cell for different short-circuit levels.

V_{OC} open-circuit voltage (V)

V_{PV} terminal voltage for PV cell (V)

The current–voltage characteristics of the PV array ($N_p = 30$, $N_s = 112$) are obtained according to the different short-circuit current values using the PV output equation, shown in Figure 1.4.

Using the current–voltage curves, the current–power curves can be obtained, which are depicted in Figure 1.5.

Figure 1.6a presents how the current of the series connected cells stays the same while the voltage increases. Meanwhile, under the same voltage, the current of the parallel connected cells is increased as shown in Figure 1.6b.

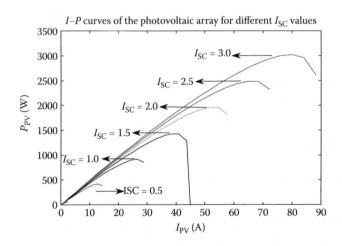

FIGURE 1.5   Current–power curves of the PV Cell for different short-circuit current values.

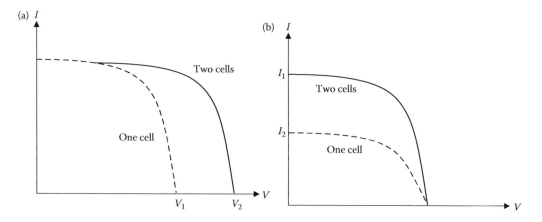

FIGURE 1.6 (a) Series and (b) parallel connection of identical cells.

1.3 PV Models and Equivalent Circuits

1.3.1 Single-Diode and Dual-Diode Models

A PV model can be expressed by the equivalent circuit shown in Figure 1.7. This model is also known as a single-diode model.

In this model, open-circuit voltage and short-circuit current are the key parameters. The short-circuit current depends on illumination, while the open-circuit voltage is affected by the material and temperature. In this model, V_T is the temperature voltage expressed as $V_T = kT/q$, which is 25.7 mV at 25°C. The ideality factor α generally varies between 1 and 5 for this model. The equations defining this model are

$$I_D = I_0 \left[e^{V_{PV}/\alpha V_T} - 1 \right],\tag{1.2}$$

$$I_{PV} = I_{SC} - I_D,\tag{1.3}$$

and

$$V_{PV} = \alpha V_T \ln \left[\frac{I_{SC} - I_{PV}}{I_0} + 1 \right].\tag{1.4}$$

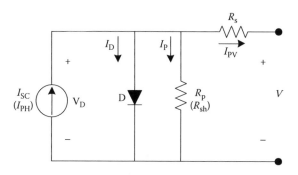

FIGURE 1.7 Single-diode model of solar cell equivalent circuit.

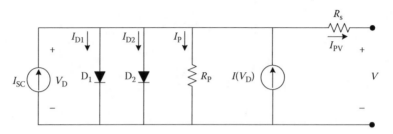

FIGURE 1.8 Dual-diode model.

The *I–V* characteristic of the solar cell can be alternatively defined by [21]

$$I_{PV} = I_{PH} - I_D,$$

$$= I_{PH} - I_0 \left[\exp \left(\frac{q(V + R_s I_{PV})}{\alpha k T} \right) - 1 \right] - \frac{V_{PV} + R_s I_{PV}}{R_p}, \qquad (1.5)$$

where I_{PH} is photocurrent (A), I_D is diode current (A), R_s is series resistance, and R_p is parallel resistance.

An alternative model for PV cells is the dual-diode model, presented in Figure 1.8. In the dual-diode model, extra degrees of freedom are provided for accuracy. However, the first model is widely used for PV applications since it is sufficient to represent PV characteristics and dynamics. Although it has more accuracy, this dual-diode model is not widely used due to its complexity.

1.3.2 Single-Diode Model without Parallel Resistance

Generally, the parallel resistance shown in the single-diode model is high enough to behave like an open circuit. Therefore, neglecting this resistance will not sacrifice the accuracy of the model significantly. A solar cell is usually represented by the equivalent one-diode model, shown in Figure 1.9 [22].

The net output current I_{PV} is the difference between the photocurrent I_{PH} and the normal diode current I_D:

$$I_{PV} = I_{PH} - I_D = I_{PH} - I_0 \left[\exp \frac{q(V_{PV} + I_{PV} R_s)}{\alpha k T} - 1 \right]. \qquad (1.6)$$

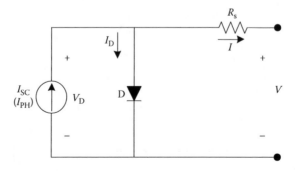

FIGURE 1.9 Model for a single solar cell.

For this model, the idealizing factor α is suggested to be 1.3 in normal operation. Based on experimental studies and theoretical analyses, the value of α can be determined for best matching the real PV characteristics. This value is generally between 1 and 2 for different applications and operations.

Figure 1.10 shows the I–V characteristic of the solar cell for a certain ambient irradiation G_a at a certain fixed cell temperature T_c.

If the cell's terminals are connected to a variable resistance R, the operating point is determined by the intersection of the I–V characteristic of the solar cell and the load I–V characteristic curves (Figure 1.10). For a resistive load, the load characteristic is a straight line with a slope of $I/V = 1/R$. It should be pointed out that the power delivered to the load depends on the resistance value of the load. However, if the load resistance is small, the cell operates in the MN region of the curve, where the cell behaves as a constant current source, almost equal to the short-circuit current. On the other hand, if the load resistance is large, the cell operates on the PS region of the curve, where the cell behaves more similar to a constant voltage source, which is almost equal to the open-circuit voltage.

For this single-diode model, the solar cell can be characterized by the following fundamental parameters of short-circuit current, open-circuit voltage, maximum power, and maximum efficiency.

By taking the output current as zero, the open-circuit voltage can be obtained as

$$V_{OC} = \frac{\alpha k T}{q} \ln\left(\frac{I_{PH}}{I_0}\right) = V_T \ln\left(\frac{I_{PH}}{I_0}\right). \tag{1.7}$$

Maximum efficiency is the ratio between the maximum power (MPP) and the incident light power:

$$\eta = \frac{P_{max}}{P_{in}} = \frac{I_{max} V_{max}}{A G_a}, \tag{1.8}$$

where G_a is ambient irradiation and A is cell area.

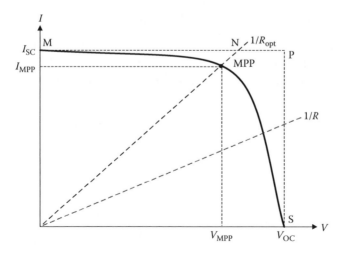

FIGURE 1.10 A typical I–V curve for a solar cell.

Fill factor is the ratio of the maximum power that can be delivered to the load and the product of I_{SC} and V_{OC}:

$$FF = \frac{P_{max}}{V_{OC}I_{SC}} = \frac{I_{max}V_{max}}{V_{OC}I_{SC}}. \tag{1.9}$$

This factor is a measure of the real *I–V* characteristics. It is higher than 0.7 for cells in good condition. This factor decreases as the cell temperature increases.

Several cells can be connected to form a module. The *I–V* relation of a module can be expressed as Equation 1.10. In this equation, M notation refers to the module variables (module voltage and current) and the PV module's current under arbitrary operating condition:

$$I_{PV}^M = I_{SC}^M \left[1 - \exp \left(\frac{V_{PV}^M - V_{OC}^M + R_S^M \, I_{PV}^M}{N_s V_T} \right) \right], \tag{1.10}$$

where the short-circuit current of the module is $I_{SC}^M = N_p I_{SC}$, the open-circuit voltage of the module is $V_{OC}^M = N_s V_{OC}^C$, and the equivalent serial resistance of the module is $R_S^M = N_s / N_p R_S$.

1.3.3 Single-Diode Model without Resistances

For simplification, the series output resistance can also be neglected since its value is usually very small. In this way, a simplified equivalent circuit shown in Figure 1.11 and the given equations can represent a PV cell.

In module structure, the mathematical model of the single-diode model without resistances can be expressed as follows:

$$I_{PV} = I_{SC} - I_D$$

$$= N_p I_{SC} - N_p I_0 \left[e^{q V_{PV}/\alpha k T N_s} - 1 \right]. \tag{1.11}$$

Therefore,

$$V_{PV} = \frac{N_s \alpha k T}{q} \ln \left[\frac{N_p I_{SC} - I_{PV}}{N_p I_0} + 1 \right]. \tag{1.12}$$

The series and parallel resistances are ignored in this model. The change of irradiance causes a change in the operating temperature of the PV array; however, this effect can be neglected in the model, since the change ratio is much slower than the other effects [22].

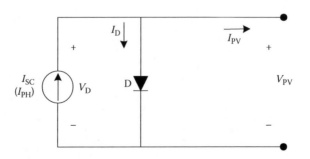

FIGURE 1.11 PV cell equivalent circuit.

1.3.4 PV Model Performance under Nominal and Standard Conditions

Practically, the performance of a module or another PV device is determined by exposing it to known conditions. The module characteristics supplied by the manufacturer are usually determined under special conditions, for example in nominal or standard conditions, shown in Table 1.1.

Under standard conditions, the following parameters are measured:

- Short-circuit current for the module $T_{SC,0}^M$
- Open-circuit voltage for the module $V_{OC,0}^M$
- Maximum power for the module $P_{max,0}^M$.

Under nominal conditions, the following parameters are delivered:

- Ambient irradiation $G_{a,ref}$
- Ambient temperature $T_{a,ref}$
- Temperature of the cell T_{ref}^C.

The PV module current can be measured under certain operating points (V^M, T_a, G_a), using the procedure shown in Table 1.2. The algorithm steps are as follows:

1. PV manufacturers' catalogues provide information about the standard conditions of the module:
   a. Maximum power $P_{max,0}^M$
   b. Short-circuit current $I_{SC,0}^M$
   c. Open-circuit voltage $V_{OC,0}^M$
   d. Number of series cells N_{SM}
   e. Number of parallel cells N_{PM}.
2. Cell data for standard conditions: $P_{max,0}^C$, $V_{OC,0}^C$, $I_{SC,0}^C$, R_S^C as described in Table 1.1.
3. Under provided operating conditions (V^M, T_a, G_a), the characteristic parameters of the cell should be determined. Therefore the short-circuit current I_{SC}^C, which is proportional to irradiation G_a, can be calculated as

$$I_{SC}^C = C_1 G_a. \tag{1.13}$$

TABLE 1.1

Operational Nominal and Standard Conditions for PVs

Nominal Conditions	Standard Conditions
Irradiation: $G_{a,ref} = 800\,W/m^2$	Irradiation: $G_{a,0} = 1000\,W/m^2$
Ambient temperature: $T_{a,ref} = 20°C$	Cell temperature: $T_0^C = 25°C$
Wind speed: $1\,m/s$	

TABLE 1.2

PV Module Current Determination under Certain Operating Points

---

Step 1—Module data for standard conditions

$P^M_{max,0}$, $I^M_{SC,0}$, $V^M_{OC,0}$, N_{SM}, N_{PM}

Step 2—Cell parameters for standard conditions

$P^C_{max,0} = P^M_{max,0}/(N_{SM}N_{PM})$

$V^C_{OC,0} = VP^M_{OC,0}/N_{SM}$

$I^C_{SC,0} = I^M_{SC,0}/N_{PM}$

$V^C_{t,0} = mkT^C/e$

$v_{OC,0} = V^C_{OC,0}/V^C_{t,0}$

$FF = \left(V^C_{OC,0} - \ln(v_{OC,0} + 0.72)\right)/(v_{OC,0} + 1)$

$FF_0 = P^C_{max,0}/\left(V^C_{OC,0}I^C_{OC,0}\right)$

$r_s = 1 - FF/FF_0$

$R^C_S = r_s V^C_{OC,0}/I^C_{SC,0}$

Step 3—Cell parameters for operating conditions (V^M, T_a, G_a)

$C_1 = I^C_{SC,0}/G_{a,0}$ $I^C_{SC} = C_1 G_a$

$T^C = T_a + C_2 G_a$

$V^C_{OC} = V^C_{OC,0} + C_3\left(T^C - T^C_0\right)$

$V^C_t = mk\left(273 + T^C\right)/e$

Step 4—Module current for operating conditions

$I^M = N_{PM}I^C_{SC}\left[1 - \exp\left(\left(V^M - N_{SM}V^C_{OC} + I^M R^C_S N_{SM}/N_{PM}\right)/\left(N_{SM}V^C_t\right)\right)\right]$

The operating temperature of the cell T^C depends exclusively on irradiation G_a and ambient temperature T_a according to

$$T^C = T_a + C_2 G_a, \tag{1.14}$$

where the constant C_2 is

$$C_2 = \frac{T^C_{ref} - T_{a,ref}}{G_{a,ref}}. \tag{1.15}$$

If T^C_{ref} is not known, it is reasonable to approximate $C_2 = 0.03\,C\,m^2/W$. The relationship between the open-circuit voltage and the temperature of solar cells is

$$V^C_{OC} = V^C_{OC,0} + C_3(T^C - T^C_0), \tag{1.16}$$

where C_3 is generally considered to be $C_3 = -2.3\,mV/C$.

4. The final step is to determine the current of the PV module for operating conditions.

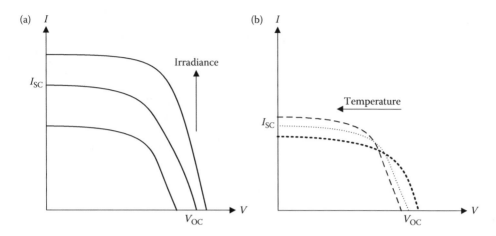

FIGURE 1.12 (a) Influence of the ambient irradiation and (b) influence of the cell temperature on the cell I–V characteristics.

1.3.5 Effects of Irradiance and Temperature on PV Characteristics

As irradiance increases, short-circuit current (I_{SC}) and open-circuit voltage (V_{OC}) of the solar cell increase. Short-circuit current is almost in a linear proportional relationship with irradiance. As temperature increases, on the other hand, open-circuit voltage decreases while short-circuit current increases. This is because temperature is a function of irradiance. In Figure 1.12a and b, the I–V characteristic of a solar cell for a certain ambient irradiation G_a and a certain cell temperature T_c is illustrated. The influence of ambient irradiation G_a and cell temperature T_c on the cell characteristics is presented in Figure 1.12.

Figure 1.12a shows that the open-circuit voltage increases logarithmically with ambient irradiation, while the short-circuit current is a linear function of ambient irradiation. The arrows show the directions of increase in irradiation and cell temperature. The influence of cell temperature on I–V characteristics is illustrated in Figure 1.12b. The dominant effect on increasing the cell's temperature is the linear decrease in the open-circuit voltage. Decreasing the cell's open-circuit voltage reduces the efficiency of the cell. The short-circuit current slightly increases with cell temperature [22].

1.4 Sun Tracking Systems

The sun changes its position from morning to night and from one season to another. A sun tracker is an apparatus that is used for orienting a solar PV panel, concentrating a solar reflector or lens toward the sun. Solar panels require a high degree of accuracy to ensure that the concentrated sunlight is directed precisely to the PV device. Solar tracking systems can substantially improve the amount of power produced by a system by enhancing morning and afternoon performance. Strong afternoon performance is particularly desirable for grid-tied PV systems, as production at this time will match the peak power demand period for the summer season. A fixed system oriented to optimize this limited time performance will have a relatively low annual production because the PV panels are fixed and do not move to track the sun.

For low-temperature solar thermal applications, trackers are not usually used. This is because tracking systems are more expensive when compared to the cost of adding more collectors. In addition, more restricted solar angles are required for winter performance, which influences the average year-round system capacity. For solar-electric applications, trackers can be relatively inexpensive when compared to the cost of a PV system. This makes them very effective for PV systems in high-efficiency panels. From the maintenance point of view, solar trackers need to be inspected and lubricated on an annual or seasonal basis.

There are several techniques to track the position of the sun and adjust the position of the panel. One of the most common techniques of tracking the sun is to use the relationship between the angle of the light source and the differential current generated in two close photodiodes due to the shadow produced by a cover over them [23–25].

The structure of the sun tracking system based on the relationship between the angle of light and the differential current in two photodiodes is demonstrated in Figure 1.13. The shadow over the photodiodes will be generated by the cover when the light source is misaligned. As a result, one photodiode is more illuminated than the other. The more illuminated photodiode generates more electrons than the other one. Therefore, the difference between the generated currents becomes larger as the misalignment angle increases. The system can use the difference between these currents to orientate the system to directly face the sun.

The sensor is composed of two modules for the "x" and "y" axes, as shown in Figure 1.14. Two photodiodes, metallic connections, resistors, resistor protectors, and a cover are the other components of the sensor modules. Figure 1.14 presents the top view of the sensor's x-axis and y-axis modules.

Dynamic range is the parameter that the sun tracking system responds to in threshold limits. The distance between both modules does not have to limit the dynamic range. If the dynamic range is increased, the sensitivity of the current differences decreases. Therefore, between the dynamic range and the sensitivity, there is a trade-off. As a solution, two

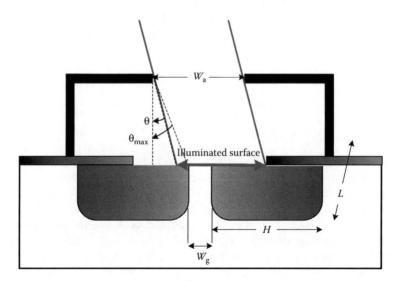

FIGURE 1.13 **(See color insert following page 80.)** The structure of the sun tracking system based on angle of the light using two photodiodes.

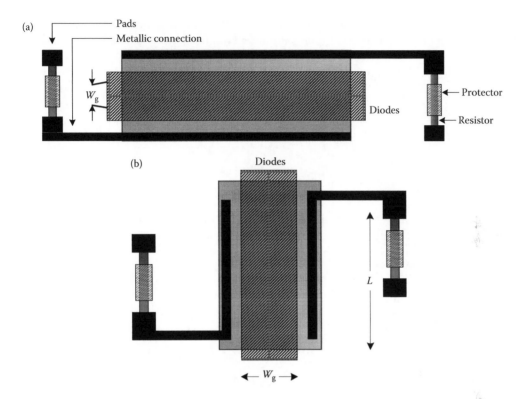

FIGURE 1.14 **(See color insert following page 80.)** (a) x-Axis sensor module structure; (b) y-axis sensor module structure.

parallel sensors can be operated to reach the desired sensitivity and high dynamic range in an optimum manner.

Currents generated by the couple sensors depend on the illuminated area. If one photodiode is more illuminated than the other, the currents generated by each photodiode become different. If the sensor is positioned with a predefined angle with respect to the source of light, this will yield different currents generated in both photodiodes as shown in Figure 1.14. In this way, the misalignment of the system is calculated. The system uses this differential signal to orientate the panel to the light source [23,24].

There is a gap between the two photodiodes, and this is necessary to avoid the combination of carriers. The PV effect is described by Equation 1.17:

$$I_L = qA \int_0^{\lambda_g} N_F(\lambda)(1 - R)S_R(\lambda)\, d\lambda. \tag{1.17}$$

In this equation, A is the illuminated area, $N_F(\lambda)$ is the number of incident photons per unit area, $S_R(\lambda)$ is the number of generated carriers to the number of incident photons with wavelength $S_R(\lambda)$, and R is photon reflection on the semiconductor surface [23,24].

The illuminated area of every photodiode can be calculated as

$$A = \left(\frac{W_a}{2} - \frac{W_g}{2} + H\, \tan\theta\right) L, \tag{1.18}$$

where W_a is the width of the aperture of the cover, W_g is the gap between the photodiodes, H is the width of the sensor, θ is the misalignment angle, and L is the depth of the sensor. The difference between two currents is

$$I_1 - I_2 = 2HL \tan \theta J_L, \qquad (1.19)$$

where I_1 and I_2 are the currents of the corresponding photodiodes and J_L is the current density generated per surface of the photodiodes. The misalignment can be calculated from

$$\theta = \tan^{-1} \left(\frac{I_1 - I_2}{2HLJ_L} \right). \qquad (1.20)$$

The angle depends on irradiation and current density. By adding currents, this variable can be normalized and its dependence on current density can be omitted:

$$I_1 + I_2 = L(W_a - W_g)J_L. \qquad (1.21)$$

Therefore, by combining Equations 1.20 and 1.22, the misalignment angle becomes

$$\theta = \tan^{-1} \left(\frac{(I_1 - I_2)(W_a - W_g)}{(I_1 + I_2)2H} \right). \qquad (1.22)$$

According to Equation 1.22, the $H/(W_a - W_g)$ ratio should be increased to increase the sensor sensitivity.

If one of the photodiodes is not illuminated at all, the maximum sensor angle occurs since one of the generated currents is zero. When there is no illumination, the photocurrent is zero and Equation 1.22 can be rewritten as

$$\theta = \tan^{-1} \left(\frac{W_a - W_g}{2H} \right). \qquad (1.23)$$

The solar sensor output should be conditioned to adapt the sensor output to the microcontroller voltage range. The microcontroller produces the inverse bias of photodiodes to orientate the panels to minimize the current difference. Usually for control purposes, current signals should be converted to voltage signals. The flowchart in Figure 1.15 presents the sensor output evaluation and motor control for the sun tracking system [23,24].

For implementation of Equation 1.22, the sensor outputs are added and subtracted by the differential and adder circuits, respectively. The reference angle (θ) is obtained by the microcontroller while actual motor positions are also fed back to the microcontroller. The reference motor positions are achieved by providing the control signals to the motor drive. Both motors and the drive circuit are powered by a power supply.

The difference signal data are compared with a maximum threshold. If the value is higher, the sun is misaligned more than the selected angle. Hence, the reverse action should be taken by the microcontroller that controls the motor drive.

A differential signal is produced by the difference between generated currents by the sensors. Movement of not only the sun but also the clouds can affect the output. The presence of clouds may cause the microcontroller to misinterpret the moving direction of the sun. This results in moving the platform in the wrong direction and consuming unnecessary energy. These malfunctions can be eliminated by using intelligent control techniques.

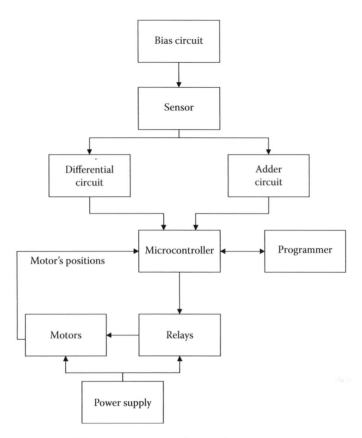

FIGURE 1.15 Block diagram of the sensor output evaluation and motor control for sun tracker.

The addition and differentiation of currents are used to identify the real status, that is, movement of the sun or a passing cloud. To detect cloudpresence, the addition of currents is used as the parameter, while the difference of currents varies with the movement of the sun. Addition of currents is not affected by a change in the position of the sun. The addition of currents is different in cloud, sunset, or sunrise conditions. In the presence of a cloud, the derivative of current addition is higher than that of sunset or sunrise situations. So, the sun tracking controller algorithm should consider current difference, current addition, current evolution slope, and current variations [23].

An algorithm is required for the determination of sun searching or sun tracking. Sun tracking is the normal operation while sun searching is required during the presence of clouds. The flowchart of the algorithm consists of several state machines as shown in Figure 1.16. The microcontroller is initialized to access the previously recorded data table to decide to run the sun search, sun tracking, or cloud algorithms. The radiation data table has the information for a typical cloudless day. The radiation, calculated from these data, is used to determine whether a cloud has passed over the sensor. The stored information makes it possible to reproduce the system movement in the case of clouds.

The sun searching or sun tracking states are a function of the radiation. The system will work in the "sun tracking" state under good meteorological conditions. This state is in charge of storing the information of the tracking movement. The system movement can be

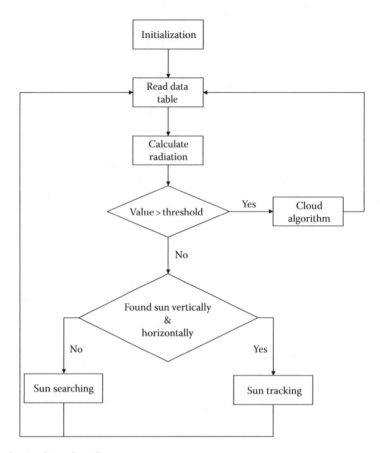

FIGURE 1.16 Sun tracking algorithm.

provided by the microcontroller that provides appropriate switching signals to the motor drive in order to reach the reference motor positions.

If a cloud is detected, the corresponding state will be "cloud algorithm." Then, the movement can be replicated by the stored information in "sun tracking" state. The system remains in this state until the radiation returns to another value. The cloud is expected to last for about 15 min. When this time is over, the system will look for the sun and maximum radiation. This temporal limitation is used to minimize the error due to the approximation. The sun's movement is characterized by azimuth and elevation angles. Depending on the time of day, the azimuth or elevation angle changes in a different way. For example, early in the morning, the elevation angle changes steeply and the azimuth angle almost remains constant. The main state selection is based on the addition voltages, because radiation is calculated from these data. Once the state is selected, the fundamental variable is the differential voltage. These data are sampled to know the sun's movement in the "sun tracking" state. The "sun search" state is in charge of orientating the sensor toward the sun if partial sunlight is available due to clouds. This action is necessary in certain conditions, for example, at dawn, and to orientate the sensor after the "cloud algorithm" state. In this state the movement of the platform is bidirectional to prevent malfunction of the system [23].

1.5 MPPT Techniques

The *I–V* characteristics of solar cells are affected by conditions of radiation and temperature. The voltage and current should be controlled to track the maximum power of PV systems. MPPT techniques are used to extract the maximum available power from solar cells. Systems composed of various PV modules located at different positions should have individual power conditioning systems to ensure the MPPT for each module [21,26].

1.5.1 Incremental Conductance–Based MPPT Technique

The incremental conductance technique is the most commonly used MPPT for PV systems [21,27–29]. The technique is based on the fact that the sum of the instantaneous conductance I/V and the incremental conductance $\Delta I/\Delta V$ is zero at the MPP, negative on the right side of the MPP, and positive on the left side of the MPP. Figure 1.17 shows the flowchart algorithm of the incremental conductance technique.

If the change in current and change in voltage is zero at the same time, no increment or decrement is required for the reference current. If there is no change in current while the voltage change is positive, the reference current should be increased. Similarly, if there is

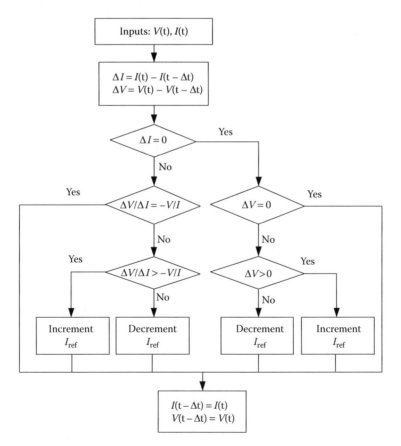

FIGURE 1.17 Incremental conductance algorithm flow-chart diagram.

no change in current while the voltage change is negative, the reference current should be decreased. If the current change is zero while $\Delta V/\Delta I = -V/I$, the PV is operating at MPP. If $\Delta V/\Delta I \neq -V/I$ and $\Delta V/\Delta I > -V/I$, the reference current should be decreased. However, if $\Delta V/\Delta I \neq -V/I$ and $\Delta V/\Delta I < -V/I$, the reference current should be increased in order to track the MPP.

Practically, due to the noise and errors, satisfying the condition of $\Delta I/\Delta V = -I/V$ may be very difficult [30]. Therefore, this condition can be satisfied with good approximation by

$$|\Delta I/\Delta V + I/V| < \varepsilon, \tag{1.24}$$

where ε is a positive small value.

Based on this algorithm, the operating point is either located in the BC interval or oscillating among the AB and CD intervals, as shown in Figure 1.18.

Selecting the step size (ΔV_{ref}), shown in Figure 1.18, is a trade-off between accurate steady tracking and dynamic response. If larger step sizes are used for quicker dynamic responses, the tracking accuracy decreases and the tracking point oscillates around the MPP. On the other hand, when small step sizes are selected, the tracking accuracy will increase. In the meantime, the time duration required to reach the MPP will increase [31].

The normalized IV, PV (power–voltage), and absolute derivative of the PV characteristics of a PV array are shown in Figure 1.19.

From these characteristics, it is seen that $|dP/dV|$ decreases as MPP is approached and increases when the operating point moves away from MPP. This relation can be given by

$$\begin{cases} dP/dV < 0 & \text{right of MPP,} \\ dP/dV = 0 & \text{at MPP,} \\ dP/dV > 0 & \text{left of MPP.} \end{cases} \tag{1.25}$$

In order to obtain the operating MPP, dP/dV should be calculated:

$$\frac{dP}{dV} = \frac{d(IV)}{dV} = I + V\frac{dI}{dV}. \tag{1.26}$$

dP/dV can be obtained by measuring only the incremental and instantaneous conductance of the PV array, that is, $\Delta I/\Delta V$ and I/V [27].

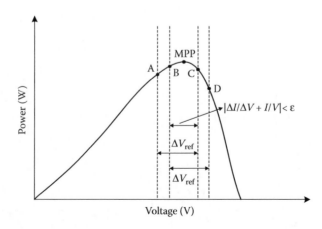

FIGURE 1.18 Operating point trajectory of incremental conductance–based MPPT.

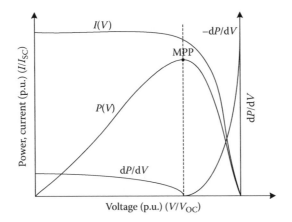

FIGURE 1.19   Normalized IV, PV, and |dP/dV| characteristics of a PV array.

1.5.2   Perturb and Observe–Based MPPT

The perturb & observe (P&O) technique is another common method of MPP tracking due to its simple structure and ease of implementation. Figure 1.20 shows a typical power–current curve of a regular PV array.

   As shown in Figure 1.20, if the operating current or, in other words, the current drawn from the PV array is perturbed in a given direction and if the power drawn from the PV array increases, the operating point becomes closer to the MPP and, thus, the operating current should be further perturbed in the same direction [32]. If the current is perturbed and this results in a decrease in the power drawn from the PV array, this means that the point of operation is moving away from the MPP and, therefore, the perturbation of the operating current should be reversed.

   The flowchart of the P&O-based MMPT technique is presented in Figure 1.21.

   In this method, an initial value for the current (I_{ref}) and current change (ΔI_{ref}) is guessed. Then the power that is associated with this current is measured from the PV panel output, that is $P_{PV}(k)$. A decrement or increment is applied to the reference current as a perturbation, which yields a change in power and the new power point becomes $P_{PV}(k+1)$. Now, the

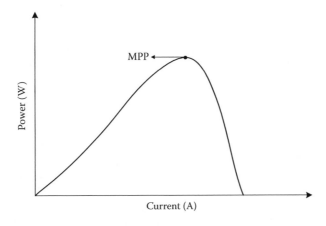

FIGURE 1.20   A typical *I–P* curve of a PV array.

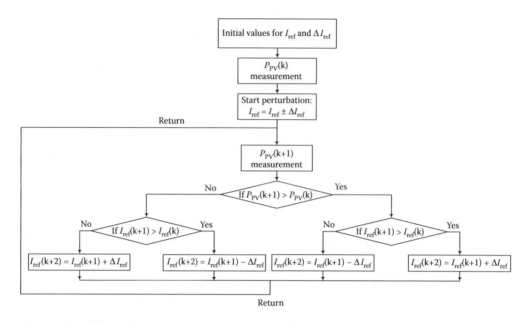

FIGURE 1.21   The flow diagram of the P&O-based MPPT method.

change in power is determined if $P_{PV}(k + 1) > P_{PV}(k)$. The change in power should be in the same direction as the change in current for further perturbation in the same direction. If they follow the opposite trend, the current reference should be reversed.

The initial value of the current can be zero or any closer value to the I_{MPP}, that is, the current at the MPP. On the other hand, choosing an appropriate incremental step (ΔI_{ref}) can quickly approach the operating point to MPP. This process should be repeated until the maximum available power from the PV array is found and extracted. The oscillation can be minimized with smaller perturbation steps. However, small step sizes cause the slow response of the MPP tracker [33]. The solution to this problem can be found using variable perturbation steps as addressed in [32,34–36]. In this method, the perturbation size should become smaller while approaching the MPP.

1.5.3   MPPT Controller–Based on Linearized *I–V* Characteristics

The *I–V* characteristic is a function of voltage, insulation level, and temperature [37–39]. From these characteristics, some important properties for the design of an MPPT controller can be explained as follows:

1. The PV array consists of two operation segments. In an *I–V* characteristic curve, one of the segments is the constant-voltage segment and the other is the constant-current segment. Therefore, the *I–V* characteristics can be approximated as a linear function in both segments as

$$I_P = -mV_p + b, \tag{1.27}$$

where m is the output conductance of the PV array. m is small in the constant-current segment; therefore the PV array exhibits highly negative output impedance. On the

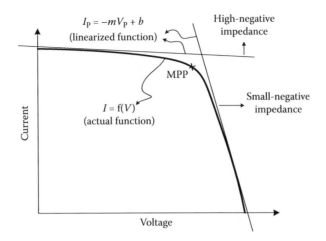

FIGURE 1.22   A typical *I–V* curve with linearized functions.

other hand, it exhibits small negative output impedance in the constant-voltage segment. The same linearization can be done for both segments with different m and b coefficients. Figure 1.22 represents a typical *I–V* curve with the linear approximations of two segments.

2. The MPP occurs at the knee of the characteristic curve, that is, $V_P = V_{PM}$. The slope is positive in the constant-current segment ($V_P < V_{PM}$), while the slope is negative in the constant-voltage segment ($V_P > V_{PM}$). Using Equation 1.27, dP_{PV}/dV_P becomes

$$\frac{dP_{PV}}{dV_P} = I_P - mV_P. \tag{1.28}$$

3. In order to move the operating point toward the zero slope point, I_P should decrease for positive slope and increase for negative slope if the PV array is controlled by current. Therefore, the MPP can be tracked by moving the operating point toward the zero slope point. For current controlling purposes, a DC/DC converter can be employed, that is, a boost converter. From the input side of the boost converter

$$I_P = I_{cap} + I_C = C_i \frac{dV_P}{dt} + I_C. \tag{1.29}$$

I_P is equal to the converter current (I_C) in steady state. Therefore, the operating point can be moved toward the MPP by adjusting I_C.

4. An MPPT controller to track the MPP using current-mode control is proposed in Figure 1.23. This MPPT controller and the current controller are based on the above properties and on Equation 1.29.

In this controller, first the PV array voltage and current are measured, which is followed by power and slope calculations. In steady state, PV current is equal to converter current. Therefore, based on the sign of the slope (dP_{PV}/dV_P), the reference current I_{MPP} is increased or decreased for moving the operating point toward the zero slope point.

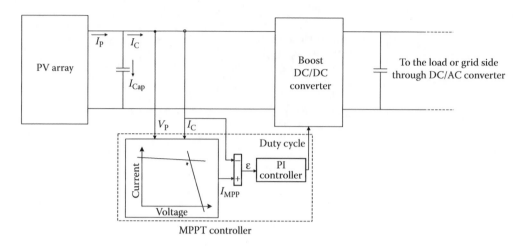

FIGURE 1.23 Linearized *I–V* characteristics–based MPPT controller.

1.5.4 Fractional Open-Circuit Voltage-Based MPPT

The approximately linear relationship between V_{MPP} and V_{OC} of the PV array, under varying irradiance and temperature levels, is the basis for the fractional V_{OC} method [40–47]:

$$V_{MPP} \approx k_1 \, V_{OC}, \tag{1.30}$$

where k_1 is a constant dependent on the characteristics of the PV array. However, it has to be computed beforehand by empirically determining V_{MPP} and V_{OC} for the specific PV array at different irradiance and temperature levels. The factor k_1 is usually between 0.71 and 0.78 [33].

Using Equation 1.32 and measuring V_{OC} from a no-loaded PV array, V_{MPP} can be calculated with the known k_1. The output terminals of the PV array should be disconnected from the power converter. This results in a temporary loss of power, which is the main drawback of this technique. To overcome this drawback, pilot cells can be used to measure V_{OC} [42]. These pilot cells should have the same irradiation and temperature as well as the same characteristics with the main PV array for better approximation of the open-circuit voltage. *P–N* junction diodes generate a voltage that is approximately 75% of V_{OC} [46]. Thus, there is no need to measure V_{OC}. A closed-loop voltage control can be implemented after the MPPT DC/DC converter for voltage regulation of the inverter input.

Figure 1.24 shows the implementation of the open-circuit voltage-based MPPT technique. V_{MPP} can be obtained using the open-circuit voltage measurements from pilot cells and Equation 1.32. Then, the measured voltage (V^*) can be compared to this value. The duty cycle is determined by the PI controller and applied to the power electronic switch through gate drives. Thus, the DC/DC converter forces the PV output voltage to reach V_{MPP}.

The PV array technically never operates at the MPP since Equation 1.32 is an approximation. This approximation can be adequate, depending on the application of the PV system. The technique is easy to implement and cheap because it does not require a complicated control system; however, it is not a real MPPT technique. Also, k_1 is not valid under partial shading conditions and it should be updated by sweeping the PV array voltage [47]. Thus, to use this method under shaded conditions, the implementation becomes complicated and incurs more power loss.

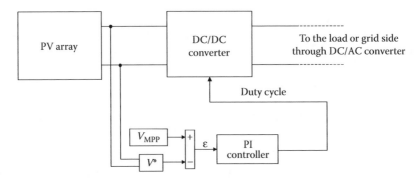

FIGURE 1.24 Implementation of fractional open-circuit-based MPPT.

1.5.5 Fractional Short-Circuit Current-Based MPPT

Similar to the open-circuit voltage, the short-circuit current of the PV array is approximately proportional to I_{MPP}, the current that corresponds to the MPP of the PV array:

$$I_{MPP} \approx k_2\, I_{SC}, \tag{1.31}$$

where k_2 is the linear proportion constant. k_2 is dependent on the characteristics of the PV array. The constant k_2 generally varies between 0.78 and 0.92.

Figure 1.25 shows an implementation example of the short-circuit current-based MPPT technique. I_{MPP} can be obtained using the short-circuit current measurements from pilot cells and Equation 1.33. The measured current can be subtracted from this value in order to obtain the error, which is fed to the PI controller. The duty cycle is determined by the PI controller and used in the power electronic switch through gate drives. Thus, the DC/DC converter forces the current to be drawn from the PV output to reach I_{MPP}.

Measuring I_{SC} during operation is very difficult since the PV array should be short-circuited. Using a current sensor and an additional switch in the power converter, the PV array can be short-circuited to measure I_{SC} [48]. This increases the number of components and cost. In addition, it causes additional power losses due to short circuit. Additional pilot cells can be used for short-circuit current measurement, which has the same characteristics as the main PV array.

MPP is never perfectly matched, as suggested by Equation 1.36, since it is an approximation of the current of the MPP. The PV array can be periodically swept from open circuit

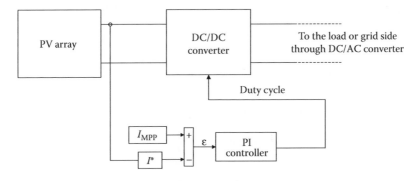

FIGURE 1.25 Implementation of fractional short-circuit-based MPPT.

to short circuit to update k_2 to ensure the proper MPPT in the presence of multiple local maxima [47]. Generally a digital signal processor (DSP) is used to implement fractional I_{SC} for MPPT purposes for PV systems. Alternatively, an easier current feedback control loop can be used to implement fractional I_{SC} [48].

1.5.6   Fuzzy Logic Control–Based MPPT

Due to developments in microcontroller and DSP technologies, fuzzy logic control [49–58] has received increased interest in MPPT applications. According to Ref. [57], fuzzy logic controllers have the advantages of working on systems with nonlinearities, not needing an accurate dynamic model and working with imprecise inputs.

Fuzzy logic control is based on three stages. The fuzzification stage converts input variables into linguistic variables based on a membership function as shown in Figure 1.26. In this case, there are five fuzzy levels, which are NB (negative big), NS (negative small), ZE (zero), PS (positive small), and PB (positive big). To increase the accuracy, a greater number of fuzzy levels can be used. a and b are based on the range of values of the numerical variable in Figure 1.26. In the membership function, some specific fuzzy levels can be designed as unsymmetrical to make them more dominant, in other words to give them more importance [49,53,57,58].

The error E and its variation (ΔE) are inputs to the fuzzy logic-based MPPT controller. E and ΔE can be calculated based on the users' preferences. Since dP/dV gets closer to zero at the MPP, the approximation of Equation 1.34 can be used [59]:

$$E(n) = \frac{P(n) - P(n-1)}{V(n) - V(n-1)} \tag{1.32}$$

and

$$\Delta E(n) = E(n) - E(n-1). \tag{1.33}$$

Alternatively, the error signal can be calculated as

$$e = \frac{I}{V} + \frac{dI}{dV}. \tag{1.34}$$

The error variations in Equation 1.35 can also be applied to Equation 1.36. The error expressed in Equation 1.36 is the sum of the instantaneous and incremental conductance

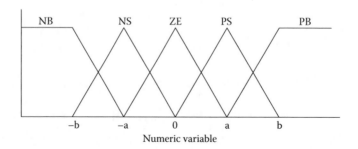

FIGURE 1.26   Membership function for inputs and outputs of fuzzy logic controller.

TABLE 1.3

Fuzzy Rule Base Table

ΔE	NB	NS	ZE	PS	PB
E					
NB	ZE	ZE	NB	NB	NB
NS	ZE	ZE	NS	NS	NS
ZE	NS	ZE	ZE	ZE	PS
PS	PS	PS	PS	ZE	ZE
PB	PB	PB	PB	ZE	ZE

and goes to zero while becoming closer to the MPP. Generally, the output of the fuzzy logic controller is the change in duty ratio ΔD of the power converter. This change in the duty ratio can be looked up in a look-up table such as Table 1.3 [50], right after E and ΔE are calculated and converted to the linguistic variables.

Different combinations of the error, E, and its variations, ΔE, can be used as the linguistic variables assigned to ΔD. For a boost converter, Table 1.3 can be used for this purpose. For example, if the operating point is far to the right of the MPP, and E is NB and ΔE is ZE, then a large decrease is required in the duty ratio to decrease the voltage, that is ΔD should be NB to reach the MPP.

The fuzzy logic controller output is converted from a linguistic variable to a numerical variable using a membership function as shown in Figure 1.27 in the defuzzification stage. By defuzzification, the controller produces an analog output signal, which can be converted to a digital signal and controls the power converter of the MPPT system.

Figure 1.27 shows an example of implementation of the fuzzy logic controller-based MPPT. Voltage and power are measured to calculate E and ΔE in Equations 1.34 and 1.35. Then, these values are evaluated by a fuzzy rule base table similar to Table 1.3. The output of the fuzzy rule base table is the required change in duty cycle. In the defuzzification stage, the numerical value of the duty cycle is determined via the conversion from linguistic values. Finally, through an analog-to-digital (A/D) converter and a gate driver, the necessary switching signal is applied to the power converter of the MPPT.

Under varying atmospheric conditions, the fuzzy logic controllers show good performance in MPPT applications. On the other hand, the effectiveness of the fuzzy logic controller depends on the accuracy of the calculation of error and its variations and the

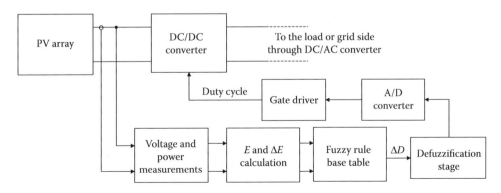

FIGURE 1.27 Implementation of fuzzy logic controller–based MPPT.

rule base table developed by the user. For better efficiency, the membership functions and rule base table can be continuously updated or tuned to achieve the optimum performance similar to an adaptive fuzzy logic controller [55]. In this way, fast convergence to the MPP and minimal fluctuation around MPP can be achieved [52]. In addition, the tracking performance depends on the type of membership function [57].

1.5.7 Neural Network–Based MPPT

Another intelligent MPPT control technique is the neural network (NN) [60–65]. NN algorithms are usually implemented by microcontrollers or DSPs.

NNs usually consist of three layers: input, hidden, and output layers. Figure 1.28 shows these layers. The NN is developed by the applier, considering the number of nodes in each layer. The NN controller's inputs are generally atmospheric parameters and PV array parameters, such as irradiance, temperature, V_{OC}, and I_{SC}. Processing these inputs, the NN controller determines the duty cycle of the power converter as the output [60,62].

The performance of the NN-based MPPT controller depends on how well the NN is trained and on the algorithms used within the hidden layer. The connections between the nodes are all weighted by gain coefficients. As shown in Figure 1.28, the connection of nodes i and j has a weight of w_{ij}. The weights between the nodes should be carefully determined in the training process in order to have a high-performance NN-based MPPT. Moreover, the inputs and outputs of the NN should be recorded over all seasons to have a sufficient training pattern. On the other hand, the NN should be specifically trained for the PV array, due to the different characteristics of different PV arrays. Because of degradation, the characteristics of the PV array also change with time and therefore the controller needs to be trained and updated in order to track the accurate MPP.

An implementation example of the NN-based MPPT method is depicted in Figure 1.29.

A pretrained NN should be trained using an experimentally obtained or calculation-based inputs versus output data table. Then, the NN controller should accurately determine the output with respect to the instantaneously measured inputs in the training data set.

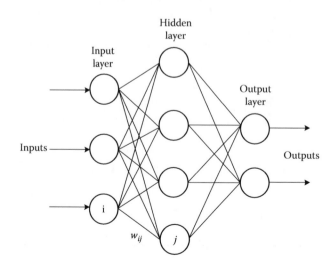

FIGURE 1.28 Example of NN structure.

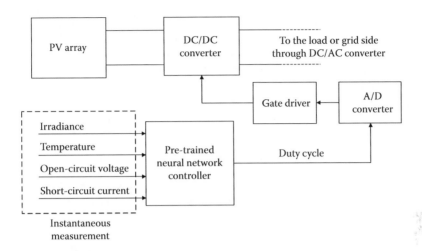

FIGURE 1.29   Implementation of NN-based MPPT.

Then, the duty cycle produced by NN is processed by necessary units in order to provide appropriate switching signals to the power converter of the PV array.

1.5.8   Ripple Correlation Control–Based MPPT

There are voltage and current ripples on the PV array due to the switching events of the connected power electronic converter. Therefore, the power output of the PV array may also contain ripples. These ripples are used in ripple correlation control (RCC) [66] to perform MPPT. RCC correlates the time derivative of the time-varying PV array power $\dot{p}$ with the time derivative of the time-varying PV array current $\dot{i}$ or voltage $\dot{v}$ to drive the power gradient to zero, in order to track the MPP.

The operating point is below the MPP ($V < V_{MPP}$ or $I < I_{MPP}$) if v or i is increasing ($\dot{v} > 0$ or $\dot{i} > 0$) and p is increasing ($\dot{p} > 0$). Conversely, the operating point is above the MPP ($V > V_{MPP}$ or $I > I_{MPP}$) if v or i is increasing and p is decreasing ($\dot{p} < 0$). According to these observations, it can be expressed that $\dot{p}\dot{v}$ or $\dot{p}\dot{i}$ are negative to the right of the MPP, positive to the left of the MPP, and zero at the MPP.

The inductor current increases if the duty ratio of the boost converter is increased; however, it decreases the PV array voltage [66]. This inductor current is also the PV array output current. Therefore, the duty ratio control input can be expressed as

$$d(t) = -k_3 \int \dot{p}\dot{v}\, dt, \qquad (1.35)$$

$$d(t) = k_3 \int \dot{p}\dot{i}\, dt, \qquad (1.36)$$

where k_3 is a positive constant. In RCC technique, if the duty ratio is controlled by Equation 1.37 or 1.38, MPP will be continuously tracked.

The derivatives of current, voltage, and power in Equations 1.37 and 1.38 are generally difficult to calculate. Therefore, alternatively, AC-coupled measurements of the PV array current and voltage can be used. It will be easier to calculate the derivatives of AC-coupled current and voltage, since they have the necessary phase information. In another way, the high-pass filters with higher cut-off frequencies than the ripple frequency can be used to estimate the derivatives. The inductor voltage can be used to calculate the derivative given in Equation 1.38 since it is proportional to the current derivative. The internal resistance of the inductor and the core loss do not have a significant effect, because the time constant of the inductor is larger than the switching interval of a power converter.

Due to the phase shift caused by the intrinsic capacitance of the PV array at high switching frequencies, Equation 1.37 may not result in effective MPPT of the system. However, correlating power and voltage as in Equation 1.37 is barely affected by the intrinsic capacitance.

Even under varying irradiance levels, RCC can accurately track the MPP with a fast response. The switching frequency of the power converter and the gain of the RCC circuit are factors limiting the time response of the MPPT. Implementation of RCC is straightforward, since it does not require any characteristics of the PV array.

Many research groups have investigated the application of RCC techniques to track MPP. The product of the signs of the time derivatives of power and duty ratio is used for integration [67]. A hysteresis-based version of RCC is purposed in [68,69]. A low-frequency fickle signal to disturb the PV array's power is utilized in [70]. In this method, a 90° phase shift in voltage or current with respect to the power at the MPP is taken into account. The use of that extra low-frequency signal instead of an inherent converter ripple is the difference of this RCC method [70].

An example of RCC-based MPPT method implementation is presented in Figure 1.30. In Figure 1.30, the voltage and current measured from the PV array output are processed to calculate the power and the derivative of the power and the voltage. Multiplication of the power and voltage derivatives is integrated and multiplied by the scale factor k_3. In this way, the duty cycle value is determined. Alternatively, integration of multiplication of the power and current derivatives could be used to determine the duty cycle, multiplying by positive scale factor k_3 based on Equation 1.38.

1.5.9 Current Sweep–Based MPPT

A sweep waveform for the PV array current can be obtained by using the *I–V* characteristic of the PV array. In this method, the sweep waveform is updated at fixed time intervals [71]. From the characteristic curve at the same intervals, V_{MPP} can be computed.

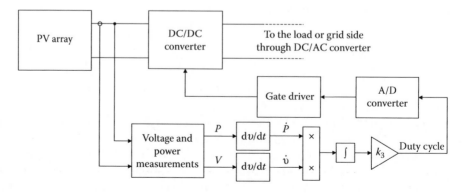

FIGURE 1.30 Implementation of RCC-based MPPT technique.

A function should be selected for the current sweep waveform such that the derivative of this function is directly proportional to the original function itself as

$$f(t) = k_4 \frac{df(t)}{dt},\qquad(1.37)$$

where k_4 is a constant. Therefore, the PV array power can be written as

$$p(t) = v(t)i(t) = v(t)f(t).\qquad(1.38)$$

At the MPP,

$$\frac{dp(t)}{dt} = v(t)\frac{df(t)}{dt} + f(t)\frac{dv(t)}{dt} = 0.\qquad(1.39)$$

Using Equations 1.41 and 1.43,

$$\frac{dp(t)}{dt} = \left[v(t) + k_4\frac{dv(t)}{dt}\right]\frac{df(t)}{dt} = 0.\qquad(1.40)$$

The solution of the differential equation in Equation 1.39 is

$$f(t) = C\,e^{t/k_4}.\qquad(1.41)$$

C is chosen to be equal to I_{max}, which is the maximum PV array current. k_4 should be negative, resulting in a decreasing exponential function with time constant $\tau = -k_4$. The new form of Equation 1.43 becomes

$$f(t) = I_{max}\,e^{-t/\tau}.\qquad(1.42)$$

By discharging some current through a capacitor, I_{max} in Equation 1.44 can be obtained. In Equation 1.42 since the derivative of Equation 1.41 is not zero, Equation 1.42 can be rewritten as

$$\frac{dp(t)}{di(t)} = v(t) + k_4\frac{dv(t)}{dt} = 0.\qquad(1.43)$$

Once V_{MPP} is computed by the current sweep method, the accuracy of the calculated MPP can be double checked through Equation 1.45.

An implementation example of sweep-based MPPT is shown in Figure 1.31. The MPP block uses the panel voltage to determine the reference voltage at MPP as given in Equation 1.45.

1.5.10 DC Link Capacitor Droop Control–Based MPPT

Topology for DC link capacitor droop control is shown in Figure 1.32.

In some cases the PV systems need to be connected to AC systems. In these cases, a specifically designed MPPT technique can be used, which is called DC-link capacitor droop control [72,73].

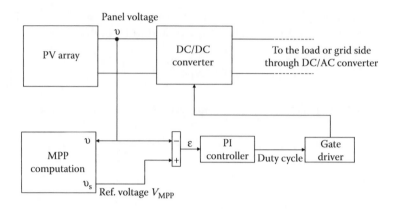

FIGURE 1.31   Implementation of current sweep-based MPPT.

The duty ratio of an ideal boost converter, in steady state operation mode, is given by

$$d = 1 - \frac{V}{V_{\text{link}}}, \qquad (1.44)$$

where V is the voltage across the input voltage of the power converter and V_{link} is the voltage across the DC link. If V_{link} is fixed, the extracted power of the PV array can be controlled by varying the input current of the inverter. The voltage V_{link} can be kept constant as long as the power required by the inverter does not exceed the maximum available power from the PV array while the current is increasing; otherwise V_{link} droops. The current control command of the inverter (I_{peak}) is at its maximum value and the PV array operates at the MPP right before that point. To achieve the MPP, d is optimized to bring I_{peak} to its maximum value in order to prevent V_{link} from drooping. This droop can be prevented by feeding the AC system line current back.

The computation of PV array power is not required in this method. However, when compared to the methods that detect the power directly, this method has lower accuracy [72], since the response of the DC-voltage control loop of the inverter directly affects its response.

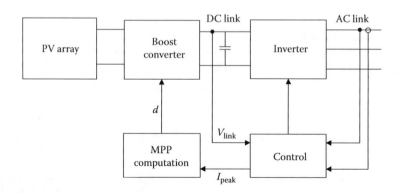

FIGURE 1.32   DC link capacitor droop control.

1.6   Shading Effects on PV Cells

The shading effect occurs when a PV array or part of it is not fully irradiated due to the effect of an obstacle. Some of the cells of the PV module can work in reverse bias acting as loads instead of power generators when the module is partially shaded. If the reverse bias exceeds the breakdown voltage of the shaded solar cell, it can act as an open circuit and damage the whole string [31,74,75].

To reduce the effects of shading, most commercial PV modules include internal bypass diodes. However, the number of diodes in the PV module is much less than the number of cells connected in series in the module [31]. Consequently, the risk of open circuits is reduced; however, the total power output of the PV module is reduced, since when one of the bypass diodes is conducting, a number of solar cells get out of order.

Figure 1.33 shows two sets of PV arrays. In the first array, each PV cell has its own bypass diode connected in parallel.

The first configuration is more resistible to the shading conditions and is available to produce more power when compared to the second configuration shown in Figure 1.33. When an array is subject to partial shading, a string of cells gets out of order in the second configuration while only the affected cells get out of order in the first configuration. Figure 1.34 shows the current–voltage curves for a fully illuminated array and two different configurations described in Figure 1.34.

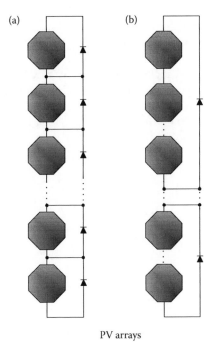

PV arrays

FIGURE 1.33   (a) Bypass diodes connected across each cell and (b) bypass diodes connected across a string of cells.

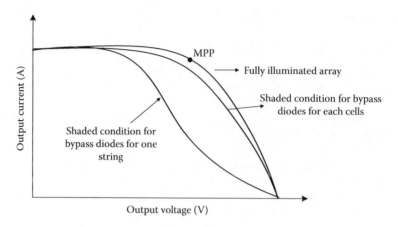

FIGURE 1.34   *I–V* characteristics for PV strings in shaded condition.

To calculate the shading effect, a shading factor (S) can be introduced as [75]

$$S = \frac{A_S}{A_C} = 1 - \frac{\overline{E}}{E_0}, \tag{1.45}$$

where A_S is the shaded area, A_C is the total area of the PV module, and $\overline{E}$ is the illumination average on the unshaded cell E_0.

The *I–V* curves of a PV cell under different shading factors behave just like the PV cells under different irradiation conditions as given in Figure 1.35.

Moreover, the power–voltage curve is also affected in the shaded condition. When local shading occurs, two different power peak values for two different voltage values exist. This is shown in Figure 1.36.

Many MPPT methods track a local peak and may not find the global value. This can be important under the local shading conditions. Therefore, the shading effect should be considered in calculating the MPP.

The shading effect can be included in the single-diode model without resistances. Equation 1.48 for the single-diode model described in Section 2.3.3 can be modified for

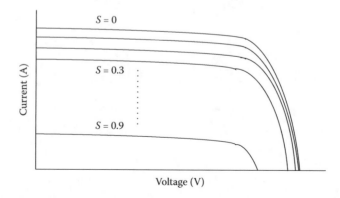

FIGURE 1.35   *I–V* characteristics under different shading conditions.

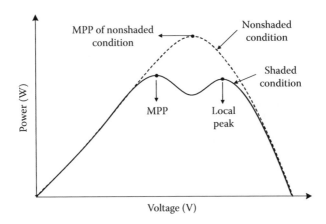

FIGURE 1.36 MPP relocation under shaded condition.

a single cell as

$$I = I_{PH} - I_S \left(e^{V/mV_T} - 1 \right),$$ (1.46)

where the photocurrent (I_{PH}) is in a linear relation with the solar irradiation, and short-circuit current (I_{SC}) is identical to I_{PH} [75]. The same linear relation is valid for shading effect (S) as

$$I_{PH} \approx I_{SC}(S) = I_{SC0}(1 - S),$$ (1.47)

where I_{SC0} is the short-circuit current of the fully illuminated cell.

For $I = 0$, Equation 1.48 can be rewritten as

$$V_{OC} = mV_T \ln \left(\frac{I_{SC}}{I_S} + 1 \right).$$ (1.48)

Therefore, V_{OC} can be written as

$$V_{OC} = mV_T \ln \left(\frac{I_{SC0}}{I_S} + 1 \right).$$ (1.49)

The unshaded cell's open-circuit voltage is

$$V_{OC}(S) = V_{OC0} + mV_T \ln(1 - S).$$ (1.50)

The shading issue affects the MPP of the PV module. Thus, the current and voltage relation at the MPP can be written as

$$I_{MPP} = I(V_{MPP}) = I_{SC} - I_S \left(e^{V_{MPP}/mV_T} - 1 \right).$$ (1.51)

By derivation of Equation 1.53, the voltage at the MPP can be obtained as

$$\frac{dP(V_{MPP})}{dV} = I(V_{MPP}) + V_{MPP} \frac{dI(V_{MPP})}{dV} = 0,$$ (1.52)

which yields

$$V_{\text{MPP}} = V_{\text{OC}} - mV_{\text{T}} \ln\left(1 + \frac{V_{\text{MPP}}}{mV_{\text{T}}}\right). \tag{1.53}$$

The nonlinear equation expressed in Equation 1.55 can be approximated using the Taylor series expansion for the logarithmic part as

$$f(V_{\text{MPP}}) = mV_{\text{T}} \ln\left(1 + \frac{V_{\text{MPP}}}{mV_{\text{T}}}\right) = f(x_0) + \frac{df(x_0)}{dV_{\text{MPP}}}(V_{\text{MPP}} - x_0) + R_1(V_{\text{MPP}}). \tag{1.54}$$

$R_1(V_{\text{MPP}})$ representing terms with high-order derivatives can be neglected. For the Taylor series approximation point V_{MPP}, defining $x_0 = cmV_{\text{T}}$ ($c = $ constant) yields

$$f(V_{\text{MPP}}) = mV_{\text{T}} \ln(1+c) - mV_{\text{T}}\frac{1/mV_{\text{T}}}{1 + V_{\text{MPP}}/mV_{\text{T}}}(V_{\text{MPP}} - cmV_{\text{T}}),$$

$$f(V_{\text{MPP}}) = mV_{\text{T}} \ln(1+c) - \frac{c}{1+c}mV_{\text{T}} + \frac{1}{1+c}V_{\text{MPP}}. \tag{1.55}$$

The linearized function can be an accurate approximation of the original function for c is between [10,1000] [75]. Therefore, the explicit solution would be

$$V_{\text{MPP}} = c_1 V_{\text{OC}} - c_2 mV_{\text{T}}. \tag{1.56}$$

The c_1 and c_2 constants can be expressed as

$$c_1 = \frac{1+c}{2+c} \quad \text{and} \quad c_2 = c_1 \ln(1+c) - \frac{c}{2+c}. \tag{1.57}$$

The voltage at the MPP can be obtained using

$$V_{\text{MPP}} = c_1 V_{\text{OC0}} - c_2 mV_{\text{T}}. \tag{1.58}$$

Then the MPP voltage as a function of shading factor becomes

$$V_{\text{MPP}}(S) = V_{\text{MPP0}} + c_1 mV_{\text{T}} \ln(1-S). \tag{1.59}$$

For high values of c, it can be assumed that $c_1 \approx 1$ and $I_{\text{SC0}} >> I_{\text{S}}$, and

$$I_{\text{MPP}} = \left(I_{\text{SC0}} - I_{\text{S}} e^{V_{\text{MPP0}}/mV_{\text{T}}}\right)(1-S). \tag{1.60}$$

Substituting $I_{\text{MPP0}} = I_{\text{SC0}} - I_{\text{S}} e^{V_{\text{MPP0}}/mV_{\text{T}}}$, the current at MPP ($I_{\text{MPP}}$) as a function of S would be

$$I_{\text{MPP}}(S) = I_{\text{MPP0}}(1-S). \tag{1.61}$$

Without any shading, the fill factor can be described as

$$\text{FF}_0 = \frac{V_{\text{MPP0}} I_{\text{MPP0}}}{V_{\text{OC0}} I_{\text{SC0}}}. \tag{1.62}$$

Under the shading conditions, a new fill factor can be defined as a function of S such that

$$FF(S) = FF_0 \frac{1 + c_3 \, \ln(1 - S)}{1 + c_4 \, \ln(1 - S)}, \tag{1.63}$$

where c_3 and c_4 constants are functions of I_{SC0} as

$$c_3(I_{SC0}) = \frac{1}{\ln(I_{SC0}/I_S) - c_2/c_1} \quad \text{and} \quad c_4(I_{SC0}) = \frac{1}{\ln(I_{SC0}/I_S)}. \tag{1.64}$$

Using maximum power, $P_{MPP0} = V_{MPP0} \, I_{MPP0}$, and c_3 the MPP under shaded conditions can be written as

$$P_{MPP}(S) = P_{MPP0}(1 - S)(1 + c_3 \, \ln(1 - S)). \tag{1.65}$$

The efficiency of the solar cell is also affected by the shading effects. The efficiency can be calculated using

$$\eta = \frac{P_{MPP}}{A_C \overline{E}}, \tag{1.66}$$

where A_C is the total cell area and $\overline{E}$ is the average illumination. P_{MPP} in Equation 1.68 can be replaced with Equation 1.69 and, using $\overline{E} = E_0(1 - S)$, the general efficiency equation $\eta_0 = P_{MPP0}/A_S E_0$ as a function of shading factor becomes

$$\eta(S) = \eta_0 (1 + c_3 \, \ln(1 - S)), \tag{1.67}$$

where E_0 is the illumination of unshaded cell and A_S is the shaded area. Figure 1.37 gives the MPP, fill factor, and efficiency under various shading factor values. The efficiency and fill factor suddenly drop for shading factors larger than a certain point. The power performance of the shaded PV cell almost linearly decreases as the shading factor increases [75].

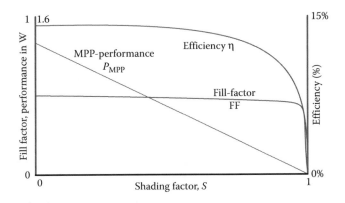

FIGURE 1.37 (See color insert following page 80.) P_{MPP}, FF, and η variations versus the shading factor S.

1.7 Power Electronic Interfaces for PV Systems

Power electronic interfaces are used either to convert the DC energy to AC energy to supply AC loads or connection to the grid or to control the terminal conditions of the PV module to track the MPP for maximizing the extracted energy. They also provide wide operating range, capability of operation over different daily and seasonal conditions, and reaching the highest possible efficiency [76]. There are various ways to categorize power electronic interfaces for solar systems. In this book, power electronic interfaces are categorized as power electronic interfaces for grid-connected PV systems and stand-alone PV systems.

1.7.1 Power Electronic Interfaces for Grid-Connected PV Systems

The power electronic interfaces for grid-connected PV systems can be classified into two main criteria: classification based on inverter utilization and classification based on converter stage and module configurations.

Based on inverter utilization, the topologies are

- Centralized inverter system
- String inverter system
- Multistring inverter system.

Based on the number of converter stages and the number of modules, topologies are

- Two-stage single module
- Single-stage multimodule
- Single-stage multilevel
- Two-stage multimodule.

These classifications are shown in Figure 1.38.

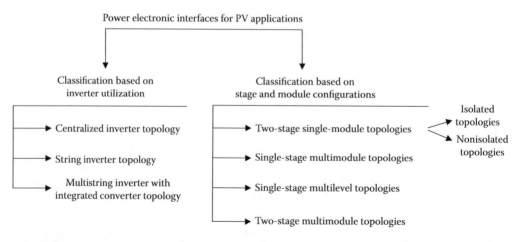

FIGURE 1.38 Power electronic interfaces classification based on inverter utilization and stage/module configurations.

1.7.1.1 Topologies Based on Inverter Utilization

1.7.1.1.1 Centralized Inverter Topology

The centralized inverter system is illustrated in Figure 1.39.

In this topology, PV modules are connected in series and parallel to achieve the required current and voltage levels. Only one inverter is used in this topology at the common DC bus. In this topology the inverter's power losses are higher than string inverter or multiinverter topologies due to the mismatch between modules and the necessity of string diodes that are connected in series. In this topology, voltage boost may not be required since the voltage of series-connected string voltages is high enough [77].

1.7.1.1.2 String Inverter Topology

String inverter topology is shown in Figure 1.40, in which the single string of modules is connected to the separate inverters for each string [78].

In this topology, if enough number of PV panels is connected in series in each string, voltage boosting may not be required. Voltage can be stepped up by a DC/DC converter at the DC side or by a transformer embedded in a high-frequency DC/DC converter. Separate MPPT can be applied to each string to increase the overall efficiency of the system [77].

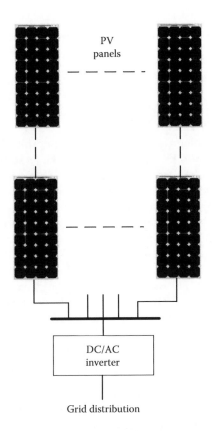

FIGURE 1.39 Conventional PV system technology using centralized inverter system topology.

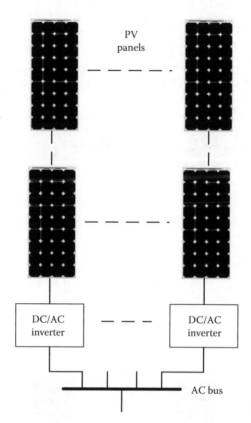

FIGURE 1.40 String inverters topology.

1.7.1.1.3   Multistring Inverter Topology

In the multistring invert topology, several strings are interfaced with their own integrated DC/DC converter to a common DC/AC inverter [79,80] as shown in Figure 1.41.

Individual PV strings can be turned on and off to use more or fewer modules. Further enlargements can be realized by adding integrated panel/converter groups. The outputs of the converters can be plugged into the existing platform, with all electrical connections in a single connector on the back plane. Therefore, this is a flexible design with high efficiency. In this topology, each PV module has its integrated power electronic interface with utility. The power loss of the system is relatively lower due to the reduced mismatch among the modules, but the constant losses in the inverter may be the same as for the string inverter. In addition, this configuration supports optimal operation of each module, which leads to an overall optimal performance [77]. This is because each PV panel has its individual DC/DC converter and maximum power levels can be achieved separately for each panel.

1.7.1.2   *Topologies Based on Module and Stage Configurations*

The power electronic conditioning circuits for solar energy systems can be transformerless, or they can utilize high-frequency transformers embedded in a DC/DC converter, which avoids bulky low-frequency transformers. The number of stages in the presented topologies refers to the number of cascaded converters/inverters in the system.

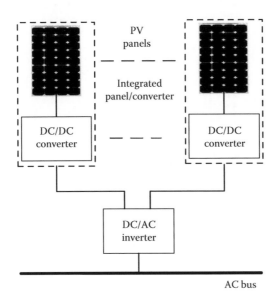

FIGURE 1.41   Multistring inverters topology.

1.7.1.2.1   Two-Stage Single-Module Topologies

The two-stage conversion systems may have many varieties. The most common two-stage topologies consist of a DC/AC grid-connected voltage source PWM inverter with a DC/DC PV-connected converter, with its associated MPPT system.

1.7.1.2.1.1   Isolated Two-Stage Single-Module Topologies   Isolated DC/DC converters consist of a transformer between the DC/AC and AC/DC conversion stage [81]. This transformer provides isolation between the PV source and load. Some typical topologies are depicted in Figure 1.42.

In the topologies shown in Figure 1.42, the output voltage of the PV panel and DC/DC converter are dc values. The two-stage DC/DC converter consists of a DC/AC inverter, a high-frequency transformer, and a rectifier. In Figure 1.42b, a capacitor is also used at the transformer input, forming an LC resonant circuit with the equivalent inductance of the transformer. This resonance circuit reduces the switching losses of the inverter. In the push–pull converter topology, a middle terminal connection is required since the inverter has only one level with less number of switching elements.

The push–pull converter shown in Figure 1.42c is modeled and simulated using Sim-PowerSystems of MATLAB®. In this system, the PV array nominal output voltage is 150 V and the switching frequency is 10 kHz. After the rectifier block, a parallel capacitive filter (1 mF) is used in order to obtain reduced oscillations at the output. The DC load is a 10 Ω resistive load.

The output current of the PV array is shown in Figure 1.43. The switching frequency is 10 kHz. A high-frequency transformer is used in the isolation stage with reduced size and cost.

The output voltage of the transformer is shown in Figure 1.44, which is a high-voltage AC voltage. It needs to be rectified prior to connection to the DC loads or DC/AC conversion.

In Figure 1.45, the output voltage of the AC/DC rectifier is shown. Usually, this output voltage varies between 0 and 150 V, which is the amplitude of AC input voltage. However,

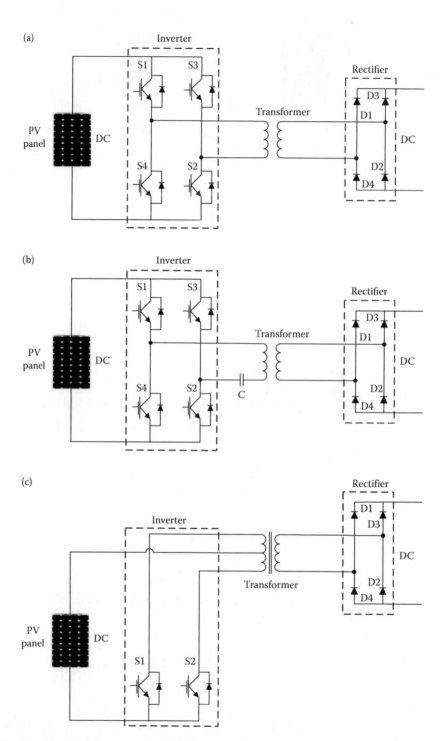

FIGURE 1.42 Isolated DC/DC converters: (a) H-bridge DC/DC converter, (b) series resonant H-bridge DC/DC converter, and (c) push–pull DC/DC converter [77].

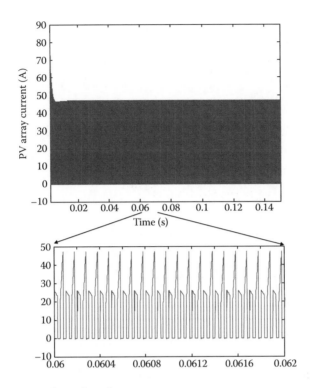

FIGURE 1.43 PV array current for push–pull converter.

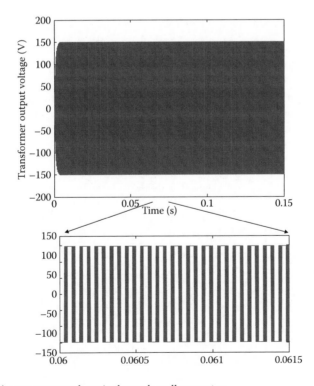

FIGURE 1.44 Transformer output voltage in the push–pull converter.

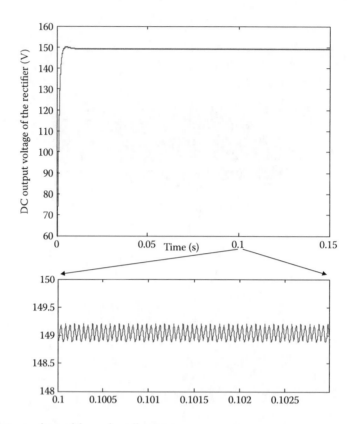

FIGURE 1.45   Output voltage of the push–pull converter.

to obtain less oscillation at the DC bus, a capacitive filter is used. Thus, the DC bus voltage is almost fixed.

The fly-back current-fed inverter shown in Figure 1.46 can be controlled to provide a rectified sine-wave output current into the inverter and to track the MPP [82]. The current into the fly-back converter is discontinuous and hence a buffer capacitor should be used to eliminate both low- and high-frequency ripples.

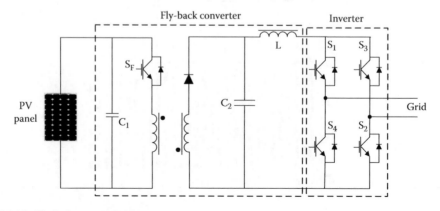

FIGURE 1.46   Fly-back current-fed inverter.

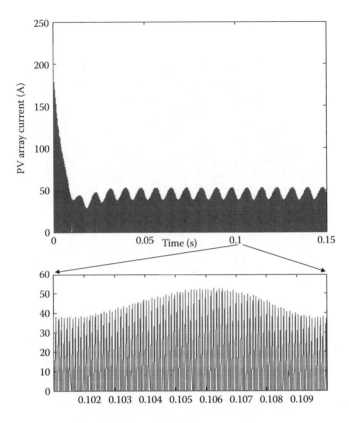

FIGURE 1.47 PV array current of the fly-back converter.

The simulation results of the fly-back converter are presented in Figures 1.47 through 1.50. The input capacitor of the simulated fly-back converter is 200 μF, the output capacitor is 1 mF, and the inductor is 0.01 mH. The nominal output voltage of the array is 150 V while the switching frequency of the converter is 10.8 kHz. The AC load resistance is 10 Ω, which is connected to the AC output terminals of the inverter. The output current of the PV array is shown in Figure 1.47.

The DC input voltage of the inverter is shown in Figure 1.48.

The DC waveform shown in Figure 1.49 is the input voltage of the inverter, or in other words it is the output voltage of the fly-back converter. This input voltage is converted to AC voltage using a pulse width modulated inverter with 10 kHz switching frequency. The AC output voltage of the inverter is shown in Figure 1.49.

The total harmonic distortion of this AC output voltage can be reduced using a series L and a parallel C filter. After defining a capacitor value (which is 1 mF in this example), the inductor value can be calculated as

$$L_f = \frac{1}{(2\pi)^2 f_{\text{cut-off}}^2 C_f}. \tag{1.68}$$

The cut-off edge frequency of the low-pass filter is 8 kHz to eliminate the disturbances and harmonics of the inverter. Therefore, higher than 8 kHz frequency components of the AC voltage will be eliminated. After implementing the passive LC filter, the output voltage of

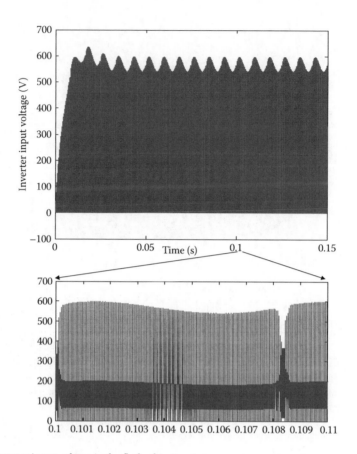

FIGURE 1.48   Inverter input voltage in the fly-back converter.

the inverter will be sinusoidal as shown in Figure 1.50. This voltage has reduced harmonic distortion and switching noise.

The topology shown in Figure 1.46 can be modified. Figure 1.51 shows a series resonant DC/DC converter plus a full bridge grid-connected inverter that is modified by adding two additional diodes [83,84]. The DCDC converter operates at 100 kHz and has a fixed voltage transfer ratio as a "DC-transformer." Switching losses are reduced by a resonant tank through zero-voltage switching. The switching losses from the converter can be reduced in this way. The MPPT is applied to the grid inverter, which uses both high and low switching frequencies. The left leg of the inverter in Figure 1.51 is controlled by a hysteresis-band controller and operates at switching frequencies between 20 and 80 kHz. The right leg of the inverter is controlled according to the polarity of the grid voltage with the grid switching frequency (60 or 50 Hz).

1.7.1.2.1.2 Nonisolated Two-Stage Single-Module Topologies   The topologies shown in Figure 1.52a and b are two-stage single-module topologies, in which a DC/DC converter is connected to a DC/AC converter for grid connection. The DC/DC converter deals with the MPP tracking and the DC/AC inverter is employed to convert the DC output to AC voltage for grid connection. These are nonisolated converters since they are transformer-less.

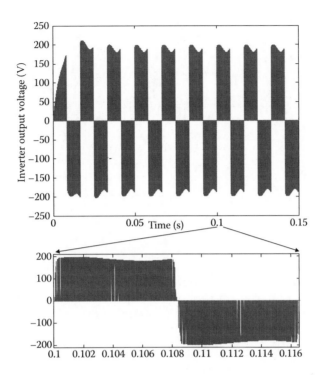

FIGURE 1.49 AC output voltage of the inverter.

In topologies shown in Figure 1.52a and b, instead of using a full bridge inverter for the DC/AC conversion stage, a half bridge inverter can also be used. In this way, the number of switching elements can be reduced and the controller can be simplified; however, for the DC bus two-series connected capacitor is required to obtain the mid-point. This mid-point of two series connected capacitors will be used as the negative terminal of the AC network of the half bridge configuration. These half bridge inverter topologies are illustrated in Figure 1.53a and b.

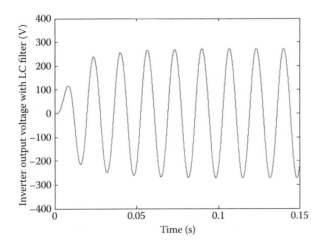

FIGURE 1.50 Inverter output voltage with LC filter.

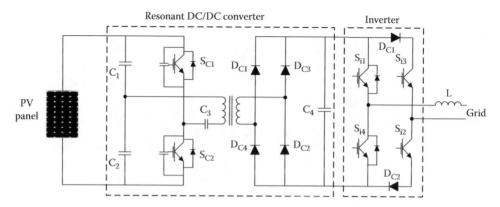

FIGURE 1.51   A series resonant DC/DC converter and its grid-connected inverter.

The current and voltage waveforms of a nonisolated two-stage converter in which a boost converter is followed by a full bridge inverter (Figure 1.52a) are demonstrated in Figures 1.54 through 1.57. The inductor of the boost converter is 0.1 mH, the capacitor of the boost converter is 5.6 mF, and the nominal output voltage of the PV array is 48 V. The switching frequency of the boost converter is 10 kHz. The 48 V output of the PV array

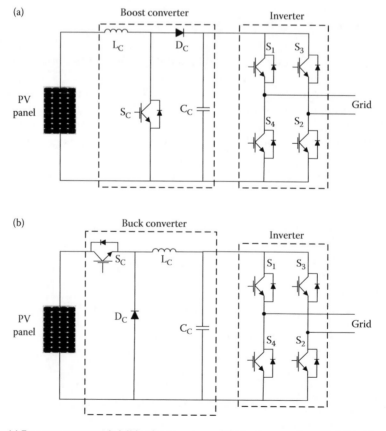

FIGURE 1.52   (a) Boost converter with full-bridge inverter and (b) buck converter with full bridge inverter.

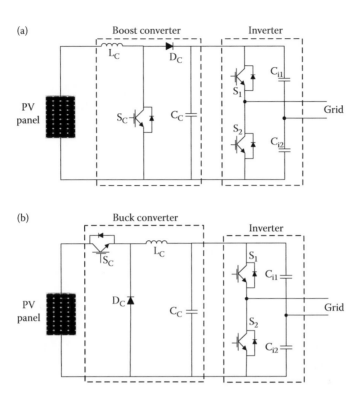

FIGURE 1.53   (a) Boost converter connected to a half bridge inverter and (b) buck converter connected to a half bridge inverter.

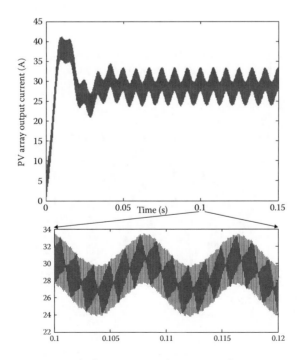

FIGURE 1.54   PV array output current for boost converter/inverter topology.

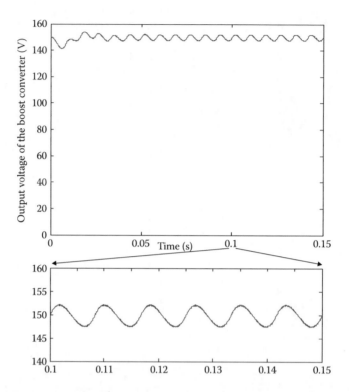

FIGURE 1.55 Boost converter output voltage or inverter input voltage for boost converter/inverter topology.

is boosted and inverted to AC voltage. The output current of the PV array is shown in Figure 1.57. The output voltage of the boost converter, which is the input voltage of the inverter, is given in Figure 1.58.

The DC/AC inversion of this DC voltage results in the AC waveform given in Figure 1.56. An LC filter is used at the inverter output in order to reduce the harmonic distortions and

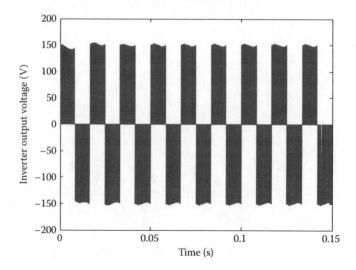

FIGURE 1.56 Inverter output voltage for boost converter/inverter topology.

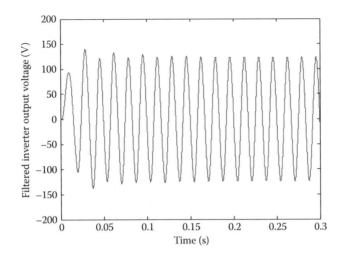

FIGURE 1.57 The filtered inverter output voltage.

switching noises. L_f is chosen as 0.4 µH and C_f is 1 mF. The sinusoidal output voltage of the inverter after filter is shown in Figure 1.57.

1.7.1.2.2 Single-Stage Multimodule Topologies

A typical single-stage inverter for multiple modules is depicted in Figure 1.58, which is the simplest grid connection topology [85]. The inverter is a standard voltage source PWM inverter, connected to the utility through an LCL filter. The input voltage, generated by the PV modules, should be higher than the peak voltage of the utility. The

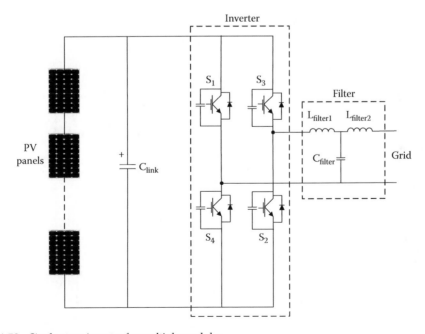

FIGURE 1.58 Single-stage inverter for multiple modules.

efficiency is about 97%. On the other hand, all the modules are connected to the same MPPT device. This may cause severe power losses during partial shadowing. In addition, a large capacitor is required for power decoupling between PV modules and the utility [86].

1.7.1.2.3   Single-Stage Multilevel Topologies

Figure 1.59 shows a multilevel converter. Each switch in this topology has an additional clamping diode protecting the switches from overvoltages. These clamping diodes avoid electromagnetic inference (EMI) influences on driver circuitries.

Multilevel inverter topologies are especially suitable for PV applications since different DC voltage levels can easily be provided with the modular structure of PV arrays [87,88]. The multilevel inverter can provide an almost sinusoidal output voltage with low harmonic distortion, at low switching frequency. In comparison with the single-level inverters, multilevel inverters produce better-quality AC voltage with less total harmonic distortion.

A half bridge diode clamped three-level inverter (HBDC) is shown in Figure 1.60. The three-level inverter can be expanded into five, seven, or even more levels by adding more modules and switches [87,88]. This allows for further reduction of the harmonic distortion. Drawbacks of this topology are the high number of required semiconductors and imbalanced loading of the different PV strings. Therefore, maximum power transfer from each individual string can be difficult to reach, especially when moderate shading occurs.

To obtain a positive voltage at the inverter's output terminals, upper switches S_1 and S_2 should turn on, while to obtain a negative voltage the lower switches S_3 and S_4 should conduct. In order to obtain zero voltage, the two intermediate switches should be turned on, that is, S_2 and S_3. The DC bus voltage should be greater than grid voltage amplitude in order to transfer power to the grid. Thus, a higher number of PV modules or individual boost DC/DC converters can be employed to achieve the desired voltage levels. In this

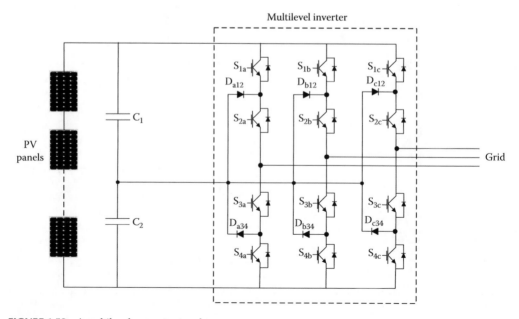

FIGURE 1.59 A multilevel converter topology.

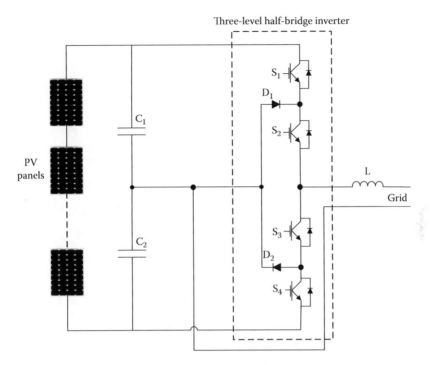

FIGURE 1.60 Grid-connected system with HBDC.

system, each of the PV strings is connected to the neutral point of the grid through the capacitors. This results in reducing the capacitive ground currents. In addition, the negative effect of these currents on electromagnetic compatibility is decreased [89,90]. However, a disadvantage of this topology is to load the DC source only during a half cycle, which increases the size of the decoupling capacitors and increases the cost of the inverter.

The output voltage is 0, $+V_{dc}$, or $-V_{dc}$ for α or 2α degrees, as shown in Figure 1.61 [91]. By controlling the α interval, the output voltage of the inverter can be controlled.

The Fourier series of the output voltage waveform can be expressed as

$$v_o(t) = \sum_{n=\text{odd}} V_n \sin(n\omega_0 t), \tag{1.69}$$

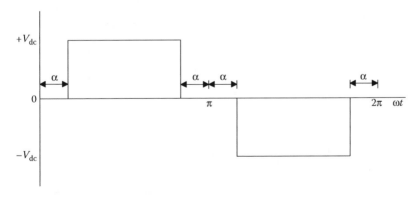

FIGURE 1.61 HBDC inverter output voltage.

where V_n is the amplitude of the nth harmonic component. If we use the half-wave period symmetry, the amplitude of the harmonic components would be

$$V_n = \frac{2}{\pi} \int\limits_{\alpha}^{\pi-a} V_{dc} \sin(n\omega_0 t)\, d(\omega_0 t) = \frac{4V_{dc}}{n\pi} \cos(n\alpha), \tag{1.70}$$

where α is the angle of zero voltage on each end of the pulses. Thus, the harmonic content can be controlled as a function of α.

The HBDC inverter for PV applications is analyzed in SimPowerSystems of MATLAB. In this model, $L = 15\,\text{mH}$ and $C = 1\,\text{mF}$. The nominal output voltages of the PV modules are assumed to be 180 V and the load resistance connected to the inverter's output is 20 Ω. For 3 s of simulation, the current of the output inductor is shown in Figure 1.62.

The output voltage of the inverter is presented in Figure 1.63.

1.7.1.2.4 Two-Stage Multimodule Topologies

In two-stage configurations, the connection of the modules and the inverter can be classified into two categories: in the first category, all modules are connected in series as shown in Figure 1.64a, which is similar to the two-stage single-module topologies. A grid-tie inverter plus a simple DC/DC converter, such as boost, buck, or buck–boost, can be used for the DC/DC conversion stage, if isolation is not required. The second category consists of a DC/DC converter for each string and a common grid-connected inverter, as shown in Figure 1.64b.

In configuration Figure 1.64b, the strings can operate at their individual MPP; therefore a better overall efficiency is expected. A PV system composed of strings with an individual DC/DC converter, a common grid-connected inverter, and the generation control circuit (GCC) is shown in Figure 1.65, which consists of two buck–boost converters with a common inductor [92]. The left leg of the inverter can control the voltage across each string individually. The upper switch along with the freewheeling diode in the lower switch and the input

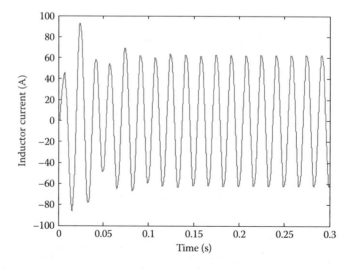

FIGURE 1.62 Output inductor current for an HBDC inverter.

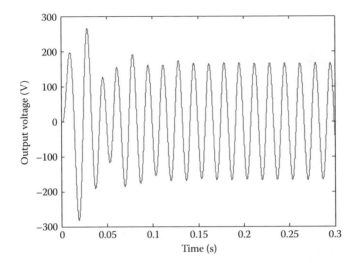

FIGURE 1.63 Output voltage for an HBDC inverter.

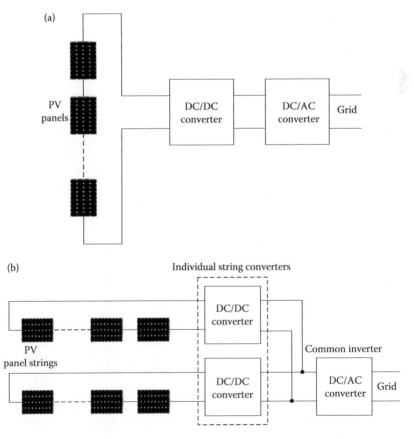

FIGURE 1.64 Configurations of dual-stage inverters for multiple PV modules: (a) modules with a common dual-stage inverter, and (b) strings with individual DC/DC converter and a common grid-connected inverter.

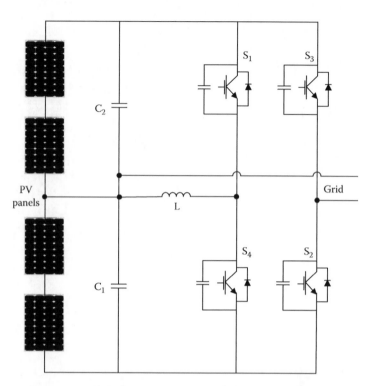

FIGURE 1.65   Utility interactive PV inverter.

inductor plus the capacitors across the strings forms a buck–boost converter. Similarly the opposite arrangement forms another buck–boost converter topology. The right leg of the inverter controls the current through the output inductor, and hence the current injected into the utility grid. The amplitude of the grid current is determined by an MPPT system. Therefore, MPPT controls the voltages across each string in order to achieve maximum power. This topology is a good solution for a multimodule system since it increases the overall efficiency of the system without adding extra components. It can also be expanded to multiple modules, by adding more chopper stages.

A modified form of the previously mentioned topology, which consists of a buck–boost converter and a half-bridge inverter, is shown in Figure 1.66.

This modified topology is an effective solution without adding extra components and it is suitable for extending it to multiple modules by adding more panel strings and DC/DC conversion stages.

The modified topology given in Figure 1.66 is analyzed for a 180 V single-panel string. The input inductor is 10 mH, the input capacitor is 1 mF, the output inductor is 15 mH, and the output capacitors are 1 mF. The load resistance at the AC bus is 10 Ω. Figure 1.67 shows the voltage of the inductor of the proposed topology, which is the inductor of the buck–boost converter.

The current of the buck/boost switch for the modified topology is shown in Figure 1.68. The output voltage of the buck–boost converter (input voltage of the DC/AC inverter stage) is shown in Figure 1.69.

The output voltage of the inverter is shown in Figure 1.70. The output voltage of the inverter can be used for local AC loads or grid connection.

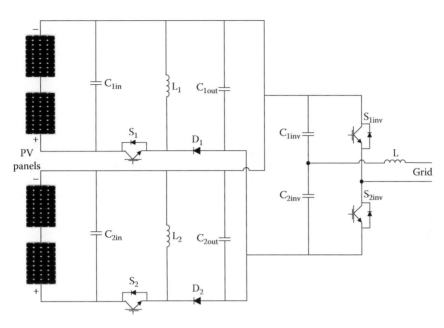

FIGURE 1.66 Modified structure of the topology given in Figure 1.65.

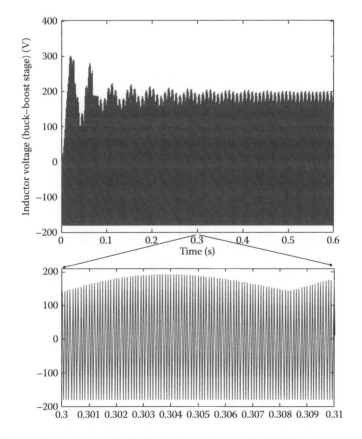

FIGURE 1.67 Voltage of the inductor at the buck–boost stage for modified GCC topology.

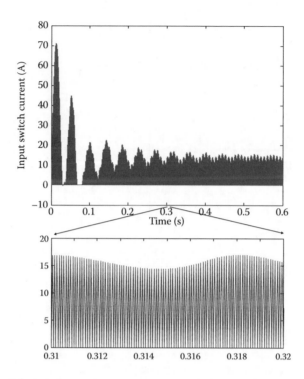

FIGURE 1.68 Input switch current.

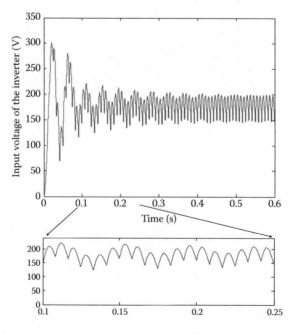

FIGURE 1.69 Input voltage of the inverter (output voltage of the DC/DC converter).

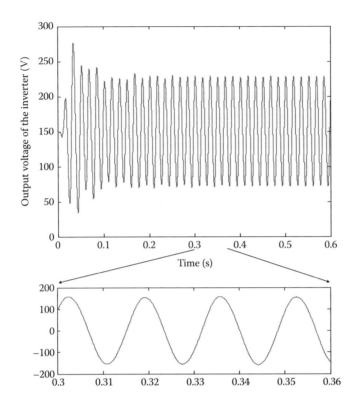

FIGURE 1.70 Inverter output voltage.

Another topology for multimodule multistring interfaces is shown in Figures 1.71 and 1.72 [80,93]. The inverter in Figure 1.71 consists of three boost converters, one for each PV string, and a common half bridge PWM inverter. The circuit can also be constructed with an isolated current- or voltage-fed push–pull or full-bridge converter [93], similar to the circuit in Figure 1.72, and a full-bridge inverter in order to connect to the utility. The voltage across each string can be controlled individually [80,93].

1.7.2 Power Electronic Interfaces for Stand-Alone PV Systems

The stand-alone PV systems are composed of a storage device and its controller for sustainable satisfaction of the load power demands [94]. The storage device with the controller should provide the power difference when the available power from the PV panel is smaller than the required power at the load bus [95]. When the available power from the PV panel is more than the required power, the PV panel should supply the load power and the excess power should be used to charge the storage device.

1.7.2.1 PV/Battery Connection: Type 1

A PV panel/battery connection topology is shown in Figure 1.73. In this topology, the DC/DC converter between the battery and the PV panel is used to capture all the available power from the PV panel. The battery pack acts as an energy buffer, charged from the PV

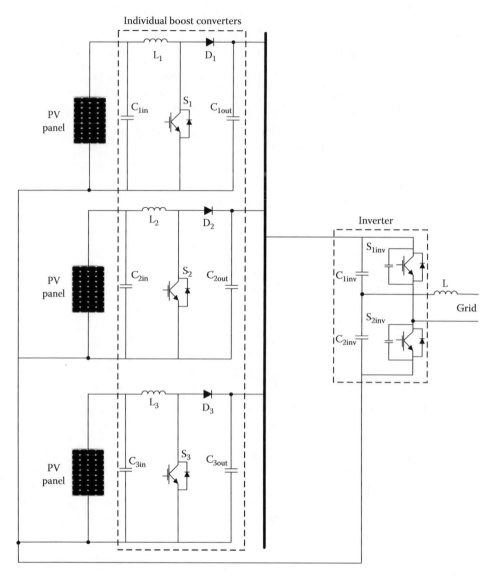

FIGURE 1.71 Topology of the power electronics of the multistring inverter.

panel and discharged through the DC/AC inverter to the load side. The charging controller determines the charging current of the battery, depending on the MPP of the PV panels at a certain time. When there is no solar radiation, the DC/DC converter disables and the stored energy within the battery supplies the load demands.

The battery size should be selected so that it can supply all the power demands during a possible no-insulation period. In addition, it could be fully charged during the insulated periods to store the energy for future use. Since the combined model produces AC electrical energy, it should be converted to AC electrical energy for domestic electrical loads. The system requires a DC/AC inverter, which is also used to match the different dynamics of the combined energy system and loads.

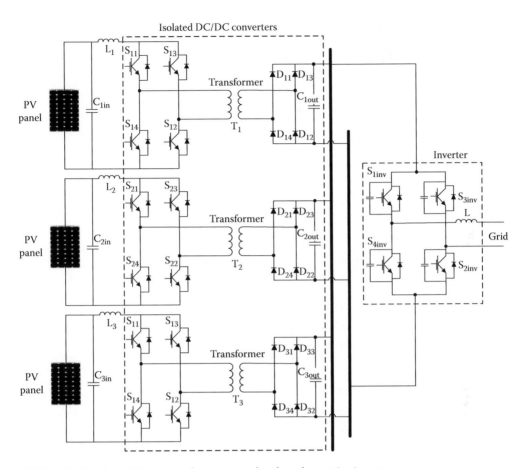

FIGURE 1.72 Topology of the power electronic interface for a three-string inverter.

1.7.2.2 PV/Battery Connection: Type 2

In the connection topology shown in Figure 1.74, the battery bank is connected to the output of the PV panel in parallel, instead of the cascaded connection. In this parallel connection, the DC/DC converter at the battery bus is never disabled, since it should be operated in either insulated or noninsulated conditions. The DC/AC inverter and the DC/DC converter should manage the load demands and capture the maximum available power from the PV panel. This topology requires a more complicated control strategy in comparison with the

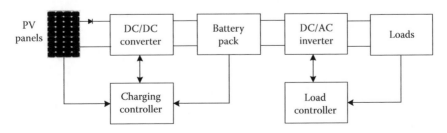

FIGURE 1.73 PV/battery connection—type 1.

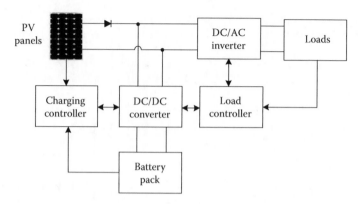

FIGURE 1.74   PV/battery connection—type 2.

connection shown in Figure 1.74, since it requires synchronized operation of the converters. In the case of sufficient solar insulation, the DC/DC converter should charge the battery. The PV panel should supply power to the load through the DC/AC inverter. In the case of no insulation, the stored energy in the battery should be transferred to the load through the DC/DC converter and the DC/AC inverter. This topology requires a bidirectional DC/DC converter to charge and discharge the battery.

1.7.2.3   PV/Battery Connection: Type 3

A modified topology of Figure 1.74 is shown in Figure 1.75, where the battery is located between the inverter and the load.

In this topology, the battery pack is connected to the AC bus through a bidirectional AC/DC converter. In this case, the DC/AC inverter should deal with the MPPT and transfer the maximum available power from the PV panel to the load side. In sufficient solar insulation conditions, the power provided by the PV panel should satisfy the load demands as

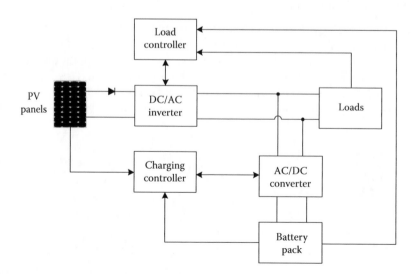

FIGURE 1.75   PV/battery connection—type 3.

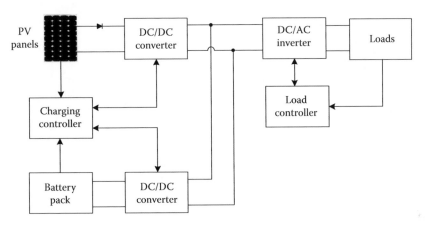

FIGURE 1.76 PV/battery connection—type 4.

well as charge the battery pack. In the no-insulation condition, the DC/AC inverter should be disabled and the bidirectional AC/DC converter should operate as a DC/AC inverter to supply sustained power to the load.

1.7.2.4 PV/Battery Connection: Type 4

In Figure 1.76, the PV and battery back are equipped with their individual DC/DC converters, which provide control flexibility. A DC/AC conversion stage is utilized to provide power for domestic AC loads. The converter of the PV panel needs to be a unidirectional power converter, while the battery pack converter should be a bidirectional converter to charge/discharge the battery.

1.7.2.5 PV/Battery Connection: Type 5

This topology incorporates multiple-input converters for PV systems. Multiple-input DC/DC converters are especially suitable for solar applications since they provide DC outputs and they need to be equipped with additional storage devices. A multiple-input DC/DC converter should be controlled so that it transfers the maximum available power from the PV panel. Based on load demand, the battery contribution can be controlled.

A multiple-input buck–boost DC/DC converter, proposed in [96], offers a solution with fewer parts while maintaining the same capability of individual DC/DC converters for different sources. Its implementation for the PV panel and the battery pack is shown in Figure 1.77.

In the topology shown in Figure 1.78, the multiple-input converter has a common inductor, diode, and output capacitor while it has different input power electronic switches for different power sources, also known as converter legs. A regular buck–boost converter has a negative output voltage and requires a transformer to obtain a positive output voltage at the DC bus. However, the topology shown in Figure 1.78 provides a positive output for the multiple-input buck–boost DC/DC converter [96–98].

The multiple-input DC/DC converter for a PV/battery system should be capable of operating the PV panel at the MPP and should control the charge/discharge of the battery pack. During night and low solar insulation periods, the leg for the PV panel input to the

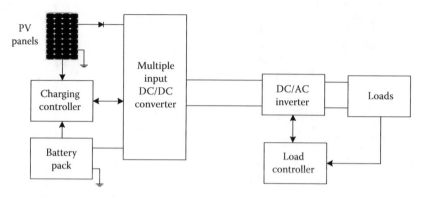

FIGURE 1.77   PV/battery connection—type 5.

converter should be disabled and the leg for the battery pack should be operated in boost mode to allow energy transfer from the battery to the load. During periods of high solar insulation, both legs can be operated for power transfer to the load side with respect to the battery state of charge and the load power demands.

1.8   Sizing the PV Panel and Battery Pack for Stand-Alone PV Applications

In this section; the procedure for sizing the PV panel and battery is discussed. Proper sizing of the PV panel and battery system is very important in order to decrease the cost, improve the performance, and increase the lifetime of stand-alone PV systems. System losses, battery charge efficiency, and PV array configuration are issues that affect system sizing.

The main sizing terms are to determine the capacity of the battery pack, that is, the ampere-hour (Ah) capacity of the battery pack, and to determine the number of parallel and series battery cells for selected voltage and capacity values as well as the required

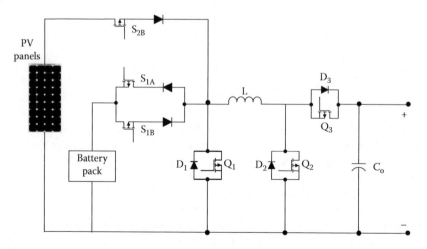

FIGURE 1.78   Positive output multiple input buck–boost converter.

number of PV cells or modules to build a sufficient PV panel. The highest load demand, system losses, and solar radiation should be taken into account for sizing.

According to IEEE Std. 1562™ [99], the PV array should be sized in order to provide sufficient energy to compensate the inefficiencies, overcome the system losses, and mainly to replace the stored energy in the battery (Ah capacity), which is consumed by the load. In order to quickly recharge the battery after low or no solar radiation periods, the PV array might be oversized. The main criterion for PV array sizing is the average daily load consumption in Ah. Since the temperature has a greater effect on the voltage of the PV array than the produced current, the effects of temperature can be neglected [100]. System sizing is based on the PV module's current [99]. On the other hand, for proper charging of the battery, the array voltage should be greater than the terminal voltage of the battery pack. The size of the system also depends on the amount of critical load. A critical system that requires more reliability and high system availability needs to be oversized. In order to appropriate PV array sizing, sun hour, load data, days of autonomy, and solar radiation should be considered.

1.8.1 Sun Hour

The equivalent number of sun hours of standard full solar irradiance at $1000 \, W/m^2$ should be obtained by using the solar radiation data on the array plane to estimate the daily module output [101]. Average available Ah/day production from the PV array can be calculated by multiplying the number of sun hours by the rated value of the module peak current.

1.8.2 Load Calculation

Proper determination of the load is one of the most important factors in sizing a stand-alone PV system. If the estimated load is greater than the actual load demand, the system should be undersized. However, if the estimated load is smaller than the actual load, the system will be oversized. By multiplying each load current to its daily duration and summing the results, it is possible to calculate equivalent daily load consumption. The Ah of the load can be calculated by multiplying the duration to the momentary current if the duration of the momentary load is known. If the duration of the momentary load is not known, the time is assumed to be 1 min and the load consumption is calculated accordingly.

For more sensitive and accurate calculation of the load data, the momentary current, running current, parasitic current, and load coincide, and the maximum and minimum load voltages should be considered [102].

1.8.3 Days of Autonomy

Autonomy is the length of time that the battery of a PV system can supply load demands without receiving energy from the PV array. Therefore, the battery should be sized to support the load during periods of low or nonsolar radiation since the array is sized to replace the Ah capacity of the battery used by the load and system losses [99,102].

Battery capacity directly determines system availability. The larger battery provides a greater number of autonomy days. However, the larger battery size brings a greater risk of sulfation, which involves losing the batteries' capability to hold the charge after being in a fully discharged mode for a long time. The ability of charge holding may be lost due to the crystallization of lead sulfate. In the risk of sulfation, a higher charging rate is required, which consequently increases the size and cost of the PV system. Contrarily, a smaller than required battery size results in more frequent and deeper charge, which reduces battery life and system availability. For critical loads and regions with lower solar irradiance, 1 or 2 weeks of autonomy should be considered. For noncritical loads and regions with higher solar irradiance, 5–7 days of autonomy is reasonable [99].

1.8.4   Solar Radiation

The solar radiation for the month with the worst-case solar radiation at a certain tilt angle should be used if the average load is constant for all months. This value is equivalent to sun hours, which is generally presented in kilowatt-hour $(kWh)/m^2$. This value is used to determine the size of the PV array. If the load is varying during different months, the battery and the array should be sized for each month. For the worst-case scenario, battery autonomy and the lowest array-to-load ratio should be used for the system design. The array-to-load ratio is the average daily PV available in Ah divided by the average daily load in Ah [99].

1.8.5   PV Array Sizing

Solar radiation, array-to-load ratio, load calculation, and system losses should be taken into account for the sizing of the PV array. These factors are explained as follows.

1.8.5.1   PV Module Selection

Some PV modules may have advantages or drawbacks when compared to others, depending on performance under various irradiance conditions and PV array size. Single and polycrystalline silicon and amorphous silicon are the most common PV modules. Various PV modules and their specifications have been explained in Section 1.1.3.

1.8.5.2   System Losses

System losses should be included in the sizing calculations, since they affect the proper sizing of the PV array. Wire and connection losses, parasitic losses (such as controller or sun tracker power requirements), battery charge–discharge efficiency, dust on the array, and inverter and line losses should be considered in the evaluation of system losses as a percentage of the system load [99,102]. Usually, these losses are about 10–20% of the rated system load. The performance of the system may reduce if the losses are not estimated accurately.

1.8.5.3   Determination of the Number of Series-Connected PV Modules

Equation 1.73 is used to calculate the number of modules in series:

$$N_s = \frac{V_{system}}{V_{module}},\qquad(1.71)$$

where N_s is the number of series-connected PV modules, V_{system} is the rated system voltage, and V_{module} is the nominal voltage per module. If the result is not an integer, the number should be rounded up to the nearest integer. Voltage drops should be taken into account in order to ensure that the module voltage is high enough to charge the battery, even in high-temperature conditions.

1.8.5.4 Determination of the Number of Parallel-Connected PV Modules

The array-to-load ratio (A:L) can be determined based on the following criteria:

- A:L can be taken as 1.1:1.2 for noncritical loads and regions with high and incessant solar radiation.
- Higher A:L values such as 1.3:1.4 or higher should be considered in the case of critical loads and regions with lower solar radiation.

The number of module strings in parallel can be calculated as

$$N_P = \frac{L_{DA} \times A{:}L}{\left((1 - SL) \times SH \times I_{mp}\right)}, \tag{1.72}$$

where N_P is the number of parallel strings, L_{DA} is the average daily load, SH represents sun hours, SL represents system losses, and I_{mp} is the module current at maximum power.

1.8.5.5 Design Verification and a Sizing Example

By increasing the number of autonomy days (DA), system availability can be increased if especially the A:L value is a relatively higher value, that is, greater than 1.3. It is a more cost-effective solution to increase the battery capacity instead of increasing the size of the PV array.

For lower values of A:L, the battery recharge time increases. Thus, the size of the array and battery should be adjusted in order to achieve the desired system availability, cost, and system life. For a sizing project, the procedure steps are given in the flowchart of Figure 1.79.

The nominal system voltage should be consistent with the voltage of the battery pack and PV modules. The days of autonomy should be determined as explained in Section 1.8.3. The total daily load can be calculated by using the power–time graph of the load. Determining the battery capacity is the next step. Battery size should be determined so that it can feed total daily load capacity during the days of autonomy (DA). System losses should be considered in Step 5 as described in Section 1.8.5.2. Then, in Step 6, the number of peak sun hours should be determined. This value can be calculated by experimental data collection or using the data available from meteorological agencies. The A:L ratio is determined in Step 7, as previously explained in Section 1.8.4. After calculating the A:L ratio, the current of the PV module at MPP and the nominal operating voltage should be gathered from the manufacturer's catalogue. Subsequently, the worst-case Ah consumption per day should be calculated. Afterwards the percentage of system losses should be converted into a decimal value and subtracted from 1. Next, system losses, sun hours, and the current at MPP should be taken into account. Finally, the number of parallel and series modules can be calculated in Steps 12 and 13. In Step 14, the total number of required modules is calculated.

As an example of the application of sizing procedure, a telecommunication system powered by a PV system is explained [99].

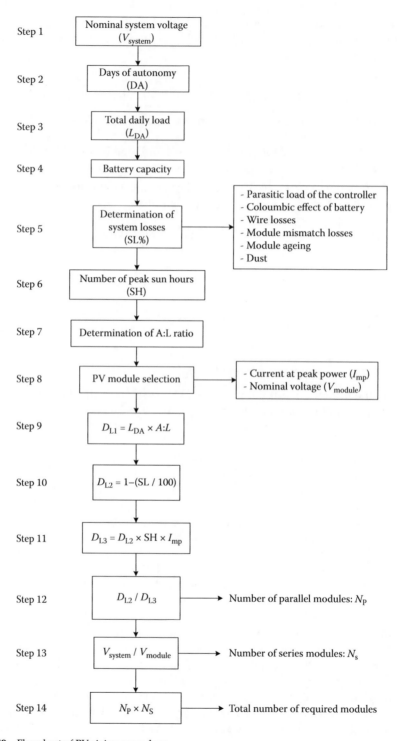

FIGURE 1.79 Flowchart of PV sizing procedure.

The nominal system voltage is selected as 48 V. Since telecommunication systems require high reliability, the days of autonomy can be selected as 2 weeks, which corresponds to the number of days that the batteries can survive feeding the system without being recharged from the PV array. The total daily load based on measurements is taken as 17.8 Ah/day for this example:

$$17.8 \, \text{Ah/day} \times 15 \, \text{day} = 267 \, \text{Ah}. \tag{1.73}$$

Selecting the battery capacity higher than this value will increase the availability of the system. On the other hand, during the days of autonomy, there might be a possibility that the actual load exceeds the total daily load. Therefore, 400 Ah of battery capacity would be enough.

The estimated system losses are given in Table 1.4. The number of peak sun hours can be found from the NREL Solar Radiation Data Manual for the specific location. In this example, the number of peak sun hours is assumed to be 4 h. For higher system availability, the A:L ratio can be selected equal to 1.7.

Mitsubishi's PV-MF110EC3 110 W solar modules are selected. This module has 24 V nominal voltage, and maximum power is obtained for current and voltage values at $I_{mp} = 6.43 \, \text{A}$ and $V_{mp} = 17.1 \, \text{V}$, respectively.

The calculation based on steps of the flowchart should be followed in order to find the number of parallel and series PV modules:

$$D_{L1} = L_{DA} \times \text{A:L} = 17.8 \times 1.7 = 30.26 \, \text{Ah/day}, \tag{1.74}$$

$$D_{L2} = 1 - \frac{\text{SL}}{100} = 1 - \frac{38}{100} = 0.62, \tag{1.75}$$

$$D_{L3} = \left(1 - \frac{\text{SL}}{100}\right) \times \text{SH} \times I_{mp} = 0.62 \times 4 \times 6.43 = 15.94 \, \text{Ah/day}, \tag{1.76}$$

where D_{L1} is the total daily load multiplied by the array to load ratio (A:L), D_{L2} is a parameter to consider all of the system losses, and D_{L3} is a parameter that corresponds to the reduced power capacity due to losses.

TABLE 1.4

Percentages of the Estimated System Losses

Type of Losses	%
Parasitic load of the controller	3
Coloumbic effect of battery	12
Wire losses	3
Module mismatch losses	3
Module aging	7
Dust	10
Total percentage of losses	38

Dividing Step 9 by Step 11 yields

$$N_P = \frac{30.26\,\text{Ah/day}}{15.94\,\text{Ah/day}} = 1.9 \approx 2, \tag{1.77}$$

which is the number of required parallel PV modules. Then for the number of series PV modules

$$N_S = \frac{V_{\text{System}}}{V_{\text{Module}}} = 48/24 = 2. \tag{1.78}$$

Therefore, the total number of PV modules required for this application is 4.

1.9 Modern Solar Energy Applications

1.9.1 Residential Applications

PV systems can be utilized to supply the power demands of a residential unit or a micro-grid neighborhood. The PV system can be implemented in the roofs of residential units. This reduces the space required for PV panels and components, thus eliminating direct structure expenses [61,103]. One of the economical advantages of this system is that, according to the power profile of the household and the tariff of the local power distributor, grid power can be bought or excess power can be sold to the utility network.

The PV panels may not be capable of sustainable satisfaction of load demands in low or noninsulation conditions. Thus, energy buffers or storage devices such as battery packs are required. When the available power from the PV panels is more than the power demand of the user, excess power can be used to charge the battery packs. Contrarily, batteries can help the PV system when the power demand of the user is more than the available power from PV panels. The system can also be connected to the grid in order to draw or inject power from/to the network in order to eliminate mismatches of generated, consumed, and stored energies.

The power profile of the resident, PV power generation characteristics, and utility electricity tariff should be considered to build an optimum and cost-effective power management strategy.

A PV system consisting of PV panels, power conditioners, and a storage device is shown in Figure 1.80 with household loads and grid connection. A DC/DC converter is required to track the MPP, and a bidirectional power electronic interface, which is able to operate as a DC/AC inverter or a AC/DC rectifier, is required in order to charge the battery pack from grid or utility and discharge the battery to the load or grid.

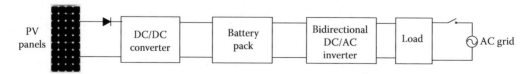

FIGURE 1.80 Power circuit of a residential PV system.

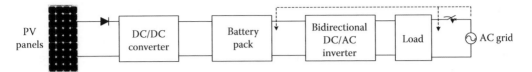

FIGURE 1.81 Operation mode 1 for the proposed residential system.

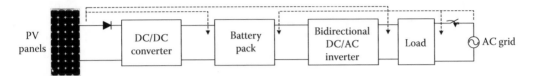

FIGURE 1.82 Operation mode 2 for the proposed residential system.

There are four operation modes for the system. These operation modes are described in Figures 1.81 through 1.84.

The following factors are considered for the operation of the proposed system [103]:

- *Power profile of the home*: The power drawn from the utility should be minimized and excess power should be injected to the network during the peak-load period of the utility. The power profile may vary from region to region, case by case, and depends on the people in the household.

- *PV panel generation characteristics*: Insulation level and solar path change with time for a roof-mounted PV panel. For best efficiency, sun tracking systems can be employed; however, this will increase the system cost. PV power availability by time should be considered for optimum power management of the proposed system.

- *Peak leveling of the network*: A typical power-leveling policy for the grid network can be to sell power to the utility during peak-load periods when the cost of a kWh is relatively more expensive. During off-peak periods of the utility, some power can be drawn for restoration in order to reshape the power level [104].

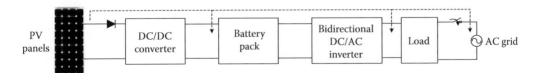

FIGURE 1.83 Operation mode 3 for the proposed residential system.

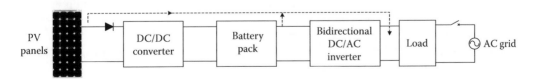

FIGURE 1.84 Operation mode 4 for the proposed residential system.

Mode 1—Off-peak load period: From midnight to sunrise, there is no available PV power. In this period, grid power is used for load demands and recharging the batteries. This mode of operation is depicted in Figure 1.81. In this mode of operation, there is no power flow from PV panels either to the battery pack or to the AC loads. Only grid power is supplied to the loads, and the battery pack is charged if needed.

Mode 2—Low-insulation period: Since the sun starts rising, PV power starts increasing; however, it is not yet sufficient to meet the load demands and charge the batteries. In this period, grid power should be used to supply the insufficient power requirements. This mode of operation is depicted in Figure 1.82. In this mode of operation, both PV panels and the grid supply power to the load. At the same time, the battery pack is charged using power from the PV panels and using grid power, if needed. However, power from PV panels has the priority to charge the battery pack since it is more economical.

Mode 3—High-insulation period: PV power is greater than load demand from late morning till middle evening. Thus, batteries are recharged in this period and excess power can be injected to the utility grid. This mode of operation is depicted in Figure 1.83. In this mode of operation, power flow direction is only from the PV panels to the battery back, load, and AC grid. No power is consumed from the battery pack or the AC grid for the loads, since the PV power is sufficient for load power requirements. It may also charge the battery pack and additionally some power can be supplied to the grid if excess power is available.

Mode 4—Discharge period: During this operation mode, from late evening till midnight, the load power is supplied by discharging the batteries as well as the available PV power at the beginning. Subsequently, the entire power requirement is satisfied by batteries as the power of the PV panel gradually decreases to zero. This mode of operation is depicted in Figure 1.84. In this mode of operation, power flow direction is from the PV panels and battery pack to the load side.

Figure 1.85 shows the power flow of the PV panel, battery pack, load, and grid in four different operation modes described in Figures 1.81 through 1.84.

It is a cost-effective solution to operate the proposed system according to the daily operation pattern, given in Figure 1.85. The battery pack is being charged during the off-peak period of the grid and the kWh cost from the grid is saved, since the battery supplies the peak-load power of the residential unit. Through bidirectional power meters, the excess power can be sold to the network, if the power policy permits.

The configuration of the proposed PV energy storage system is shown in Figure 1.86, in which the DC/DC converter is responsible for MPP tracking and the bidirectional converter of the battery is a single-phase full-bridge converter, which is capable of operating both DC/AC inverter and AC/DC converter modules.

Various parts of the proposed system, boost converter, MPPT algorithm, bidirectional DC/AC–AC/DC converter, and power conditioner circuitry, are explained in the following subsections.

1.9.1.1 Boost Converter

The boost converter between the PV panels and the battery pack is responsible for controlling and adjusting the operating point of the PV panels. The configuration of the boost converter, PV panel, and batteries is shown in Figure 1.87.

In steady-state conditions, the voltage of the C_p capacitor reaches the output voltage of the PV panel. Therefore, it does not draw any current.

$$I_{PV} = I_C. \tag{1.79}$$

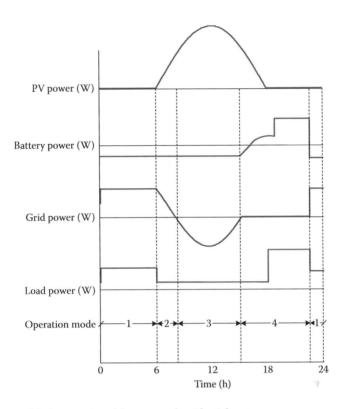

FIGURE 1.85 Pattern of daily operation of the proposed residential system.

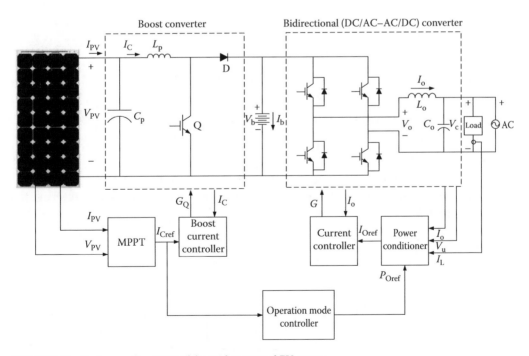

FIGURE 1.86 System configuration of the grid connected PV system.

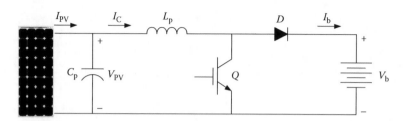

FIGURE 1.87 The boost converter circuit.

The PV panel output current (I_{PV}) can be controlled by controlling the converter current (I_C). This can be done by current mode control of the boost converter.

1.9.1.1.1 State-Space Model of the Boost Converter

Using the averaging method, two different circuits for the on and off positions of the switch can be derived and averaged. When the switch is in the "on" position, the circuit is a parallel CL circuit, and

$$L_p \frac{dI_C}{dt} = V_{PV}. \tag{1.80}$$

When the switch is off, it is an open circuit; thus the inductor current charges the battery.

$$L_p \frac{dI_C}{dt} = V_{PV} - V_b. \tag{1.81}$$

Equations 1.82 and 1.83 can be averaged based on the duty cycle, d, as

$$L_p \frac{dI_C}{dt} = V_{PV} - (1 - d)V_b = V_{PV} - V_b + dV_b. \tag{1.82}$$

The block diagram given in Figure 1.88 is the implementation of Equation 1.84.

1.9.1.1.2 Current Controller

A proportional (P) or a proportional-integral (PI)-type controller can be applied to control the current of the boost converter. The control scheme of the boost converter with a proportional control is shown in Figure 1.89.

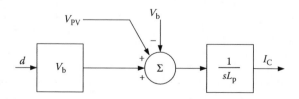

FIGURE 1.88 The implementation block diagram of Equation 1.84.

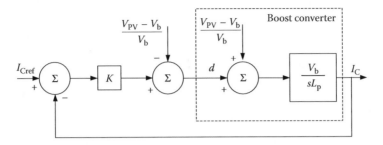

FIGURE 1.89 Current control scheme of the boost converter.

The equivalent transfer function of the system in Figure 1.89 can be written as

$$\frac{I_C}{I_{Cref}} = \frac{KV_b/L_p}{s + KV_b/L_p}. \tag{1.83}$$

The transfer function can be rewritten as a function of the bandwidth as:

$$\frac{I_C}{I_{Cref}} = \frac{\mu}{s + \mu}, \tag{1.84}$$

where μ is the bandwidth of the control loop. In order to decrease the rise time of the system, the bandwidth should be increased. Therefore, for greater values of the proportional gain (K), the system will have a faster response. However, the bandwidth of the system should be limited to be less than the switching frequency of the controller. The rise time and the bandwidth relationship can be approximated as

$$t_r \cong \frac{0.35}{\mu}. \tag{1.85}$$

1.9.1.2 MPPT Control of the PV Panel

The incremental conductance-based MPPT method, details of which were described earlier in Section 2.5.1, is used in this example.

For better performance and maximum efficiency, the increment of the conductance should be varied based on the distance between the MPP and the actual operating point. For greater distances the increment should be greater for faster response time. For smaller distances, the increment should be less for smaller steady-state error. Based on this idea, the increment or decrement of the reference current should be adjusted based on

$$\Delta I_{ref} = \pm\beta \left(\frac{\Delta V}{\Delta I} + \frac{V}{I} \right), \tag{1.86}$$

where ΔI_{ref} is the adjustment of the reference current and β is the scaling factor. The sign of the change in the reference current is determined based on whether the current operating point is on the left- or right-hand side of the MPP.

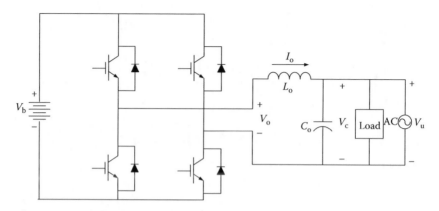

FIGURE 1.90 Bidirectional inverter/converter.

1.9.1.3 Bidirectional Inverter/Converter

The bidirectional inverter/converter should deal with the active power flow from/to batteries to/from the AC load or grid. The circuit configuration of the bidirectional converter is shown in Figure 1.90.

The power of the PV panels and batteries is transferred to the load and utility grid if the power flow direction is from the DC bus to the AC bus. If the power flow direction is from AC side to DC side, this means that the grid power is used to recharge the batteries. Meanwhile, the load is also powered by grid power.

In Figure 1.90, V_c is the capacitor voltage, which is equal to the grid voltage V_u, I_o is the current of the inductance, which is equal to the converter current, and V_o is the output voltage of the converter. PWM signals are applied to the switch gates, which are produced by comparing a triangle carrier signal (-1 to 1) with the control voltage v_{ctrl}. Therefore, the output voltage may alternate between $+V_b$ and $-V_b$.

$$V_o = v_{ctrl}V_b. \tag{1.87}$$

From this equation, the inductance current can be written as

$$L_o \frac{di_o}{dt} = V_o - V_c = v_{ctrl}V_b - V_u, \tag{1.88}$$

where

$$V_c = V_u. \tag{1.89}$$

The block diagram of Equation 1.90 is depicted in Figure 1.91.

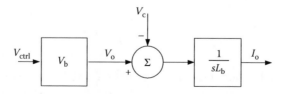

FIGURE 1.91 Block diagram implementation of Equation 1.90.

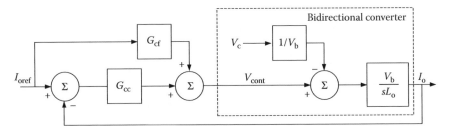

FIGURE 1.92 Block diagram of the current control scheme of the bidirectional converter.

1.9.1.3.1 Current Control for the Bidirectional Inverter/Converter

The converter current should be controlled in order to control the power transfer if the grid voltage is assumed to be constant. The voltage of the capacitance is a disturbance input for the system. To eliminate the effect of the disturbance, this value can be subtracted from the control signal. However, since the disturbance signal is alternating periodically, adding or subtracting this value may cause stability problems. Therefore, a feed-forward controller can be used to damp the effect of the disturbances. Figure 1.92 shows the block diagram of the closed-loop controller and the bidirectional controller.

The transfer function of the capacitor voltage to the inductor current can be written as

$$\frac{I_o(s)}{V_c(s)} = \frac{1/L_o}{s + G_{cc}V_b/L_o} \tag{1.90}$$

and the transfer function from the reference output current to the actual output current would be

$$\frac{I_{oref}(s)}{I_o(s)} = \frac{V_b \, (G_{cc} + sL_o/V_b)/L_o}{s + G_{cc}V_b/L_o} = \frac{s + G_{cc}V_b/L_o}{s + G_{cc}V_b/L_o} \approx 1, \tag{1.91}$$

where

$$\frac{sL_o}{V_b} = G_{cf}. \tag{1.92}$$

From these transfer functions, it can be observed that in order to decrease the disturbance effect, the G_{cc} gain should be increased. On the other hand, the transfer function from the reference current to the actual converter current is regardless of G_{cc} gain. Therefore, converter current will track the reference value successfully when the G_{cc} value is high enough.

1.9.1.4 Power Conditioner

The power conditioner is an upper loop for the current controller of the bidirectional converter. Therefore, the power conditioner is the unit determining the value of the reference current. The main task of the power conditioner is to determine the active power flow from the DC bus to the utility grid or from the utility grid to the DC bus. This reference active power value is the input for the power conditioner. While it is performing this task, it also forces the converter to supply the required harmonic currents and reactive power of residential loads. The power factor of the system becomes unity if it is measured from the grid

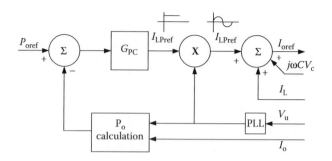

FIGURE 1.93 Block diagram of the power conditioner.

side. This means that the power conditioner works as an active filter. The block diagram of the power conditioner is shown in Figure 1.93.

The phase-locked-loop block shown in Figure 1.93 is responsible to perform the necessary measurements for grid synchronization. This block internally calculates the phase angle and the frequency of the input signal.

The output active power over a period can be calculated by

$$P_{\mathrm{o}} = \frac{1}{T} \int\limits_{t}^{t+T} i_{\mathrm{o}}(t) v_{\mathrm{u}}(t) \, \mathrm{d}t. \tag{1.93}$$

The reference output active power is compared to the actual active power. The error is multiplied by G_{pc}, which is the proportional-integral controller to calculate I_{LPref}. Then it is multiplied to a sine wave, which is in phase with the grid. Therefore, I_{LPref} is the factor for the reference active power. The load current and capacitance current values are added to this current in order to take reactive power demand and harmonics into account. As a result, the converter reference will include the active power of the user as well as the reactive power and the harmonic currents of the load. The controller for the power conditioner, G_{pc}, is a PI controller given by

$$G_{\mathrm{pc}} = \frac{K_{\mathrm{p}}s + K_i}{s}. \tag{1.94}$$

The complete system, in Figure 1.86, is simulated to observe the dynamic performances of the system. The number of series cells is 112, and the number of parallel cells is 30 in a module. In one array, the number of series modules is 6 and the number of parallel modules is 7. The short-circuit module current and the open-circuit module voltage are $I_{\mathrm{SC}} = 1.2\,\mathrm{A}$ and $V_{\mathrm{OC}} = 14\,\mathrm{V}$, respectively. The series array resistance is $R_{\mathrm{s}} = 10\,\mathrm{m\Omega}$. In the PI-based current controller of the MPPT system, the proportional gain is 20. The PV array current, voltage, and power are obtained as given in Figures 1.94 through 1.96.

Due to the I–V characteristics of the PV array, as the current drawn from the PV array increases, the voltage of the PV array decreases. The PV array power is shown in Figure 1.96, where it starts increasing from zero to the maximum as the PV array current increases.

The reference PV array current, which is determined by the MPPT model, is given in Figure 1.97. Based on Figures 1.94 and 1.97, it is seen that the reference PV array current is successfully tracked by the boost converter.

The PV array's $\mathrm{d}P/\mathrm{d}I$ ratio is shown in Figure 1.98. From Figure 1.98 it is seen that the $\mathrm{d}P/\mathrm{d}I$ ratio decreases as the operating point gets closer to the MPPT.

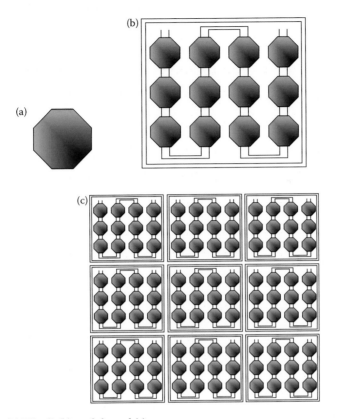

FIGURE 1.2 (a) PV cell, (b) module, and (c) array.

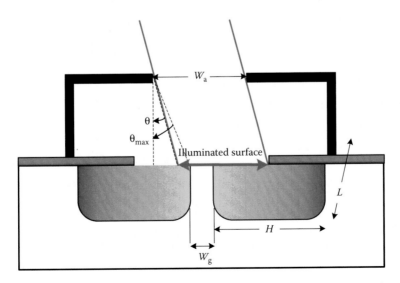

FIGURE 1.13 The structure of the sun tracking system based on angle of the light using two photodiodes.

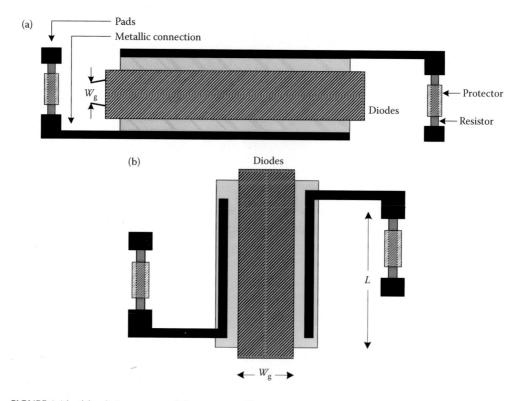

FIGURE 1.14 (a) *x*-Axis sensor module structure; (b) *y*-axis sensor module structure.

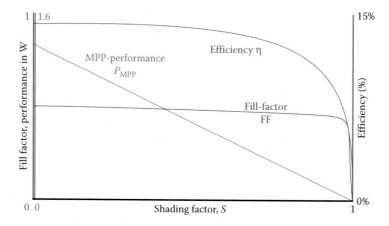

FIGURE 1.37 P_{MPP}, FF, and η variations versus the shading factor *S*.

FIGURE 1.104 SAUV II. (Courtesy of Solar Powered Autonomous Underwater Vehicle (SAUV II), Falmouth Scientific, Inc., available online: http://www.falmouth.com/DataSheets/SAUV.pdf)

FIGURE 1.110 (a) Solar-powered aircraft. (Modified from Y. Perriard, P. Ragot, and M. Markovic, *IEEE International Conference on Electric Machines and Drives*, pp. 1459–1465, May 2005.)

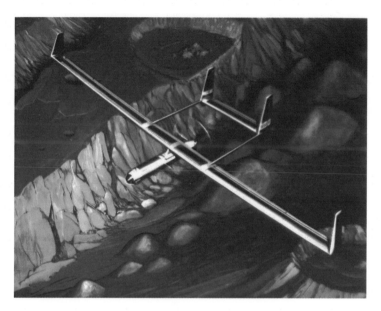

FIGURE 1.112 The solar-powered air vehicle. (Modified from K.C. Reinhardt, et al., *Proceedings of the IECEC 96 Energy Conversion Engineering Conference*, Vol. 1, pp. 41–46, August 1996.)

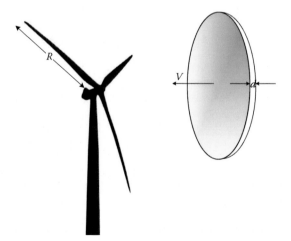

FIGURE 2.1 Kinetic energy of wind.

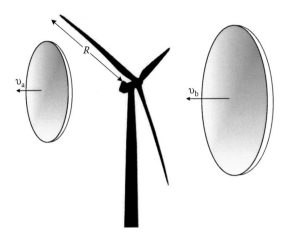

FIGURE 2.4 Wind speed before and after the turbine.

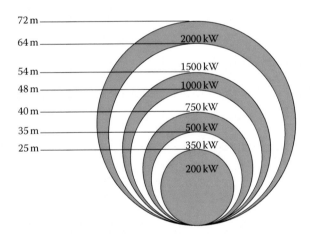

FIGURE 2.10 Turbine output power for different wind turbine diameters.

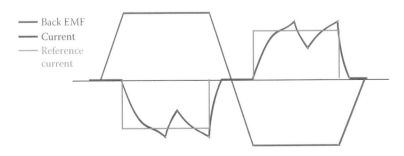

FIGURE 2.15 Back EMF, real current, and reference current waveform of a BLDC machine.

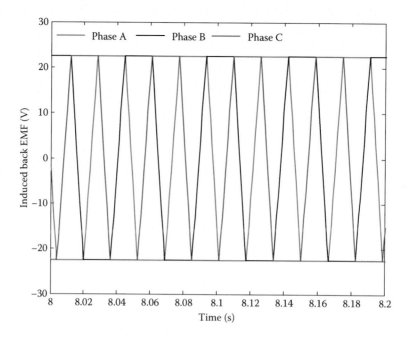

FIGURE 2.21 Back EMF induced by the BLDC phases.

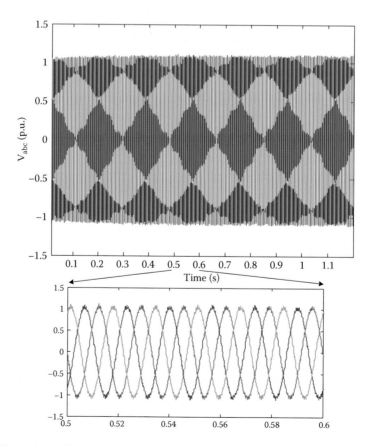

FIGURE 2.57 Three-phase voltage measured from the grid side bus.

FIGURE 3.6 La Rance tidal power station. (From Dam of the tidal power plant on the estuary of the Rance River, Bretagne, France, *image licensed under the Creative Commons Attribution 2.5 License*. With permission.)

FIGURE 3.42 A SeaGen turbine installed at Strangford Lough. (Courtesy of Seaflow and SeaGen Turbines, Marine Current Turbines Ltd., available at: http://www.marineturbines.com, March 2008.)

FIGURE 3.43 SeaGen turbine farm. (Courtesy of Seaflow and SeaGen Turbines, Marine Current Turbines Ltd., available at: http://www.marineturbines.com, March 2008.)

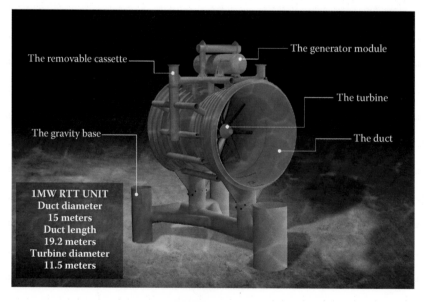

The removable cassette

The generator module

The turbine

The gravity base

The duct

1MW RTT UNIT
Duct diameter
15 meters
Duct length
19.2 meters
Turbine diameter
11.5 meters

FIGURE 3.44 A generating system with a horizontal axis turbine, Rotech Tidal Turbine. (Courtesy of Rotech Tidal Turbine (RTT), Lunar Energy™, Harnessing Tidal Power, available at: http://www.lunarenergy.co.uk/)

FIGURE 3.46 Semisubmersible turbine. (Courtesy of Semi-Submersible Turbine, TidalStream Limited *"The Platform for Tidal Energy,"* available at: http://www.tidalstream.co.uk/)

FIGURE 3.47 Position transformation of SST to its maintenance position. (a) Water is started to be pumped out of the main body, (b) swing arm started to roll, (c) rotors are rising to the water surface, (d) rotors reach a stable position on the sea surface. (Courtesy of Semi-Submersible Turbine, TidalStream Limited *"The Platform for Tidal Energy,"* available at: http://www.tidalstream.co.uk/)

FIGURE 3.48 Comparison of SST with a wind turbine for the same power rating. (Courtesy of Semi-Submersible Turbine, TidalStream Limited *"The Platform for Tidal Energy,"* available at: http://www.tidalstream.co.uk/)

FIGURE 3.50 Stingrays mounted on the seabed. (Courtesy of Engineering Business Ltd., UK, "Stingray tidal stream generator," available at: http://www.engb.com, March 2008.)

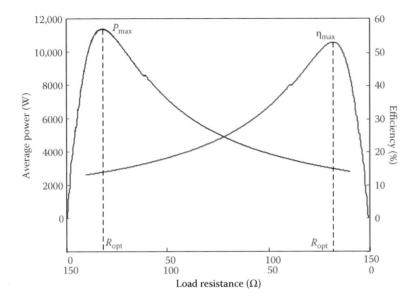

FIGURE 4.34 Average power and efficiency variation versus load resistance.

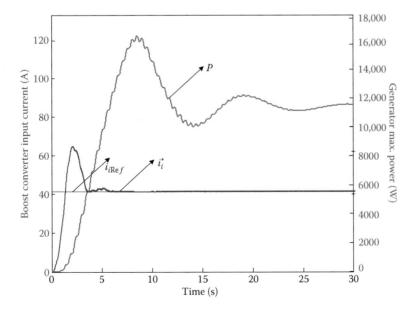

FIGURE 4.38 Reference current tracking and power output for the linear PM generator.

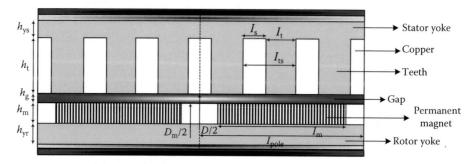

FIGURE 4.53 Cross-sectional view of two poles of the generator.

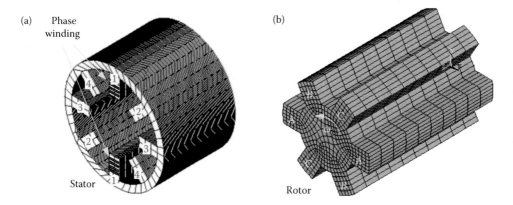

FIGURE 4.57 8/6 pole SRG cross-sectional view, (a) stator of the SRG and (b) rotor of the SRG.

FIGURE 4.72 Pelamis WEC device at sea (Ocean Power Delivery Ltd). (Courtesy of R. Henderson, *Renewable Energy*, 31 (2), 271–283, 2006.)

Wave direction

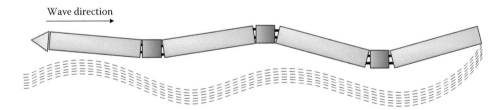

FIGURE 4.73 General layout of the Pelamis WEC device.

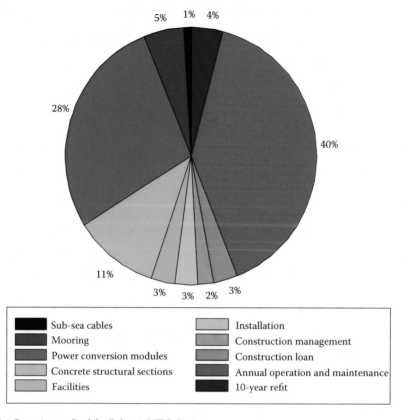

5%   1%   4%

28%

40%

11%

3%   3%   2%   3%

	Sub-sea cables		Installation
	Mooring		Construction management
	Power conversion modules		Construction loan
	Concrete structural sections		Annual operation and maintenance
	Facilities		10-year refit

FIGURE 4.76 Cost pie graph of the Pelamis WEC device.

FIGURE 4.79 Front view of the Wave Dragon with wave reflectors in the sides and the reservoir in the middle. (Courtesy of Wave Dragon ApS, available online at http://www.wavedragon.net/)

FIGURE 4.82 WSE position in storm protection mode. (Courtesy of M. Kramer, "The wave energy converter: Wave Star, a multi point absorber system," Technical Report, Aalborg University and Wave Star Energy, Bremerhaven, Denmark.)

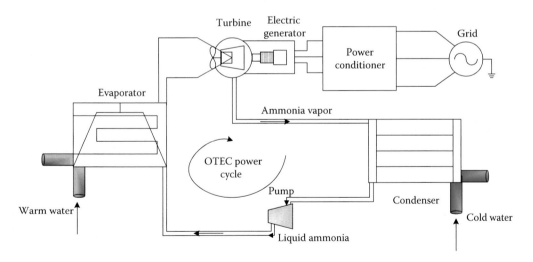

FIGURE 5.1 A general overview of an OTEC system.

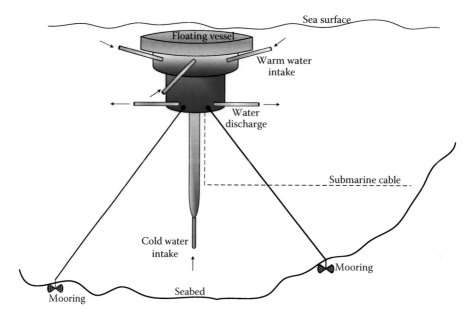

FIGURE 5.2 An offshore OTEC application.

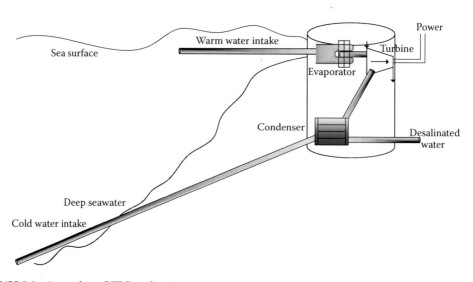

FIGURE 5.3 A nearshore OTEC application.

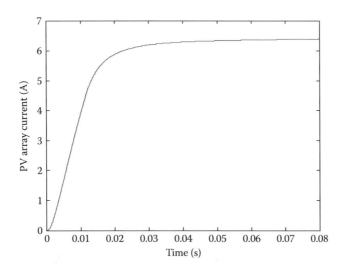

FIGURE 1.94 PV array current.

1.9.2 Electric Vehicle Applications

In this section, details regarding the electrical implementation and addition of solar generators to the electric vehicle (EV) are presented.

A typical simple configuration of a solar EV is shown in Figure 1.99 [59,105]. In this configuration, the outputs of individual PV panels are combined and connected to a common DC/DC converter for MPPT. The MPPT converter is then connected to the battery pack. The DC bus voltage is then converted to AC voltage through an inverter to control the traction machine.

In order to increase the efficiency, instead of connecting solar panels using only one power tracker, separate maximum power trackers can be used, since each solar panel may have

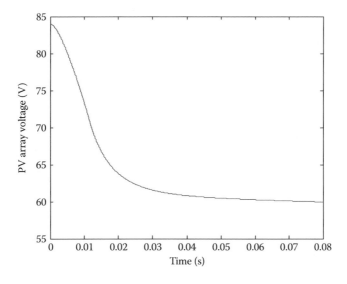

FIGURE 1.95 PV array voltage.

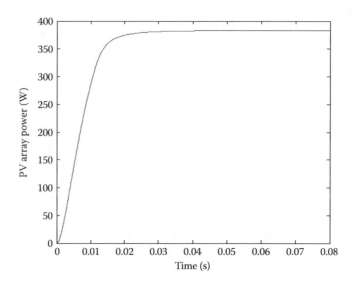

FIGURE 1.96 PV array power.

different temperature and illumination conditions with different maximum power levels (Figure 1.100).

A 60–100 V battery bank is reasonable for a small-sized EV. In fact, most manufacturers recommend that the solar panel nominal voltage should be at least 5% lesser than the battery voltage for enough boosting and charging of the batteries. This sets a range for the nominal voltage of the main panel [59].

Energy optimization and efficiency can be reached with the use of regenerative braking and ultracapacitors. A regenerative braking system is able to convert kinetic energy to electricity, and an ultracapacitor bank acts as a high-power source, improving vehicle performance and battery bank life.

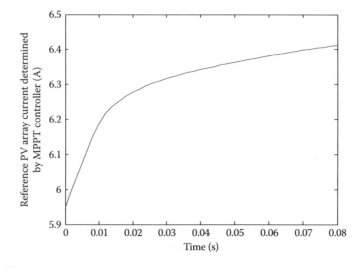

FIGURE 1.97 Reference PV array current.

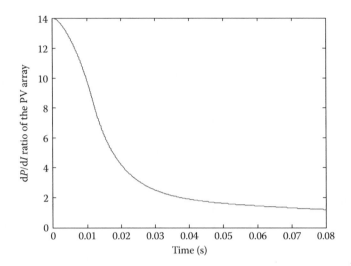

FIGURE 1.98   dP/dI ratio of the PV array.

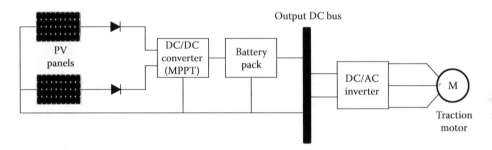

FIGURE 1.99   Block diagram of a solar EV.

The electrical power system of a solar EV is shown in Figure 1.101, which consists of two different DC voltage buses for the traction and auxiliary power requirements [106]. This topology also contains an external outlet to charge the batteries in the case of a plug-in hybrid EV. Ultracapacitors can also be used to capture the braking energy at higher efficiencies and may help overcome the transient conditions of the vehicle, such as acceleration or hill climbing, due to their faster response times. Utilizing ultracapacitors will also reduce the

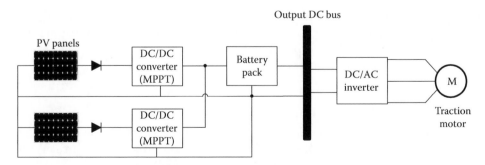

FIGURE 1.100   Distributed solar panel and power tracker configuration.

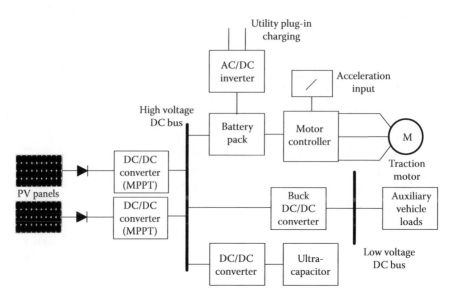

FIGURE 1.101 Block diagram of the electrical system of a solar hybrid EV.

stress on the batteries, battery size, and cost, and may help increase the lifetime of the battery pack.

In this topology, an AC machine or a DC brushless motor can be used for higher efficiency and less maintenance. The auxiliary power requirement of the vehicle (such as for headlights and backlights, wiper, air conditioner, car stereo, hydraulics, etc.) is supplied through the low-voltage DC bus. The output of the peak power trackers are connected to the high-voltage DC bus. A DC/DC converter supplies the low-voltage auxiliary devices. Batteries are connected to the high-voltage DC bus to supply sustained driver demands in the low insulation or shaded periods. An external charger for the batteries can also be considered to charge the vehicle batteries from the utility grid when the car is parked. The external charger can be a controlled AC/DC converter or an uncontrolled AC/DC converter followed by a DC/DC converter to obtain the appropriate charging voltage for the batteries.

The power is supplied from the PV panels and the battery pack. The power flow through a solar vehicle concept can be described as [107]

$$P_{\text{supplied}}(t) = P_s(t) + I_B(t)E_B[I_B(t)], \qquad (1.95)$$

where the solar power $P_s(t)$ and the battery current $I_B(t)$ are functions of time. The battery voltage $E_B[I_B(t)]$ is also a function of the battery current. This is due to the internal resistance and voltage–current characteristic of the batteries.

The power flow through the solar EV is illustrated in Figure 1.102.

1.9.2.1 Solar Miner II

The Solar Miner II [108] uses solar cells made of single-crystalline silicon, rated at 14% efficiency. Their array produces about 900 W at 120 V under full sun. The array is divided into three subarrays. Two of them produce 144 V at approximately 3.1 A, and the third subarray produces 144 V at approximately 1.5 A. Each subarray drives a DC/DC converter.

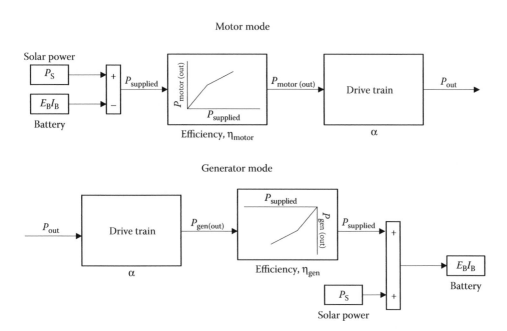

FIGURE 1.102   Power flow diagram in a solar EV.

The outputs from the power trackers are connected in parallel to a main bus at 96 V. The power trackers keep the solar array operating at its highest efficiency by conditioning the input and maintaining the MPP of the solar cells. The output controller monitors the batteries' state of charge to ensure maximum power transfer to the battery pack. A 150 kg pack of Delphi lead-acid battery pack is connected to the 96 V bus to run the motor and various electrical peripherals. The motor and the motor controller are developed by New Generation Motors, Inc.

The electrical layout of the Solar Miner II is shown in Figure 1.103.

The motor is an axial flux, three-phase, DC brushless machine that operates at a nominal 96 V. The air gap between the rotor and the stator is adjustable for torque and speed optimization. The motor can operate as a generator during regenerative braking [108].

A DC/DC step-down converter converts 96 V to 12 V for the horn, ventilation fans, gap-adjust motor, and DSP. A DSP monitors the electrical systems and computes power consumption. It also sends the data out through a radio modem to a remote computer in the support vehicle. Dual, eight-channel, 10-bit A/D converters integrated on the DSP read the Hall-effect sensor to measure current. Four 8-bit D/A converters set the speed of the motor and run the driver's LCD speedometer display. A 25 A Hall-effect sensor, model IHA-25, manufactured by F. W. Bell, measures the solar cells current being fed to the batteries. Array wires are wrapped twice through the current sensor to decrease error and double the accuracy. The DSP's 10-bit A/D converter converts this signal with a resolution of 10 mV that represents 100 mA in sensing current. Another Hall-effect sensor, rated at 100 A, measures the motor current with 500-mA resolution.

The motor controller's integrated 8-bit A/D converter samples the voltage on the battery bus at a 10 times/s rate. It has 1 V accuracy. The motor controller also measures the wheel's revolutions per minute and temperature.

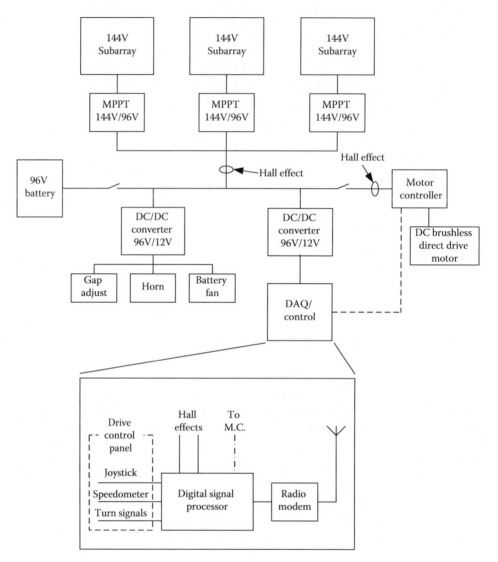

FIGURE 1.103 Layout of the Solar Miner II electrical system.

The DSP calculates the distance and speed in real time. It also calculates the power from the solar array and the power to the motor from respective currents and bus voltage. Subsequently, it calculates the total generated and used energy.

During periods of stationary charging, the solar array can be oriented toward the sun to extract maximum power.

1.9.3 Naval Applications

Solar energy has many applications in the naval and ocean industry. Solar energy is especially used for autonomous underwater or undersea vehicles (AUVs). AUVs have a unique capability in that they are able to transit in the ocean in three dimensions following a predefined path [109,110]. In this section, a few examples of naval applications powered by solar energy are introduced.

1.9.3.1   Solar-Powered AUV

In this subsection, the electrical power structure of a long-endurance solar-powered autonomous underwater vehicle (SAUV) is described [109].

The SAUV II (Figure 1.104) is a solar-powered AUV designed and developed by the U.S. Office of Naval Research. It is designed for long-endurance missions such as monitoring, surveillance, or station keeping where real-time bidirectional communications to shore are critical. The SAUV II operates continuously using solar energy to recharge its lithium ion batteries during daylight hours, to manage energy consumption over a long (up to several months) period of time. This underwater survey concept vehicle has an environmental monitoring tool that is autonomous and mobile for sampling in shallow water over large survey areas.

The SAUV II is easy to deploy and operate due to its relatively small size (5 m length × 1.1 m width × 0.5 m height; 1.0 m^2 topside solar array panel; 200 kg total overall weight in air). The vehicle can be launched from a boat ramp onshore and can independently dip to a predetermined area of interest. The vehicle may be either preprogrammed before the survey or programmed during the survey via radio frequency (RF) communication or iridium satellite phone. The mission planner is a Microsoft-Windows-based graphical user interface that renders intuitive programming and mission viewing. The mission controller is based on a PC-104-embedded single board processor that is capable of interpreting high-level commands to control the vehicle's behavior during its mission.

A solar energy subsystem test-bed has been developed to make empirical measurements of subsystem performance and to investigate efficiencies, capabilities, and limitations of the components relative to their application on AUVs [110]. The test-bed system consists of solar panels (arrays), a battery system, a charge controller, a battery monitoring system, and current and voltage sensors to monitor energy.

The measurement data are directly acquired by an on-board microprocessor, and stored and transmitted to a computer system in the laboratory by a direct line initially and later by wireless telemetry. This enables testing of the telemetry system and comparing data received via telemetry, sent by the microprocessor. PC software on the laboratory computer

FIGURE 1.104   (See color insert following page 80.) SAUV II. (Courtesy of Solar Powered Autonomous Underwater Vehicle (SAUV II), Falmouth Scientific, Inc., available online: http://www.falmouth.com/DataSheets/SAUV.pdf)

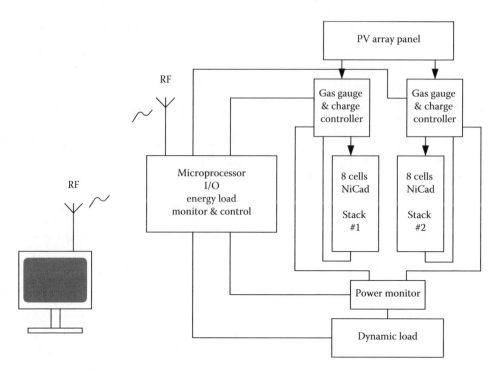

FIGURE 1.105   Solar energy system test-bed block diagram.

monitors, analyzes, and displays the data in real time. It also provides the capability to analyze the data statistically over longer periods of time in order to develop histograms to assist later planning strategies for energy utilization and battery charging. A variable and controllable load is integrated into the system, which allows for reasonable simulation of battery discharging, simulating expected AUV energy usage profiles. A block diagram of the system is shown in Figure 1.105.

The amount of energy collected by the solar arrays is measured as a function of time. The efficiency of the charge/discharge system is determined by measuring the amount of energy out of the batteries compared to the measured values in the battery. The important parameter calculated from measurements is the energy efficiency (i.e., discharge [in watt-hours]/charge [watt-hours]). Generally, the charging and discharging efficiency of typical batteries is around 90% since most of the batteries generate some heat or require some heat.

1.9.3.2 *Solar-Powered Boat,* Korona

The solar-powered boat (*Korona*) is shown in Figure 1.106 [112]. This boat has an electrical motor that is powered solely by PV cells.

The solar power boat *Korona* is driven by a three-phase asynchronous electrical motor whose energy is solely provided by $9\,m^2$ of PV cells with 10 car batteries as storage medium. The boat body and propeller are especially designed to combine optimum driving characteristics with efficient use of energy [112].

All sensors are connected to the on-board information management system (IMS) of the solar boat. This is shown in Figure 1.107. The lower part represents the energy chain, starting from the left with the PV cells (PV generator) as energy source. A DC/DC converter

FIGURE 1.106 Solar boat *Korona*. (Courtesy of R. Leiner, *Proceedings of Africon*, pp. 1–5, October 2007.)

adjusts the voltage of the cells to about 450 V, which is the voltage of the battery pack, consisting of 10 conventional car batteries. The batteries function as energy storage medium. The propeller electric machine is an asynchronous three-phase motor. A DC/AC inverter converts the DC voltage of the energy storage system to an AC voltage to drive the electric machine [112].

1.9.4 Space Applications

PV modules can provide all the power generation for space systems. Arrays of PV solar cells have been utilized to power spacecraft since 1958 and have been the primary source of power for both commercial and military spacecraft [113].

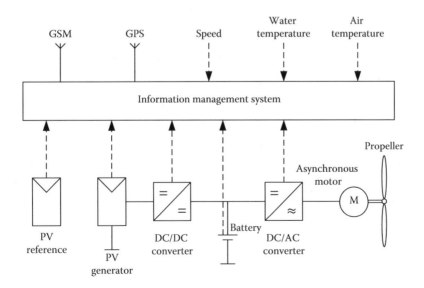

FIGURE 1.107 Block diagram of energy chain and additional inputs to the IMS.

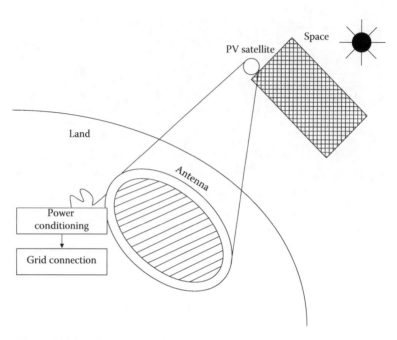

FIGURE 1.108   Concept of the solar power satellite system.

1.9.4.1   Solar Power Satellite Energy System

The concept of the solar power satellite energy system is to place giant satellites, covered with vast arrays of solar cells, in geosynchronous orbit 22,300 miles above the earth's equator [114]. Each satellite is illuminated by sunlight 24 h a day for most of the year. Because of the 23″ tilt of the earth's axis, the satellites pass either above or below the earth's shadow. It is only during the equinox period in the spring and fall that they will pass through the shadow. Therefore, they will be shadowed for less than 1% of the time during the year. The solar cells will convert sunlight to electricity, which will then be transformed to RF energy for transmission through an antenna on the satellite and beam to a receiver site on the earth. It will be reconverted to electricity by the receiving antenna, and the power will be routed into the electric distribution network. Figure 1.108 illustrates the concept of a solar power satellite energy system.

Each satellite has an output equivalent to the current capacity of a large earth-based power plant. The advantage of placing solar cells in space instead of on the earth is that the energy is available 24 h a day, and the total solar energy available to the satellite is four to five times more than the energy available anywhere on the earth. Testing has demonstrated that wireless energy transmission to the earth can be accomplished at very high efficiencies. Tests have also shown that the energy density in the RF beam can be limited to safe levels [114].

1.9.4.2   Pathfinder, the Solar Rechargeable Aircraft

The system level configuration of the electrical power system of the *Pathfinder* is demonstrated in Figure 1.109. The storing and recycling energy system utilizes the fuel cells/electrolyzer. This system provides the sustained propulsion power to aircraft motors

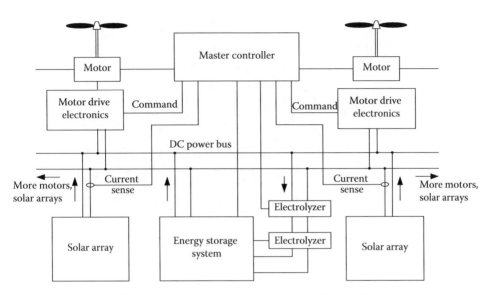

FIGURE 1.109 Schematic of the propulsion system of a solar-rechargeable aircraft.

and night flights [115]. The fuel cell/electrolyzer system is a more suitable candidate than rechargeable battery technologies due to its higher energy conversion efficiency and better cycling capabilities.

1.9.4.3 Solar-Powered Aircraft

The power train of the solar-powered aircraft presented in Figure 1.110a and b consists of solar cells on the wings, an energy storage system, energy management electronics, a brushless DC motor, and propellers [116].

As seen from Figure 1.110b, the solar arrays are connected to the main DC bus through MPPT DC/DC converters. Batteries are charged from the main DC bus. The aircraft propulsion consists of the motor drive system, which is powered from the DC bus through the DC/AC inverter. The auxiliary power requirements are also provided by the main DC bus.

1.9.4.4 Solar Airplane

The general scheme of the airplane electrical power system (power train) is presented in Figure 1.111 [117]. It is composed of solar cells placed on the wings, maximum power point trackers used to maximize the energy obtained by the sun, batteries to store energy for the night flight, DC–AC inverter and electronic management to drive the motor, and finally the motor and propeller.

In the configuration given in Figure 1.111, the solar modules are connected to the individual MPPT DC/DC converters since each module may have different MPPs due to their position differences. Hence, there are two main DC buses for each wing with two other battery packs. Two motors drive the propellers for each wing and these motors are powered by the DC/AC inverters connected to the DC buses. The main feature for the energy management is the efficiency of the power train submodules. The efficiency of the system

(a)

(b)

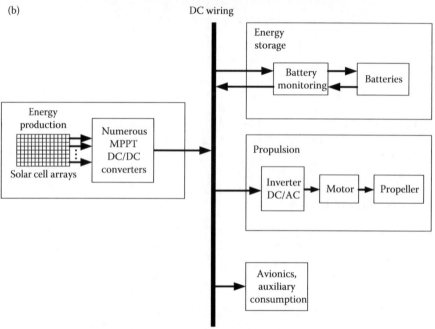

FIGURE 1.110   (See color insert following page 80 for [a].) (a) Solar-powered aircraft and (b) aircraft power train. (Modified from Y. Perriard, P. Ragot, and M. Markovic, *IEEE International Conference on Electric Machines and Drives*, pp. 1459–1465, May 2005.)

depends greatly on the various functional modes: horizontal flight at low altitude, climbing, and horizontal flight at high altitude.

1.9.4.5 *Solar-Powered Unmanned Aerial Vehicle*

The solar-powered unmanned aerial vehicle (UAV) shown in Figure 1.112 is another space application of solar energy. In this concept, the time of year (date) and hour, and the latitude at which the UAV operates determines the charge/discharge time available for the energy storage system and the total amount of solar power available for propulsion. If the available solar energy per wing area is lower, larger wing area should be considered to generate sufficient power to propel the airframe and the masses of the energy storage, fuel cell, and electrolyzer units. In the proposed system, if the power produced by solar arrays is more

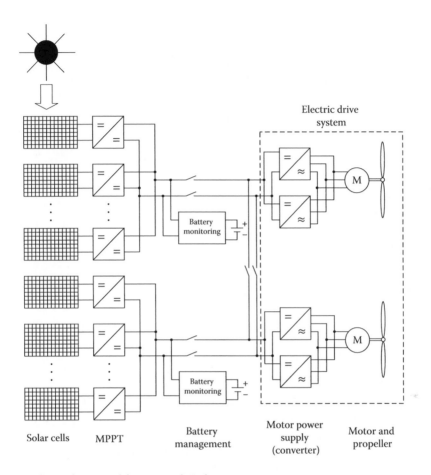

FIGURE 1.111 Electrical system of the proposed airplane.

than the required power for propulsion, the excess energy is used in the electrolyzer to produce oxygen and hydrogen for future use in the fuel cell. When the power required for propulsion is higher than the power produced by solar arrays, fuel cells also supply

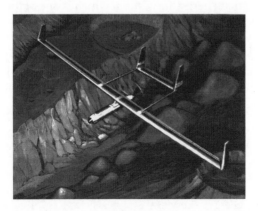

FIGURE 1.112 **(See color insert following page 80.)** The solar-powered air vehicle. (Modified from K.C. Reinhardt, et al., *Proceedings of the IECEC 96 Energy Conversion Engineering Conference*, Vol. 1, pp. 41–46, August 1996.)

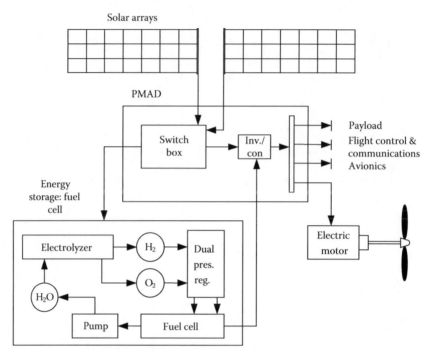

FIGURE 1.113   Power system of the solar-powered UAV.

power to the drive train. The power flow from solar arrays to the vehicle propulsion and to the electrolyzer is determined by a power management strategy. In addition, the fuel cell power is available to balance the power.

The power and propulsion system of the UAV is shown in Figure 1.113.

1.10   Summary

This chapter focuses on solar energy harvesting, which is one of the most important renewable energy sources that has gained increased attention in recent years. Solar energy is plentiful; it has the greatest availability among other energy sources. This chapter deals with *I–V* characteristics of PV systems, PV models and equivalent circuits, sun tracking systems, MPPT systems, shading effects, power electronic interfaces for grid-connected and standalone PV systems, sizing criteria for applications, and modern solar energy applications such as residential, vehicular, naval, and space applications.

References

1. R. Chapo, "Solar energy overview," Ezinearticles.com, December 2008, available at: http://ezinearticles.com
2. C.-J. Winter, R.L. Sizmann, and L.L. Vant-Hull, *Solar Power Plants, Fundamentals, Technology, Systems, Economics*, Springer, New York (ISBN 0-387-18897-5), 1991.

3. R. Sier, "Solar stirling engines," Stirling and Hot Air Engines, available at: http://www. stirlingengines.org.uk/sun/sun.html

4. D. Boehmer, "Overview of flasolar.com," Flasolar.com, available at: http://www.flasolar.com/ overview.htm

5. J. Nelson, *The Physics of Solar Cells*, Imperial College Press, UK, 2003.

6. Atomstromfreie, Greenpeace Energy, "Large-scale photovoltaic power plants range 1–50," available at: http://www.pvresources.com

7. 154 MW Victoria (Australia) Project, "Future Solar Energy Project," available at: http://www.solarsystems.com.au/154MWVictorianProject.html

8. U.S. Department of Energy, Energy Efficiency and Renewable Energy, Solar Energy Technologies Program, "PV physics," available at: http://www1.eere.energy.gov/solar

9. U.S. Department of Energy, Energy Efficiency and Renewable Energy, Solar Energy Technologies Program, "PV systems," available at: http://www1.eere.energy.gov/solar

10. U.S. Department of Energy, Energy Efficiency and Renewable Energy, Solar Energy Technologies Program, "PV devices," available at: http://www1.eere.energy.gov/solar

11. M. van Cleef, P. Lippens, and J.Call, "Superior energy yields of UNI-SOLAR® triple junction thin film silicon solar cells compared to crystalline silicon solar cells under real outdoor conditions in Western Europe," *17th European Photovoltaic Solar Energy Conference and Exhibition*, Munich, October 2001.

12. E. Bum, N. Cereghetti, D. Chianese, A. Realini, and S. Reuonieo, "PV module behavior in real conditions: Emphasis on thin film modules," LEEE-TISO, CH-Testing Centre for PV-Modules, University of Applied Sciences of Southern Switzerland (SUPSI).

13. U.S. Department of Energy, Energy Efficiency and Renewable Energy, Solar Energy Technologies Program, "Solar cell materials," available at: http://www1.eere.energy.gov/solar

14. The Encyclopedia of Alternative Energy and Sustainable Living, available at: http://www. daviddarling.info/encyclopedia/S/AE_silicon.html

15. U.S. Department of Energy, Energy efficiency and renewable energy, solar energy technologies program, "Polycrystalline and single crystalline thin films," available at: http://www1.eere.energy.gov/solar

16. Missouri Department of Natural Resources, "Missouri's solar energy resource," available at: http://www.dnr.mo.gov

17. U.S. Department of Energy, Energy efficiency and renewable energy, solar energy technologies program, "Concentrating solar power," available at: http://www1.eere.energy.gov

18. A. Lasnier and T.G. Ang, *Photovoltaic Engineering Handbook*, Adam Hilger, Bristol and New York, pp. 69–97, 1990.

19. U.S. Department of Energy, Energy efficiency and renewable energy, solar energy technologies program, "Current–voltage measurements," available at: http://www1.eere.energy.gov/solar

20. M.A.S. Masoum, H. Dehbonei, and E.F. Fuchs, "Theoretical and experimental analyses of photovoltaic systems with voltage and current-based maximum power-point tracking," *IEEE Transactions on Energy Conversion*, 17 (4), 514–522, 2002.

21. H. Koizumi and K. Kurokawa, "A novel maximum power point tracking method for PV module integrated converter," *Proceedings of the IEEE 36th Power Electronics Specialists Conference*, pp. 2081–2086, 2005.

22. A.D. Hansen, P. Sørensen, L.H. Hansen, and H. Bindner, "Models for a stand-alone PV system," Risø-R-1219(EN)/SEC-R-12, Risø National Laboratory, December 2000, Roskilde.

23. C. Aracil, J.M. Quero, L. Castañer, R. Osuna, and L.G. Franquelo, "Tracking system for solar power plants," *Proceedings of the IEEE 32nd Annual Conference on Industrial Electronics*, pp. 3024–3029, November 2006.

24. M. Dominguez, I. Ameijeiras, L. Castaner, J.M. Wuero, A. Guerrero, and L.G Franquelo, "A novel light source position sensor," Patent number P9901375.

25. D.A. Pritchard, "Sun tracking by peak power positioning for photovoltaic concentrator arrays," *IEEE Control Systems Magazine*, 3 (3), 2–8, 2003.

26. D.P. Hohm and M.E. Ropp, "Comparative study of maximum power point tracking algorithms using an experimental, programmable, maximum power point tracking test bed," *Proceedings of the IEEE 28th Photovoltaic Specialists Conference*, pp. 1699–1702, 2000.

27. K.H. Hussein, "Maximum photovoltaic power tracking: An algorithm for rapidly changing atmospheric conditions," *IEE Proceedings of the Transmission and Distribution*, 142 (1), 59–64, 1995.

28. D. O'Sullivan, H. Spruyt, and A. Crausaz, "PWM conductance control," *Proceedings of the IEEE 19th Annual Power Electronics Specialists Conference*, Vol. 1, 351–359, 1988.

29. L. Bangyin, D. Shanxu, L. Fei, and X. Pengwei, "Analysis and improvement of maximum power point tracking algorithm based on incremental conductance method for photovoltaic array," *Proceedings of the Seventh International Conference on Power Electronics and Drive Systems*, pp. 637–641, November 2007.

30. D.P. Hohm and M.E. Ropp, "Comparative study of maximum power point tracking algorithm," *Proceedings of the 28th IEEE Photovoltaic Specialists Conference*, pp. 1699–1702, September 2000.

31. S. Silvestre and A. Chouder, "Shading effects in characteristic parameters of PV modules," *Spanish Conference on Electron Devices*, pp. 116–118, January–February 2007.

32. N. Femia, G. Petrone, G. Spagnuolo, and M. Vitelli, "Optimization of perturb and observe maximum power point tracking method," *IEEE Transactions on Power Electronics*, 20 (4), 963–973, 2005.

33. T. Esram, and P.L. Chapman, "Comparison of photovoltaic array maximum power point tracking techniques," *IEEE Transactions on Energy Conversion*, 22 (2), 439–449, 2007.

34. W. Xiao and W.G. Dunford, "A modified adaptive hill climbing MPPT method for photovoltaic power systems," *Proceedings of the 35th Annual IEEE Power Electronics Specialists Conference*, pp. 1957–1963, 2004.

35. A. Al-Amoudi and L. Zhang, "Optimal control of a grid-connected PV system for maximum power point tracking and unity power factor," *Proceedings of the Seventh International Conference on Power Electronics and Variable Speed Drives*, pp. 80–85, 1998.

36. C.-C. Hua and J.-R. Lin, "Fully digital control of distributed photovoltaic power systems," *Proceedings of the IEEE International Symposium on Industrial Electronics*, pp. 1–6, 2001.

37. J.H.R. Enslin and D.B. Snyman, "Simplified feed-forward control of the maximum power point in PV installations," *Proceedings of the IEEE Industrial Electronics Conference*, Vol. 1, pp. 548–553, 1992.

38. A.S. Kislovski and R. Redl, "Maximum-power-tracking using positive feedback," *Proceedings of the IEEE 25th Power Electronics Specialists Conference*, Vol. 2, pp. 1065–1068, 1994.

39. C.Y. Won, D.H. Kim, S.C. Kim, W.S. Kim, and H.S. Kim, "A new maximum power point tracker of photovoltaic arrays using fuzzy controller," *Proceedings of the IEEE Power Electronics Specialists Conference*, pp. 396–403, 1994.

40. J.J. Schoeman and J.D. van Wyk, "A simplified maximal power controller for terrestrial photovoltaic panel arrays," *Proceedings of the 13th Annual IEEE Power Electronics Specialists Conference*, pp. 361–367, 1982.

41. M. Buresch, *Photovoltaic Energy Systems*, McGraw-Hill, New York, 1983.

42. G.W. Hart, H.M. Branz, and C.H. Cox, "Experimental tests of open loop maximum power-point tracking techniques," *Solar Cells*, 13, 185–195, 1984.

43. D.J. Patterson, "Electrical system design for a solar powered vehicle," *Proceedings of the 21st Annual IEEE Power Electronics Specialists Conference*, pp. 618–622, 1990.

44. M.A.S. Masoum, H. Dehbonei, and E.F. Fuchs, "Theoretical and experimental analyses of photovoltaic systems with voltage and current-based maximum power-point tracking," *IEEE Transactions on Energy Conversions*, 17 (4), 514–522, 2002.

45. H.-J. Noh, D.-Y. Lee, and D.-S. Hyun, "An improved MPPT converter with current compensation method for small scaled PV applications," *Proceedings of the IEEE 28th Annual Conference of Industrial Electronics Society (IECON 2002)*, pp. 1113–1118, 2002.

46. K. Kobayashi, H. Matsuo, and Y. Sekine, "A novel optimum operating point tracker of the solar cell power supply system," *Proceedings of the 35th Annual IEEE Power Electronics Specialists Conference*, pp. 2147–2151, 2004.

47. B. Bekker and H.J. Beukes, "Finding an optimal PV panel maximum power point tracking method," *Proceedings of the Seventh AFRICON Conference*, Africa, pp. 1125–1129, 2004.

48. S. Yuvarajan and S. Xu, "Photo-voltaic power converter with a simple maximum-power-point-tracker," *Proceedings of the 2003 International Symposium on Circuits Systems*, pp. III-399–III-402, 2003.

49. R.M. Hiloowala and A.M. Sharaf, "A rule-based fuzzy logic controller for a PWM inverter in photo-voltaic energy conversion scheme," *Proceedings of the IEEE Industry Applications Society Annual Meeting*, pp. 762–769, 1992.

50. C.-Y. Won, D.-H. Kim, S.-C. Kim, W.-S. Kim, and H.-S. Kim, "A new maximum power point tracker of photovoltaic arrays using fuzzy controller," *Proceedings of the 25th Annual IEEE Power Electronics Specialists Conference*, pp. 396–403, 1994.

51. T. Senjyu and K. Uezato, "Maximum power point tracker using fuzzy control for photovoltaic arrays," *Proceedings of the IEEE International Conference on Industrial Technologies*, pp. 143–147, 1994.

52. G.-J. Yu, M.-W. Jung, J. Song, I.-S. Cha, and I.-H. Hwang, "Maximum power point tracking with temperature compensation of photovoltaic for air conditioning system with fuzzy controller," *Proceedings of the IEEE Photovoltaic Specialists Conference*, pp. 1429–1432, 1996.

53. M.G. Simoes, N.N. Franceschetti, and M. Friedhofer, "A fuzzy logic based photovoltaic peak power tracking control," *Proceedings of the IEEE Internationl Symposium on Industrial Electronics*, pp. 300–305, 1998.

54. A.M.A. Mahmoud, H.M. Mashaly, S.A. Kandil, H. El Khashab, and M.N.F. Nashed, "Fuzzy logic implementation for photovoltaic maximum power tracking," *Proceedings of the 9th IEEE International Workshop on Robot Human Interactive Communications*, pp. 155–160, 2000.

55. N. Patcharaprakiti and S. Premrudeepreechacharn, "Maximum power point tracking using adaptive fuzzy logic control for grid connected photovoltaic system," *IEEE Power Engineering Society Winter Meeting*, pp. 372–377, 2002.

56. B.M.Wilamowski and X. Li, "Fuzzy system based maximum power point tracking for PV system," *Proceedings of the 28th Annual Conference on IEEE Industrial Electronics Society*, pp. 3280–3284, 2002.

57. M. Veerachary, T. Senjyu, and K. Uezato, "Neural-network-based maximum-power-point tracking of coupled-inductor interleaved-boostconverter-supplied PV system using fuzzy controller," *IEEE Transactions on Industrial Electronics*, 50 (4), 749–758, 2003.

58. N. Khaehintung, K. Pramotung, B. Tuvirat, and P. Sirisuk, "RISC microcontroller built-in fuzzy logic controller of maximum power point tracking for solar-powered light-flasher applications," *Proceedings of the 30th Annual Conference on IEEE Industrial Electronics Society*, pp. 2673–2678, 2004.

59. A. Abusleme, J. Dixon, and D. Soto, "Improved performance of a battery powered electric car, using photovoltaic cells," *IEEE Bologna PowerTech Conference*, Italy, 2003.

60. T. Hiyama, S. Kouzuma, and T. Imakubo, "Identification of optimal operating point of PV modules using neural network for real time maximum power tracking control," *IEEE Transactions on Energy Conversions*, 10 (2), 360–367, 1995.

61. K. Ro and S. Rahman, "Two-loop controller for maximizing performance of a grid-connected photovoltaic-fuel cell hybrid power plant," *IEEE Transactions on Energy Conversion*, 13 (3), 276–281, 1998.

62. A. Hussein, K. Hirasawa, J. Hu, and J. Murata, "The dynamic performance of photovoltaic supplied dc motor fed from DC–DC converter and controlled by neural networks," *Proceedings of the International Joint Conference on Neural Networks*, pp. 607–612, 2002.

63. X. Sun, W. Wu, X. Li, and Q. Zhao, "A research on photovoltaic energy controlling system with maximum power point tracking," *Proceedings of the Power Conversion Conference*, pp. 822–826, 2002.

64. K. Samangkool and S. Premrudeepreechacharn, "Maximum power point tracking using neural networks for grid-connected system," *International Conference on Future Power Systems*, pp. 1–4, November 2005.

65. L. Zhang, Y. Bai, and A. Al-Amoudi, "GA-RBF neural network based maximum power point tracking for grid-connected photovoltaic systems," *Proceedings of the International Conference on Power Electronics, Machines and Drives*, pp. 18–23, 2002.

66. P. Midya, P.T. Krein, R.J. Turnbull, R. Reppa, and J. Kimball, "Dynamic maximum power point tracker for photovoltaic applications," *Proceedings of the 27th Annual IEEE Power Electronics Specialists Conference*, pp. 1710–1716, 1996.

67. V. Arcidiacono, S. Corsi, and L. Lambri, "Maximum power point tracker for photovoltaic power plants," *Proceedings of the IEEE Photovoltaic Specialists Conference*, pp. 507–512, 1982.

68. Y.H. Lim and D.C. Hamill, "Simple maximum power point tracker for photovoltaic arrays," *Electronics Letters*, 36, 997–999, 2000.

69. "Synthesis, simulation and experimental verification of a maximum power point tracker from nonlinear dynamics," *Proceedings of the 32nd Annual IEEE Power Electronics Specialists Conference*, pp. 199–204, 2001.

70. L. Stamenic, M. Greig, E. Smiley, and R. Stojanovic, "Maximum power point tracking for building integrated photovoltaic ventilation systems," *Proceedings of the IEEE Photovoltaic Specialists Conference*, pp. 1517–1520, 2000.

71. M. Bodur and M. Ermis, "Maximum power point tracking for low power photovoltaic solar panels," *Proceedings of the Seventh Mediterranean Electrotechnical Conference*, pp. 758–761, 1994.

72. T. Kitano, M. Matsui, and D.-H. Xu, "Power sensor-less MPPT control scheme utilizing power balance at DC link-system design to ensure stability and response," *Proceedings of the Seventh Annual Conference on IEEE Industrial Electronics Society*, pp. 1309–1314, 2001.

73. M. Matsui, T. Kitano, D.-H. Xu, and Z.-Q. Yang, "A new maximum photovoltaic power tracking control scheme based on power equilibrium at DC link," *Conference on Record of the 1999 IEEE Industry Applications Conference*, pp. 804–809, 1999.

74. F. Batrinu, G. Chicco, J. Schlabbach, and F. Spertino, "Impacts of grid connected photovoltaic plant operation on the harmonic distortion," *Proceedings of the IEEE Mediterranean Electrotechnical Conference*, MELECON 2006, pp. 861–864, May 2006.

75. V. Quaschning and R. Hanitsch, "Influence of shading on electrical parameters of solar cells," *Conference on Record of the 25th IEEE Photovoltaic Specialists Conference*, pp. 1287–1290, May 1996.

76. S.B. Kjaer, J.K. Pedersen, and F. Blaabjerg, "Power inverter topologies for photovoltaic modules—a review," *Proceedings of the IAS'02 Conference*, Vol. 2, pp. 782–788, 2002.

77. F. Blaabjerg, Z. Chen, and S.B. Kjaer, "Power electronics as efficient interface in dispersed power generation systems," *IEEE Transactions on Power Electronics*, 19 (5), 1184–1194, 2004.

78. M. Meinhardt and G. Cramer, "Past, present and future of grid connected photovoltaic- and hybrid-power-systems," *Proceedings of the IEEE Power Engineering Society Summer Meeting*, Vol. 2, pp. 1283–1288, 2000.

79. T. Shimizu, M. Hirakata, T. Kamezawa, and H. Watanabe, "Generation control circuit for photovoltaic modules," *IEEE Transactions on Power Electronics*, 16 (3), 293–300, 2001.

80. M. Meinhardt and D. Wimmer, "Multistring-converter: The next step in evolution of string-converter technology," *Proceedings of the EPE'01 Conference*, Graz, Austria, 2001.

81. B.K. Bose, P.M. Szczeny, and R.L. Steigerwald, "Microcomputer control of a residential photovoltaic power conditioning system," *IEEE Transactions on Industry Applications*, 21 (5), 1182–1191, 1985.

82. T. Shimizu, K. Wada, and N. Nakamura, "A flyback-type single-phase utility interactive inverter with low-frequency ripple current reduction on the dc input for an ac photovoltaic module system," *Proceedings of the PESC'02 Conference*, Vol. 3, pp. 1483–1488, 2002.

83. S. Saha and V.P. Sundarsingh, "Novel grid-connected photovoltaic inverter," *IEE Proceedings on Generation, Transmission, and Distribution*, 143 (2), 219–224, 1996.

84. A. Lohner, T. Meyer, and A. Nagel, "A new panel-integratable inverter concept for grid-connected photovoltaic systems," *Proceedings of the ISIE'96 Conference*, Vol. 2, pp. 827–831, 1996.

85. B. Lindgren, "Topology for decentralized solar energy inverters with a low voltage ac-bus," *European Conference on Power Electronics and Applications*, 1999.

86. S.B. Kjaer and F. Blaabjerg, "A novel single-stage inverter for AC-module with reduced low-frequency ripple penetration," *Proceedings of the Tenth European Conference on Power Electronics Applications*, pp. 2–4, 2003.
87. M. Calais, V.G. Agelidis, L.J. Borle, and M.S. Dymond, "A transformerless five level cascaded inverter based single-phase photovoltaic system," *Proceedings of the PESC'00 Conference*, Vol. 1, pp. 224–229, 2000.
88. M. Calais and V.G. Agelidis, "Multilevel converters for single-phase grid connected photo voltaic systems—an overview," *Proceedings of the ISIE'98 Conference*, Vol. 1, pp. 224–229, 1998.
89. S. Kjaer, J. Pedersen, and F. Blaabjerg, "A review of single-phase grid-connected inverters for photovoltaic modules," *IEEE Transactions on Industry Applications*, 41 (5), 1292–1306, 2005.
90. O. Lopez, R. Teodorescu, and J. Doval-Gandoy, "Multilevel transformerless topologies for single-phase grid-connected converters," *IEEE Industrial Electronics, IECON 2006*, pp. 5191–5196, November 2006.
91. D.W. Hart, *Introduction to Power Electronics*, Prentice-Hall, Englewood Cliffs, NJ,1996.
92. T. Shimizu, O. Hashimoto, and G. Kimura, "A novel high-performance utility-interactive photovoltaic inverter system," *IEEE Transactions on Power Electronics*, 18, 704–711, 2003.
93. C. Dorofte, "Comparative analysis of four dc/dc converters for photovoltaic grid interconnection & design of a dc/dc converter for photovoltaic grid interconnection," Technical Report, Aalborg University, Aalborg, Denmark, 2001.
94. S.J. Chiang, "Design and implementation of multi-functional battery energy storage systems," PhD dissertation, Department of Electronics and Engineering, National Tsing Hua University, Hsin-Chu, Taiwan, R.O.C., 1994.
95. C.M. Liaw, T.H. Chen, S.J. Chiang, C.M. Lee, and C.T. Wang, "Small battery energy storage system," *IEE Proceedings on Electric Power Applications*, 140 (1), 7–17, 1993.
96. N.D. Benavides and P.L. Chapman, "Power budgeting of a multiple-input buck–boost converter," *IEEE Transactions on Power Electronics*, 20,1303–1309, 2005.
97. O.H.A. Shirazi, O. Onar, and A. Khaligh, "A novel telecom power system," *Proceedings of the International Telecommunication Conference, INTELEC 2008*, SanDiego, 1–8.
98. A. Khaligh, "A multiple-input dc-dc buck–boost converter topology," *Proceedings of the Applied Power Electronics Conference and Exposition*, Austin, TX, February 2008.
99. IEEE Std 1562™-2008, IEEE guide for array and battery sizing in stand-alone photovoltaic (PV) systems.
100. NREL/CP-411-20379: Emery, Keith, et al., "Temperature and irradiance behavior of photovoltaic devices." *Photovoltaic and Reliability Workshop*, Laxmi Mrig, Editor. September 1995.
101. NREL/TP-463-5607: Marion, William, and Wilcox, "Solar radiation data manual for flat-plate and concentrating collectors," April 1994.
102. IEEE Std 1013™-2007, IEEE recommended practice for sizing lead–acid batteries for stand-alone photovoltaic (PV) systems, pp. 1–54, July 2007, New York.
103. S.J. Chiang, K.T. Chang, and C.Y. Yen, "Residential photovoltaic energy storage system," *IEEE Transactions on Industrial Electronics*, 45 (3), 385–394, 1998.
104. A. Pivec, B.M. Radimer, and E.A. Hyman, "Utility operation of battery energy storage at the BEST facility," *IEEE Transactions on Energy Conversion*, EC:1-3, 47–54, 1986.
105. R.J. King, "Photovoltaic applications for electric vehicles," *IEEE 21st Photovoltaic Specialists Conference*, Vol. 2, pp. 977–981, May 1990.
106. P. Singh, P.V. Glahn, and W. Koffke, "The design and construction of a solar electric commuter car," *Photovoltaic Specialists Conference*, Vol. 1, pp. 712–716, 1991.
107. M.W. Daniels and P.R. Kumar, "The optimal use of the solar powered automobile," *IEEE Control Systems Magazine*, 19 (3), 12–22, 1999.
108. L. McCarthy, J. Pieper, A. Rues, and C.H. Wu, "Performance monitoring in UMR's solar car," *IEEE Instrumentation and Measurement Magazine*, 3 (3), 19–23, 2000.
109. D. Crimmins, C. Deacutis, E. Hinchey, M. Chintala, G. Cicchetti, and D. Blidberg, "Use of a long endurance solar powered autonomous underwater vehicle (SAUV II) to measure dissolved

oxygen concentrations in Greenwich Bay, Rhode Island, U.S.A," *Oceans 2005—Europe*, 2 (20–23), 896–901, June 2005.

110. R. Blidberg, J. Jalbert, and M.D. Ageev, "The AUSI/IMTP solar powered autonomous undersea vehicle," *OCEANS '98 Conference Proceedings*, Vol. 1, pp. 363–368, 1998.

111. Solar Powered Autonomous Underwater Vehicle (SAUV II), Falmouth Scientific, Inc., available online: http://www.falmouth.com/DataSheets/SAUV.pdf

112. R. Leiner, "Research solar power boat—data management and online visualization," *Proceedings of Africon*, pp. 1–5, October 2007.

113. D.C. Senft, "Opportunities in photovoltaics for space power generation," *Proceedings of the IEEE Photovoltaic Specialist Conference*, 3–7 pp. 536–541, January 2005.

114. R.H. Nansen, "Wireless power transmission: The key to solar power satellites," *IEEE Aerospace and Electronic Systems Magazine*, 11 (1), 33–39, 1996.

115. N.J. Colella and G.S. Wenneker, "Pathfinder. Developing a solar rechargeable aircraft," *IEEE Potentials*, 15 (1), 18–23, 1996.

116. Y. Perriard, P. Ragot, and M. Markovic, "Round the world flight with a solar aircraft: Complex system optimization process," *IEEE International Conference on Electric Machines and Drives*, pp. 1459–1465, May 2005.

117. P. Ragot, M. Markovic, and Y. Perriard, "Optimization of electric motor for a solar airplane application," *IEEE Transactions on Industry Applications*, 42 (4), 1053–1061, 2006.

118. K.C. Reinhardt, T.R. Lamp, J.W. Geis, and A.J. Colozza, "Solar-powered unmanned aerial vehicles," *Proceedings of the IECEC 96 Energy Conversion Engineering Conference*, Vol. 1, pp. 41–46, August 1996.

2

Wind Energy Harvesting

2.1 Introduction

Wind is the airflow that consists of many gases in the atmosphere of the earth. Rotation of the earth, uneven heating of the atmosphere, and the irregularities of the ground surface are the main factors that create winds. Motion energy of the wind flow is used by humans for many purposes such as water pumping, grain milling, and generating electricity. Windmills that are used for electricity generation are called wind turbines in order to distinguish them from the traditional mechanical wind power applications.

Wind is a sustainable energy source since it is renewable, widely distributed, and plentiful. In addition, it contributes to reducing the greenhouse gas emissions since it can be used as an alternative to fossil-fuel-based power generation [1]. Wind turbines capture the kinetic energy of winds and convert it into a usable form of energy. The kinetic energy of winds rotates the blades of a wind turbine. The blades are connected to a shaft. The shaft is coupled to an electric generator. The generator converts the mechanical power into electrical power.

Even though wind turbines currently provide only 1% of the worldwide power supply, wind energy is one of the fastest growing renewable energy technologies all over the world. In countries such as Denmark, Spain, Portugal, and Germany, wind power accounts for approximately 19%, 9%, 9%, and 6% of the required electric power, respectively. From 2000 to 2007, the global wind power generation increased to approximately five times of its previously recorded capacity [2].

Wind energy conversion systems (WECS) involve many fields of various disciplines such as kinematics, mechanics, aerodynamics, meteorology, power electronics, power systems, as well as topics covered by structural and civil engineering. The main focus of this book is on electrical power interfaces and grid-connected topologies for WECS.

This chapter firstly focuses on some basic features of winds. A brief history of wind energy is presented in Section 2.3. Fundamentals of wind energy harvesting, basic parts of wind turbines, and the wind turbine types are described in Sections 2.4 and 2.5. These are followed by an introduction to different types of electric machines for wind turbine applications. This section also deals with many power electronic interfaces for wind power applications. The final section provides discussions on wind energy harvesting research and development.

2.2 Winds

The wind is the phenomenon of air moving from the equatorial regions toward the poles, as light warm air rises toward the atmosphere, while heavier cool air descends toward the earth's surface [3]. Therefore, cooler air moves from the North Pole toward the Equator and warms up on its way, while already warm air rises toward the North Pole and gets cooler and heavier, until it starts sinking back down toward the poles. Another phenomenon that is affecting global winds is caused by the "Coriolis force," which makes all winds on the northern hemisphere divert to the right and all winds from the southern hemisphere divert to the left [3].

Both of the above-mentioned phenomena affect global winds that exist on the earth's surface. Hence, as the wind rises from the Equator, there will be a low-pressure area close to ground level attracting winds from the North and South. At the poles, there will be high pressure due to the cooling of the air. In order to find the most suitable sites for wind turbines, it is crucial to study the geological data of the area since the wind's speed and direction are highly influenced by the local topology. Surface roughness and obstacles not only will affect the speed of the wind, but also affect its direction and overall power.

During the daytime, land masses are heated by the sun faster than water bodies due to their higher specific densities, which allows them to have better heat transfer capabilities. Thus, during the day, air rises from the land, flows out to the sea, and creates a low-pressure area at ground level, which attracts the cool air from the sea. At nightfall, there is often a calm period when the land and sea temperatures are equal. During the night, the high pressure is inland, and the wind blows in the opposite direction.

In order to efficiently capture wind energy, several key parameters need to be considered: air density, area of the blades, wind speed, and rotor area. The force of the wind is stronger at higher air densities. Wind force generates torque, which causes the blades of the wind turbine to rotate. Therefore, the kinetic energy of the wind depends on air density; therefore heavier (denser) winds carry more kinetic energy. At normal atmospheric pressure and at 15°C (59°F), the weight of the air is 1.225 kg/m^3, but if the humidity increases, the density decreases slightly. Air density is also influenced by temperature; therefore warmer winds are less dense than cold ones, so at high altitudes the air is less dense [3].

In addition, the area of the blades (air-swept area), that is, the diameter of the blade, plays an important role in the captured wind energy. The longer the blade, the bigger the rotor area of the wind turbine, and therefore, more wind can be captured under the same conditions [3].

The other parameter is the wind speed. It is expected that wind kinetic energy rises as wind speed increases [3].

The kinetic energy of the wind can be expressed as

$$E_k = \frac{1}{2}mv^2 = \frac{1}{2}\rho Vv^2 = \frac{1}{2}\rho A d v^2 = \frac{1}{2}\rho R^2 \pi d v^2, \tag{2.1}$$

where E_k is the wind kinetic energy, m is the wind mass, v is the wind speed, ρ is the air density, A is the rotor area, R is the blade length, and d is the thickness of the "air disc" shown in Figure 2.1.

Hence, the overall power of wind (P) is [3]

$$P = \frac{E_k}{t} = \frac{1}{2}\rho R^2 \pi \frac{d}{t} v^2 = \frac{1}{2}\rho R^2 \pi v^3, \tag{2.2}$$

$$P = \frac{1}{2}\rho R^2 \pi v^3. \tag{2.3}$$

The power content of the wind varies with the cube (the third power) of the average wind speed (Figure 2.2).

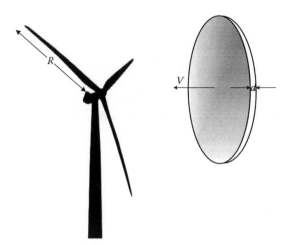

FIGURE 2.1 **(See color insert following page 80.)** Kinetic energy of wind.

Wind speed is one of the most important parameters in determining the available wind power. Therefore, it is important to accurately monitor and measure it. Wind speed distribution can be measured by a cup anemometer. The cup anemometer works on the principle of simple measurement of the number of revolutions per minute. Data acquisition systems are used to store data for a long period, for instance one year [3].

Information collected from anemometers is useful for measuring the wind rose, which shows the information of wind speed distributions in various directions. Depending on the number of sectors, various wind rose resolutions exist. Usually the compass is divided into 12 sectors, one for each 30° of the horizon, but more precise anemometers have 24 or more sectors. Figure 2.3 shows an example of a wind rose.

Based on Figure 2.3, winds coming from the north have the greatest speed, while winds from the southwest are calmer [3]. A wind rose can have several different useful data such as wind frequency, mean wind speed, and mean cube of the wind speed, which are important

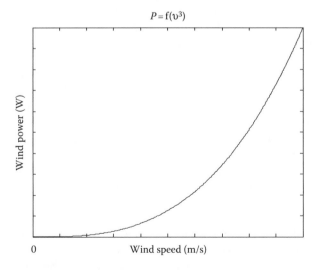

FIGURE 2.2 Specific wind power due to wind speed variation.

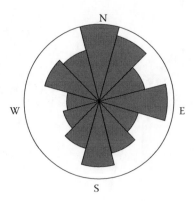

FIGURE 2.3   Wind rose example.

to calculate potential wind power [3]. Terrain specifications are the other factors affecting the winds that are discussed in Section 2.3.1 [4,5].

2.3   History of Wind Energy Harvesting

Wind turbines, as machines, were used by ancient civilizations (Persians, Romans, etc.) mostly as vertical axle windmills with several blades. The primary use was to grind corn and wheat and for water irrigation systems [6]. Later on, windmills were developed mostly in the Netherlands, Denmark, and Scotland for grinding mills. The first windmill that produced electricity was constructed by Professor James Blyth in Scotland, Glasgow in 1887, at Anderson's College [7]. A great research contribution was made in Russia, in 1931. A modern horizontal axis wind generator with 100 kW of power was mounted on a 30 m (100 ft) tower connected to the local 6.3 kV distribution system.

Ten years later, in Castleton, Vermont, USA, the world's first megawatt-size wind turbine supplied the local electrical distribution system [8]. The first wind machines were DC-type electrical machines. Later in the 1950s, in Denmark, research and development led to the development of two-bladed wind turbines, which utilized AC electrical machines instead of DC machines. At the same time, a Danish group of researchers brought the concept of three-bladed wind turbine. After the first oil crisis in 1973, interest in wind energy increased in several countries such as Germany, Sweden, the UK, and the USA [8].

In the early 1980s, the Tvind generator, a 2 MW machine, was constructed as a fairly revolutionary machine at that time. The machine had 54 m rotor diameter running at variable speed with a synchronous generator, and indirect grid connection using power electronic interfaces. This was the first time that power electronics was used to control wind-generated power [8]. In this period, the cost per kWh of electricity dropped by 50% due to the appearance of the 55 kW Nordtank wind turbine generators. Thousands of 55 kW Nordtanks were delivered to the wind farms in California in the early 1980s, especially in Palm Springs, California [3].

Later developments in wind turbines were geared toward improving the 2 MW machine, creating wind farms as a group of wind turbines, establishing offshore sites for improved wind power, and establishing new control techniques to increase the overall efficiency of the wind generators.

2.4 Fundamentals of Wind Energy Harvesting

2.4.1 Wind Turbine Siting

Wind turbines can be grid connected or independently operated from isolated locations. The two critical factors in finding the most suitable locations for wind turbines are wind speed and the quality of wind. This information is provided by wind roses. The most suitable sites for wind turbines are turbulence-free locations since turbulence makes wind turbines less efficient and affects overall stability of the turbine.

Wind turbulence is influenced by the surface of the earth. Depending on the roughness of the terrain, the wind can be more or less turbulent. It is only at 3000 ft and above ground level that the wind speed is not affected by friction against the earth's surface and therefore there is no turbulence [9].

Wind power can be categorized into seven classes [4] according to the wind speed (m/s) and wind power density (W/m^2). These wind power classes are shown in Table 2.1. It should be noted that in this table each wind power class corresponds to two power densities. For example, wind power class 4 represents the wind power density range between 200 and 250 W/m^2.

The wind map provides useful information about the classes in different geographic locations. Based on the wind's annual average speed and power, the wind power classes are determined for specific locations and those wind classes are indicated on wind maps. These maps also provide information regarding the wind's potential at specific locations and these maps can be used for wind turbine siting determinations.

Another important factor for wind turbine siting is roughness of terrain. There are roughness classes, which explain the relation between wind speeds and landscape conditions. Roughness class varies from 0 to 4, where class 0 represents water surfaces and open terrains with smooth surfaces, and class 4 represents very large cities with tall buildings and skyscrapers. Two key parameters regarding the different geographical locations affect the wind turbine siting: the friction coefficient and the roughness classification [5]. The friction coefficients are based on different terrain characteristics and are shown in Table 2.2.

TABLE 2.1

Wind Power Classes

Wind Power Class	At a Height of 10 m (33 ft)		At a Height of 50 m (164 ft)	
	Wind Power Density (W/m^2)	Wind Speed (m/s)	Wind Power Density (W/m^2)	Wind Speed (m/s)
1	0	0	0	0
1–2	100	4.4	200	5.6
2–3	150	5.1	300	6.4
3–4	200	5.6	400	7.0
4–5	250	6.0	500	7.5
5–6	300	6.4	600	8.0
6–7	400	7.0	800	8.8
7	1000	9.4	2000	11.9

TABLE 2.2

Friction Coefficient

Terrain Characteristics	Friction Coefficient (α)
Smooth hard ground, calm weather	0.10
Tall grass on level ground	0.15
High cops, hedges, and shrubs	0.20
Wooded countryside, many trees	0.25
Small town with trees and shrubs	0.30
Large city with tall buildings	0.40

The reduced wind speed rate with respect to the friction coefficient and terrain height can be expressed as

$$\left(\frac{v}{v_0}\right) = \left(\frac{H}{H_0}\right)^{\alpha}, \tag{2.4}$$

where v_0 is the original wind speed associated with the terrain height of H_0 and v is the reduced wind speed at a terrain height of H.

Roughness classes for different roughness lengths are provided in Table 2.3 with specific description of the terrains.

The reduced wind speed rate for different terrain heights and roughness lengths can be expressed as

$$\left(\frac{v}{v_0}\right) = \frac{\ln(H/z)}{\ln(H_0/z)}, \tag{2.5}$$

where z is the length of the roughness.

Wind speed increases exponentially with altitude. Therefore, the speed of wind at 10 m will be totally different from its speed at 100 and 200 m heights. The formula that gives exact information on actual wind speed at a certain height level above the earth's surface

TABLE 2.3

Roughness Classes

Roughness Class	Description	Roughness Length [z (m)]
0	Water surface	0.0002
1	Open areas with a few windbreaks	0.03
2	Farm land with some windbreaks more than 1 km apart	0.1
3	Urban districts and farm land with many windbreaks	0.4
4	Dense urban or forest	1.6

is presented in Equation 2.6:

$$v = v_{\text{ref}} \frac{\ln(z/z_0)}{\ln(z_{\text{ref}}/z_0)},$$ (2.6)

where v is the wind speed at height z above ground level, z is the height above ground level for the desired velocity (v), z_0 is the roughness length in the current wind direction, and z_{ref} is the height where the exact wind speed, v_{ref}, is known.

The ideal wind turbine site is a location with smooth wind, constant speed, and without turbulence, which is only possible in locations with roughness class 0. As mentioned earlier, roughness class 0 represents water surfaces and open terrains with smooth surfaces, where wind speeds are constant and there is no turbulence because there are no obstacles. Due to very low roughness, the wind speed at 30 m above water surface is not different from its speed at 50 m height. Moreover, due to the low turbulence, the lifetime of the turbine will be extended. However, there are great variations in wind speed and direction, depending on the weather and local surface conditions, with the greatest variations happening during the daytime.

Because of low turbulence, lots of wind turbines are situated offshore, in sites about 5–10 km in the sea. Wind shade (turbulence due to the obstacles in lands like forests, hills, etc.) does not affect wind turbines that are far in the sea. Therefore, based on their siting, there are two main types of wind turbines: onshore (site is on the land) and offshore (site is in the sea).

The generated electricity from onshore wind turbines can be transferred and connected to the grid through transmission lines. Offshore turbines are exposed to less turbulent winds (no wind shades); however, it is difficult to place transmission lines from the sea to the required inland locations.

2.4.2 Wind Turbine Power

2.4.2.1 Betz Law

Betz's law demonstrates the theoretical maximum power that can be extracted from the wind. The wind turbine extracts energy from the kinetic energy of the wind. Higher wind speeds result in higher extracted energy. It should be noted that the wind speed after passing through the turbine is much lower than before reaching the turbine (before energy is extracted). This means that there are two wind speeds: one before the wind approaches (in front of) the turbine and the other after (behind) the turbine. Figure 2.4 shows both speeds of the wind; after the turbine, the wind has a decreased speed.

The decreased wind speed, after the turbine, provides information on the amount of possible extracted energy from the wind. The extracted power from the wind can be calculated using Equation 2.7.

$$P_{\text{extract}} = \frac{E_k}{t} = \frac{1}{2}\rho R^2 \pi \frac{d}{t}(v_b^2 - v_a^2) = \frac{1}{2}\rho R^2 \pi \frac{v_a + v_b}{2}(v_b^2 - v_a^2),$$ (2.7)

where P_{extract} is the maximum extracted power from the wind, v_a and v_b are wind speeds after and before passing through the turbine, ρ is the air density, and R is the radius of the blades.

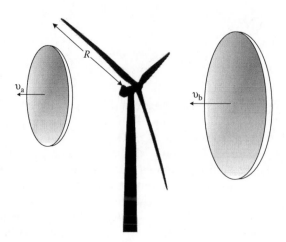

FIGURE 2.4 **(See color insert following page 80.)** Wind speed before and after the turbine.

The relation of total amount of power P_{total} to the extracted one $P_{\text{extracted}}$ can be calculated as

$$P_{\text{total}} = \frac{1}{2}\rho R^2 \pi v_b^3, \tag{2.8}$$

$$P_{\text{extract}} = \frac{1}{2}\rho R^2 \pi \frac{v_a + v_b}{2}(v_b^2 - v_a^2), \tag{2.9}$$

$$\frac{P_{\text{extract}}}{P_{\text{total}}} = \frac{(v_a + v_b)/2(v_b^2 - v_a^2)}{v_b^3} = \frac{1}{2}\left(1 - \frac{v_a^2}{v_b^2}\right)\left(1 + \frac{v_a}{v_b}\right). \tag{2.10}$$

For maximum power extraction, the ratio of the wind speed after and before the turbine can be calculated using

$$\frac{\mathrm{d}(P_{\text{extract}}/P_{\text{total}})}{\mathrm{d}(v_a/v_b)} = 0,$$

$$\frac{\mathrm{d}(P_{\text{extract}}/P_{\text{total}})}{\mathrm{d}(v_a/v_b)} = \frac{1}{2}\left[-3\left(\frac{v_a}{v_b}\right)^2 - 2\left(\frac{v_a}{v_b}\right) + 1\right] = 0. \tag{2.11}$$

Solving Equation 2.11 for v_a/v_b yields

$$\frac{v_a}{v_b} = \frac{1}{3}. \tag{2.12}$$

As a result, Equation 2.10 reaches its maximum value for $(v_a/v_b) = 1/3$.

$$\left.\frac{P_{\text{extract}}}{P_{\text{total}}}\right|_{\frac{v_a}{v_b} = \frac{1}{3}} \approx 59.3\%. \tag{2.13}$$

Equation 2.13 shows that the maximum extracted power from the wind is 59.3% of the total available power. In other words, it is not possible to extract all 100% of wind energy since the wind speed after the turbine cannot be 0. The effect of v_a/v_b ratio on the $P_{\text{extract}}/P_{\text{total}}$

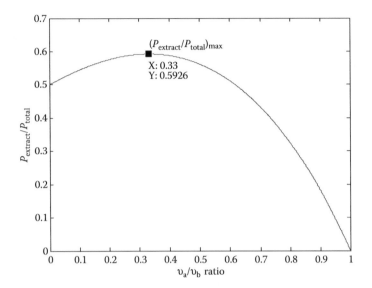

FIGURE 2.5 Maximum theoretical extracted power from wind.

is shown in Figure 2.5. The maximum extracted power is approximately 2/3 of total wind power.

Betz's law indicates that the maximum theoretical extracted wind power is 59%. However, in practice, the real efficiency of the wind turbine is slightly different.

2.4.2.2 Power Curve

The power curve is a diagram that shows the specific output power of a turbine at various wind speeds, which is important in turbine design. Power curves are made by a series of measurements for one turbine with different wind speeds. In order to obtain accurate power curves, the wind should be nonturbulent. Figure 2.6 shows that the output power decreases for wind speeds above 20 m/s.

The "cut-in" and "cut-out" wind speeds can be found from the power curve. The "cut-in" shows the minimum wind speed needed to start the turbine (which depends on turbine

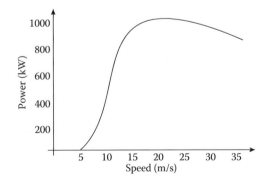

FIGURE 2.6 The power curve for a wind turbine.

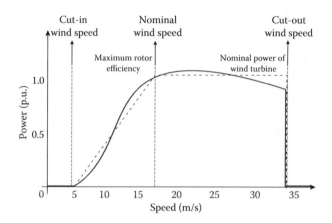

FIGURE 2.7   Power curve characteristics.

design) and to generate output power. Usually it is 3 m/s for smaller turbines and 5–6 m/s for bigger ones.

The "cut-out" wind speed represents the speed point where the turbine should stop rotating due to the potential damage that can be done if the wind speed increases more than that.

The power curve can be divided into a few regions. These regions are presented in Figure 2.7 and are separated by cut-in, nominal, and cut-out speeds. Once the rotor of the turbine starts spinning, it can be assumed that the amount of produced output power is linearly increasing with the speed.

After the wind speed reaches nominal speed (for which the turbine is designed), the output power remains relatively constant; however, an increase in output power above the nominal value might damage the system. Therefore, despite increasing the wind speed, the speed of the turbine should be controlled to be constant. This is achievable using several controlling techniques, such as pitch angle control or controlling the load of the turbine.

After the wind reaches the cut-out speed, in order to protect the system from any kind of damage, the turbine should be stopped by setting the pitch angle such that it does not face the upcoming wind. Usually, at very high speeds, the stability of the blades and cables is weak; thus the system should immediately stop.

2.4.2.3   Power Coefficient

The power coefficient is the ratio of the electrical output power of the wind turbine to the total (potential) wind power, which depends on the wind speed. The power coefficient curve of a wind turbine is shown in Figure 2.8. It can be observed from Figures 2.8 and 2.9 that both curves have a maximum for the same wind speeds. Hence, the power coefficient curve always has a maximum point for maximum possible turbine power, which is usually for the most frequent wind speeds.

The power that can be captured from the wind with πR^2 effective area of the blades is given by

$$P_{\text{total}} = \frac{1}{2} \rho c_{\text{p}} R^2 \pi v_{\text{b}}^3. \tag{2.14}$$

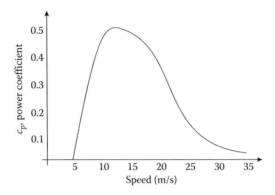

FIGURE 2.8 Power coefficient curve of a wind turbine.

The power coefficient c_p depends on the specific design of the wind turbine (especially the particular aerodynamic structure of the blades). Each wind turbine has its own power coefficient c_p that depends mainly on the tip speed ratio λ. As shown in Equation 2.15, the tip speed ratio depends on the geometry of the wind turbine, its rotational speed, and the length of the blades:

$$\lambda = \frac{Rw}{v}, \qquad (2.15)$$

where w is the rotor speed, R is the length of the blades, and v denotes the wind speed.

In addition, the power coefficient depends on the pitch angle, which is defined as the angle between the blade surface and the plane of the wind rotor, as shown in Figure 2.9. There is a great difference between an onshore and offshore wind turbine's power coefficient. For onshore turbines, for example, the blades are designed such that the optimal tip speed is limited to roughly 50–70 m/s, because the blade tips cause excessive acoustical noise and can cause damage at higher speeds. On the other hand, the noise does not play an important role in offshore turbines and higher speeds lead to slightly higher optimal values of c_p [10].

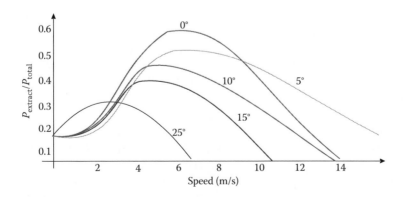

FIGURE 2.9 Speed versus power with respect to different pitch angle values.

2.5 Wind Turbine Systems

Basically, a wind turbine consists of three main parts: a turbine, a nacelle, and a tower. The wind turbine presented in Ref. [11] demonstrates some other parts inside the tower and nacelle.

2.5.1 Basic Parts of Wind Turbines

The blades are the main components that capture the kinetic energy of the wind and help the turbine rotate. The rotor is the part that couples the generator to the rotating part of the turbine, directly or through gearboxes. The pitch is the ability to change the wind-facing angle of the blades in order to maintain a constant wind turbine speed if the wind speed changes. With different pitch angles, the effective blade surface facing the wind direction can be controlled. For nominal speed, the pitch angle can be set to zero so that the blades fully face the upcoming wind. Above the nominal speed, the pitch angle can be increased; thus the effective blade surface is decreased and eventually a constant speed can be maintained. The brake is the mechanical speed reducer that prevents the generator speed from increasing above the maximum value. Even though the pitch for the blades can be helpful for speed reduction, the brakes have a faster response than that of the pitch control. The low-speed shaft is connected to a gearbox with a high turn ratio that provides faster rotating speed to the generator during low wind speed conditions. The gearbox is the component that allows the wind turbine shaft to be coupled to the generator shaft. The generator is the mechanical-to-electrical energy conversion unit of the system that is driven by the mechanical power of the turbine. The electrical output of the generator is connected to the grid or load through power electronic converters. The controller, which is the brain of the system, is responsible for the torque and the speed control of the generator, determines the pitch angle, controls the yaw motor to face the wind direction, and controls the power electronic interfaces. The anemometer is a measurement device to assess the wind speed. The wind vane is an elevated device that shows the direction of the wind. This instrument may also be operated together with the yaw mechanism to measure the wind direction. The nacelle is the enclosure of the system and all of the generating components such as the generator, drive train, and the like are placed inside of the nacelle. The high-speed shaft drives the generator without a gearbox and it is essentially a gearbox with a turn ratio that is smaller than that of the low-speed shaft. It is effective during high-wind-speed conditions. The yaw drive ensures that the rotor is aligned with the wind direction. It helps produce the maximal amount of energy at all times by keeping the turbine facing the wind as its direction changes. The yaw motor is the device that provides mechanical rotation to the yaw drive for the yaw mechanism's operation. The tower supports the body of the turbine and the other components. Some of the main components are described in detail in the following subsections.

2.5.1.1 The Tower

The main purpose of the tower is to support the nacelle and resist vibration due to the wind speed variations. The cables that connect the generator (on top of the tower, inside the nacelle) and transmission line (down, in the basement of the tower) are inside the tower. The tower is the main component that carries most of the other components such as the turbine, nacelle, blades, generator, and so on.

The height of the tower is different for offshore and onshore turbines. The higher towers are more appropriate for wind energy harvesting, since winds contain less turbulence in higher altitudes. However, stability issues limit the height of the tower. Onshore wind systems have higher towers than offshore turbines, because the land has higher roughness than the water surface [12]. On the water surface, there are almost no obstacles; hence the low tower length is sufficient to capture the wind [12]. In onshore applications, there may be some objects around the tower that may block the wind speed. In areas with high roughness, high turbine towers are required to avoid the effect of wind blocking objects such as buildings, mountains, hills, trees, and so on.

2.5.1.2 Yaw Mechanism

The yaw mechanism is composed of the yaw motor and the yaw drive. The "yaw" mechanism turns the whole nacelle toward the wind direction in order to face the wind directly. Regardless of the direction of the wind, the yaw mechanism can help the turbine face the wind by changing the direction of the nacelle and the blades. During the rotation of the nacelle, there is a possibility of twisting the cables inside of the tower. The cables will become more and more twisted if the turbine keeps turning in the same direction, which can happen if the wind keeps changing in the same direction. The wind turbine is therefore equipped with a cable twist counter, which notifies the controller that it is time to straighten the cables.

2.5.1.3 The Nacelle

The gearbox, generator, and the control electronics are all located inside of the nacelle. The nacelle is connected to the tower through the yaw mechanism. Inside the nacelle, two shafts connect the rotor of the turbine to the rotor of the electrical generator through the gearbox. The gearbox is the mechanical energy converter that connects the low-speed shaft of the turbine to the high-speed shaft of the electrical machine.

The control electronics inside the nacelle record the wind speed, direction data, rotor speed, and generator load, and then determine the control parameters of the wind operation system. If the wind changes direction, the controller will send a command to the yaw system to turn the whole nacelle and turbine to face the wind.

The electrical generator is the main part of the nacelle. It is the heaviest part and produces electrical energy, which is transferred through the cables to the grid. There are different types of generators that are used for wind turbines, and depending on the type of generator, wind turbines can operate with either fixed or variable speeds. Fixed speed (FS) turbines use synchronous machines, and operate at an FS that depends on the grid's frequency. These machines are not the best solution for the wind turbines, because the wind always changes its speed. Variable speed turbines use DC machines, brushless DC (BLDC) machines, and induction machines. DC machines are not commonly used due to the maintenance problems with the brushes. Induction and BLDC machines are more suitable for wind applications. Various generator types for wind energy harvesting are explained in Section 2.6.

2.5.1.4 The Turbine

The turbine, also called "low-speed rotor," usually has two to six blades. The most common number of blades is three since they can be positioned symmetrically, maintain the system's lightness, and ensure the stability of the wind power system (WPS). Two-blade turbines have high stresses in cut-in speed; therefore, the speed and power of the wind are insufficient for starting the rotation of the turbine and higher minimum wind speed values are required at the beginning. The radius of the blades is directly proportional to the amount of captured energy from the wind; hence an increased blade radius would result in a higher amount of captured energy.

The blades are aerodynamic and they are made of a composite material such as carbon or Plexiglas, and are designed to be as light as possible. Blades use lift and drag forces caused by wind; therefore, by capturing these forces, the whole turbine will rotate. The blades can rotate around their longitudinal axis to control the amount of captured wind energy. This is called "pitch control." If the wind speed increases, the pitch control can be used to change the effective blade surface, hence keeping the turbine power constant. The pitch angle control is usually used for wind speeds above the nominal speed.

2.6 Wind Turbines

Wind turbines can be classified based on several criteria. One classification is based on the position of rotational axis and the other one is based on the size of the wind turbine.

2.6.1 Wind Turbines Based on Axis Position

Based on axis position, wind turbines are classified as the horizontal axis and vertical axis turbines.

Horizontal axis wind turbines (HAWTs) are more common than vertical axis wind turbines (VAWTs). The horizontal axis turbines have a horizontally positioned shaft, which helps ease the conversion of the wind's linear energy into a rotational one.

VAWTs have a few advantages over the horizontal axis wind turbines. VAWTs' electrical machines and gearbox can be installed at the bottom of the tower, on the ground, whereas in HAWTs, these components have to be installed at the top of the tower, which requires additional stabilizing structure for the system. Another advantage of the VAWTs is that they do not need the yaw mechanism since the generator does not depend on the wind direction. The most famous design of VAWT is Darrieus type of turbine.

There are a few disadvantages that limit the utilization of the VAWTs. Due to the design of blades, the sweep area of VAWT is much smaller. Wind speed is low near the surface and usually turbulent; hence these wind turbines harvest less energy than horizontal axis ones. Additionally, VAWTs are not self-starting machines and must be started in motoring mode, and then switched to generating mode.

2.6.2 Wind Turbines Based on Power Capacity

Another classification criterion for wind turbines is based on their installed power capacity. Based on the installed power capacity, wind turbines are divided into *small, medium, and large wind turbines* (Figure 2.10).

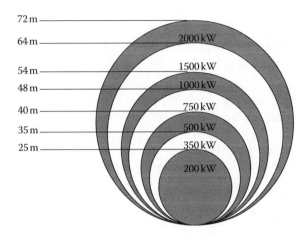

FIGURE 2.10 **(See color insert following page 80.)** Turbine output power for different wind turbine diameters.

The output power rating of *small wind turbines* is less than 20 kW. Small wind turbines are used for residential applications, supplying households with electricity, and they are designed for low cut-in wind speeds (generally 3–4 m/s). They are also suitable for remote places, where the grid is far away and the power transmission is difficult. Small wind generators provide isolated power systems for a household's load, and are usually connected to batteries, as shown in Figure 2.11. It is projected that small wind turbines would contribute to 3% of U.S. electrical consumption by 2020 [12].

Medium wind turbines usually provide between 20 and 300 kW installed power. They are usually used to supply either remote loads that need more electrical power or commercial buildings. Medium wind turbines usually have a blade diameter of 7–20 m, and the tower is not higher than 40 m. They are almost never connected to a battery system. They are directly connected to the load through DC/AC power electronic inverters.

Large wind turbines are of MW power range. These turbines incorporate complex systems, and wind farms typically consist of several to hundreds of those large turbines. One of the world's biggest wind turbines is located in Emden, Germany [13], built by the German company Enercon as an offshore turbine. The price of 1 kW installed power from a large wind turbine is significantly less than the price of 1 kW from a small turbine. Currently, the installed price of a large turbine is about $500/kW, and the price of energy is around 30–40 cents/kWh, depending on the site and turbine size. Enercon turbines have a power output of 5 MW with a rotor blade diameter of 126 m, sweeping an area of over 12,000 m^2 [13].

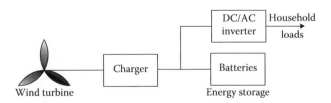

FIGURE 2.11 Connection scheme of a typical small wind turbine.

2.7   Different Electrical Machines in Wind Turbines

There are various types of electrical machines that are used in wind turbines. There is no clear criterion for choosing a particular machine to work as a wind generator. The wind generator can be chosen based on the installed power, site of the turbine, load type, and simplicity of control. BLDC generators, permanent magnet synchronous generators (PMSGs), induction generators, and synchronous generators are the machine types that are used in wind turbine application.

Squirrel cage induction or BLDC generators are generally used for small wind turbines in household applications. Doubly fed induction generators (DFIGs) are usually used for megawatt size turbines. Synchronous machines and permanent magnet synchronous machines (PMSMs) are the other machines that are used for various wind turbine applications.

2.7.1   BLDC Machines

Advancements over the course of the past 15–20 years in the development of the BLDC machines have made them very popular for multiple applications. In addition, the development of fast semiconductor switches, cost-effective DSP processors, and other microcontrollers have benefited the improvement of the motor/generator drives.

BLDC machines are widely used in small wind turbines (up to 15 kW) due to their control simplicity, compactness, lightness, ease of cooling, low noise levels, and low maintenance. Due to the existence of a magnetic source inside the BLDCs, they are the most efficient electric machines. Recent introduction of high-energy density magnets (rare-earth magnets) has allowed the achievement of very high flux densities in these machines bringing compactness. There is no current circulation in the rotor for the magnetic field; therefore, the rotor of a BLDC generator does not heat up. The absence of brushes, mechanical commutators, and slip rings suppresses the need for regular maintenance and suppresses the risk of failure associated with these elements. Moreover, there is no noise associated with the mechanical contact. The switching frequency of the driving converter is high enough so that the harmonics are not audible [14,15].

Due to its mechanical performance, the BLDC generator drive system can provide additional increase in power density with the advanced control techniques.

Since the BLDC generator has permanent magnets (PMs) inside the machine, it can be classified as a kind of PM machine. Due to the existence of PMs, the brushes and commutator, which supply magnetic flux in a regular DC machine, are not required. Eddy current losses are also reduced since the stator has laminated steel [16–19]. The cross-sectional view of a three-phase BLDC generator with four poles is shown in Figure 2.12. A trapezoidal electromotive force (EMF) is induced in the stator winding as the rotor rotates.

The waveform of the induced EMF from the stator winding is shown in Figure 2.13. The concentric winding of the machine and rectangular distribution of the magnetic flux in the air gap generate this nonsinusoidal EMF [17,20]. Due to this waveform, a BLDC generator has approximately 15% higher power density in comparison to a PMSG, which has a sinusoidal winding configuration and sinusoidal magnetic flux distribution in the air gap [21]. This is due to the fact that the effective time-integral of a trapezoidal waveform, which is proportional to the output power of the generator, is higher than that of a sinusoidal waveform with the same amplitude.

FIGURE 2.12 BLDC generator cross-sectional view (inner rotor type).

The BLDC machines can be controlled using various strategies and techniques. The BLDC model using differential equations for all three-phase voltages, torque, and position is [21] as follows:

$$e_{an}(t) = Ri_a(t) + L\frac{di_a(t)}{dt} + v_{an}(t), \tag{2.16}$$

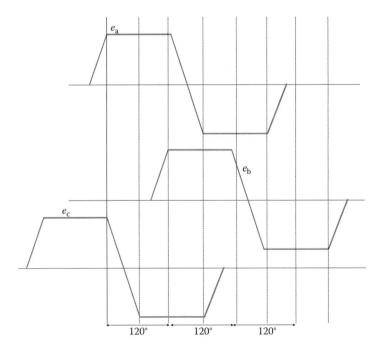

FIGURE 2.13 Induced EMF of a three-phase BLDC generator.

$$e_{bn}(t) = Ri_b(t) + L\frac{di_b(t)}{dt} + v_{bn}(t), \tag{2.17}$$

$$e_{cn}(t) = Ri_c(t) + L\frac{di_c(t)}{dt} + v_{cn}(t), \tag{2.18}$$

$$T_{cm}(t) = T_L(t) + J\frac{d\omega(t)}{dt} + b\omega(t), \tag{2.19}$$

$$\omega(t) = \frac{d}{dt}\theta(t). \tag{2.20}$$

These differential equations present the model of the BLDC. The electromagnetic force per phase, $e(t)$, is defined by Equations 2.16 through 2.18, where $i(t)$ stands for phase current, L stands for phase inductance, and v is the phase voltage. Equation 2.19 shows the electrical torque, where J is the moment of inertia and b represents the friction losses.

Figure 2.14 shows the equivalent circuit of the BLDC generator connected to a diode rectifier. This is the simplest way of using the BLDC machine for wind applications because no switch is necessary to control the phase current. The full bridge rectifies the induced voltages of variable frequencies caused by the variable wind speeds. Basically, the waveform of the induced EMF is converted by the diode rectifier to DC voltage, regardless of the input shape. Usually these types of wind turbines are connected to batteries; therefore, the rectified electrical power is used to charge the battery.

The inductor L_s ($L_s = L_a = L_b = L_c$) is the synchronous reactance of the BLDC generator. The DC-link voltage is

$$V_d = 1.35\ V - \frac{3}{\pi}\omega L_s I_d, \tag{2.21}$$

where V represents line-to-line induced voltage in the generator and the second part of the equation shows the voltage drop on the DC-link caused by the synchronous reactance, and I_d is the phase current. The DC-link voltage and machine current are inversely proportional; therefore, the DC-link voltage can be used to control the machine current and power [12].

The back EMF waveform of the diode rectifier input is presented in Figure 2.15. The phase current should be in the opposite direction of the back EMF waveform in order to get maximum generated power [20,21]. In phasor form, the power of a BLDC machine is

$$P = |\bar{E}|\,|\bar{I}|\cos(\alpha), \tag{2.22}$$

where P is the power of the BLDC generator, E is the average induced back EMF, I is the current, and α is the angle between the EMF and current vectors. In order to obtain

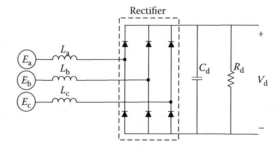

FIGURE 2.14 Diode rectifier connected to a BLDC generator.

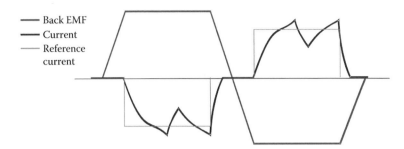

FIGURE 2.15 **(See color insert following page 80.)** Back EMF, real current, and reference current waveform of a BLDC machine.

maximum power, the α angle can either be $0°$ or $180°$, and $\cos(\alpha) = 1$. However, in the case where α is equal to $0°$, the machine is operating in motoring mode. Therefore, the α angle should be $180°$.

Since the real-time waveform of the induced EMF in the BLDC generator is not a sinusoidal waveform, it contains harmonics. The trapezoidal waveform is generated by the machine design. All the harmonics have to be included, and in the single-phase BLDC generator, the average power can be expressed as [17–20]

$$P_{avg} = \frac{1}{T} \int_0^T e(t)i(t)\, dt, \qquad (2.23)$$

$$P_{avg} = \frac{1}{T} \int_0^T [e_a(t)i_a(t) + e_b(t)i_b(t) + e_c(t)i_c(t)]\, dt, \qquad (2.24)$$

$$P_{avg} = \frac{1}{T} \int_0^T e_a(t)i_a(t)\, dt + \frac{1}{T} \int_0^T e_b(t)i_b(t)\, dt + \frac{1}{T} \int_0^T e_c(t)i_c(t)\, dt, \qquad (2.25)$$

$$P_{avg} = \frac{3}{T} \int_0^T e_a(t)i_a(t)\, dt. \qquad (2.26)$$

The phase voltage and phase current can be represented by

$$e_a(t) = E_{a1} \sin(\omega t) + E_{a2} \sin(2\omega t) + E_{a3} \sin(3\omega t) + \cdots, \qquad (2.27)$$

$$e_b(t) = E_{b1} \sin\left(\omega t - \frac{2\pi}{3}\right) + E_{b2} \sin 2\left(\omega t - \frac{2\pi}{3}\right) + E_{b3} \sin 3\left(\omega t - \frac{2\pi}{3}\right) + \cdots, \qquad (2.28)$$

$$e_c(t) = E_{c1} \sin\left(\omega t - \frac{4\pi}{3}\right) + E_{c2} \sin 2\left(\omega t - \frac{4\pi}{3}\right) + E_{c3} \sin\left(3\omega t - \frac{4\pi}{3}\right) + \cdots, \qquad (2.29)$$

$$i_a(t) = I_{a1} \sin(\omega t - \varphi_1) + I_{a2} \sin 2(\omega t - \varphi_2) + I_{a4} \sin 4(\omega t - \varphi_4) + \cdots, \qquad (2.30)$$

FIGURE 2.16 Back EMF and voltage phasor diagram.

$$i_b(t) = I_{b1} \sin\left(\omega t - \varphi_1 - \frac{2\pi}{3}\right) + I_{b2} \sin 2\left(\omega t - \varphi_2 - \frac{2\pi}{3}\right) + I_{b4} \sin 4\left(\omega t - \varphi_4 - \frac{2\pi}{3}\right) + \cdots,$$

$$(2.31)$$

$$i_c(t) = I_{c1} \sin\left(\omega t - \varphi_1 - \frac{4\pi}{3}\right) + I_{c2} \sin 2\left(\omega t - \varphi_2 - \frac{4\pi}{3}\right) + I_{c4} \sin 4\left(\omega t - \varphi_4 - \frac{4\pi}{3}\right) + \cdots.$$

$$(2.32)$$

where φ_n represents the phase difference between each voltage and current harmonic. In order to be able to control the phase current and power factor ($\cos \varphi$), the full bridge diode rectifier has to be replaced with a controlled rectifier with active switches. As shown in Figure 2.16, φ increases with the speed (ω), as the reactance increases with the speed.

$$V = E - (j\omega L_s + R)I_d. \tag{2.33}$$

Therefore, the phase angle, φ_n, can be kept at zero by controlling the phase current. The average power corresponding to the zero phase angle can be expressed as

$$P_{avg} = \frac{1}{T} \int_0^T e(t)i(t) \, dt = \frac{1}{2}(E_1 I_1 + E_2 I_2 + E_3 I_3 + \cdots). \tag{2.34}$$

In this way, the output power can be maximized. Figure 2.17a shows the schematic of the controlled rectifier used for BLDC phase current control. Phase current waveform is shown in Figure 2.17b. Usually a hysteresis regulator is used to control current.

A well-known BLDC generator in the wind industry is the 3.7 kW Skystream turbine [22]. Technical specification of the Skystream 3.7 kW wind turbine is presented in Table 2.4.

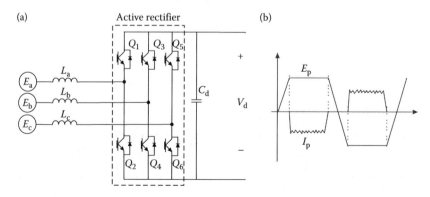

FIGURE 2.17 (a) BLDC generator connected to an active rectifier. (b) Phase voltage–current waveform.

TABLE 2.4

Skystream Technical Specifications

Rated power	1.8 kW (2.4 kW max.)	Alternator	Slotless PM brushless
Weight	170 lb (77 kg)	Yaw control	Passive
Rotor diameter	12 ft (3.72 m)	Grid feeding	120/240 VAC 50–60 Hz
Swept area	115.7 ft² (10.87 m²)	Braking system	Electronic stall regulation
Rated speed	50–330 rpm	Cut-in wind speed	3.5 m/s
Shutdown speed	370 rpm	Rated wind speed	9 m/s
Tip speed	9.7–63 m/s	Survival wind speed	63 m/s

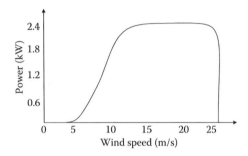

FIGURE 2.18 Performance graph of a Skystream small wind turbine with a BLDC generator.

Figure 2.18 shows the performance graph of the Skystream small wind turbine with a BLDC generator. This relationship of power to wind speed should be taken into consideration by small wind turbine designers and users.

Figure 2.19 displays the basic scheme of a small wind turbine used for household applications. The battery unit is used as a back-up system and requires an inverter unit in order to convert DC current into 120 V AC/60 Hz.

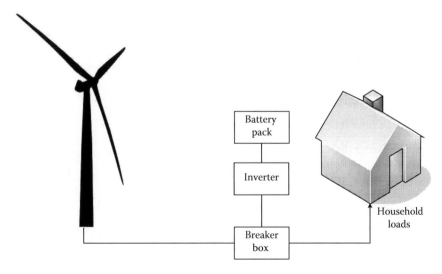

FIGURE 2.19 Basic schematic of a small wind turbine for household application. (Modified from W.D. Jones, *IEEE Spectrum*, 43 (10), 2006.)

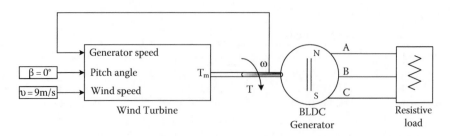

FIGURE 2.20   Block diagram of a wind turbine with a BLDC generator.

In order to investigate the behavior of a wind turbine with a BLDC generator, a small-scale model, shown in Figure 2.20, is analyzed. The wind turbine produces mechanical torque that drives the BLDC generator. In this example, the pitch angle is 0; therefore, there is no pitch angle controller and the wind speed is constant at 9 m/s. The BLDC generator electrical parameters are: stator resistance $R_s = 0.2\,\Omega$, stator inductance $L_s = 8.5\,mH$, flux induced by magnets $\varphi = 0.175\,Wb$, and back EMF flat area is 120°, where the flat area represents the duration when the wave has a constant amplitude in Figure 2.21. The mechanical parameters of the generator are: inertia $J = 0.05\,kg\,m^2$, friction factor $F = 0.005\,N\,ms$, and number of poles $p = 4$.

The trapezoidal back EMF induced by the BLDC generator is shown for phases A, B, and C in Figure 2.21. This figure shows that each phase voltage has 120° of flat area and there are 120° between each phase voltage.

The wind turbine generates 10 N m mechanical torque that drives the generator. This mechanical torque yields the electromagnetic-induced torque as shown in Figure 2.22. The

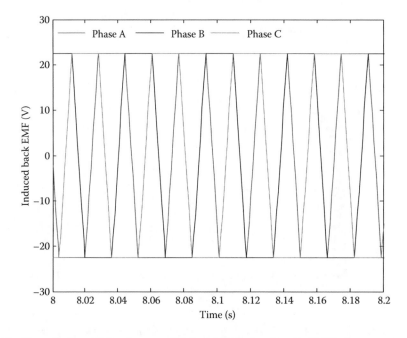

FIGURE 2.21   **(See color insert following page 80.)** Back EMF induced by the BLDC phases.

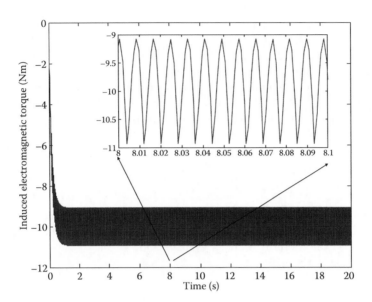

FIGURE 2.22 Electromagnetic torque induced by the BLDC generator.

generator shaft speed in radians per second is given in Figure 2.23. The main disadvantage of the BLDC generator is the nonsinusoidal-induced back EMF and nonsinusoidal current including harmonic components. Therefore, there are some ripples in the induced electromagnetic torque due to the distorted current waveform.

The generator speed is proportional to the induced torque of the machine, since the machine speed is zero at the beginning and it gradually increases as the induced torque increases.

2.7.2 Permanent Magnet Synchronous Machines

This type of machine can be used in both fixed and variable speed applications. The PMSG is very efficient and suitable for wind turbine applications.

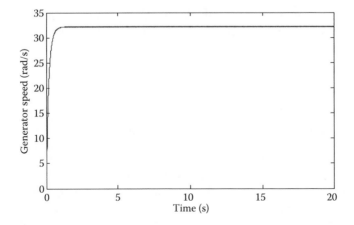

FIGURE 2.23 BLDC generator speed.

PMSGs allow direct-drive (DD) energy conversion for wind applications. DD energy conversion helps eliminate the gearbox between the turbine and generator; thus these systems are less expensive and require less maintenance [23,24]. However, the lower speed determined by the turbine shaft is the operating speed for the generator.

The analytical model of a small PM synchronous machine can be represented using the electrical equivalent equations of the machine and its mechanical design specifications. The rms value of the fundamental component of the excitation voltage induced in a phase winding of the machine [21,25] can be written as

$$E_f = \frac{2\pi}{\sqrt{2}} f N_{ph} K_{w1} \varphi_p,$$ (2.35)

where f is the frequency, N_{ph} is the number of turns per phase, K_{W1} is the fundamental harmonic winding factor, and φ_p is the flux per pole due to the fundamental space harmonic component of the excitation flux density distribution.

The flux per pole is defined as

$$\varphi_p = B_{l\,max} \frac{lD}{p},$$ (2.36)

where $B_{l\,max}$ is the peak value of the fundamental space harmonic component of the excitation flux density distribution, l is the effective length of the stator core, D is the stator inner diameter, and p is the number of pole pairs.

The flat value (when the flux density reached its maximum amplitude and remains constant) of the rectangular air gap flux density distribution produced by the PMs, B_g, is related to the maximum value of the flux density, $B_{l\,max}$ [26], by

$$B_{l\,max} = k_f B_g,$$ (2.37)

where k_f is the form factor of the excitation field. The form factor can be expressed as a function of pole-arc to polepitch ratio, which is represented by α.

$$k_f = \frac{4}{\pi} \sin\left(\frac{\alpha\pi}{2}\right).$$ (2.38)

The flat value of the excitation flux density distribution is related to the remnant flux density. Thus, the relative permeability of the PMs can be represented as [23]

$$B_g = \frac{l_m/\mu_r}{(l_m/\mu_r)(1/C_\phi) + K_c l_g (1 + p_{rl})} B_r,$$ (2.39)

where l_m is the radial thickness of the PM, μ_r is the relative permeability of the PM, C_ϕ is the flux focusing factor, K_c is Carter's coefficient, l_g is the mechanical air gap clearance, and p_{rl} is the normalized rotor leakage permeance. The p_{rl} generally varies between 0.05 and 0.2 [25]. The flux focusing factor is related to the detailed structure of the machine and is dependent on the radial height of the PMs [27]. Carter's coefficient is used to compensate for the slot's effect [28]. Slot effect is the loss due to the reduced magnetic flux in the air gap. Magnetic flux weakening in the air gap is due to slot dimensions and air gap geometry.

The relative permeability of the PM material is close to unity. Therefore, the effective air gap can be considered constant and relatively large in a PM machine with magnets mounted on rotor surface. Thus, the d- and q-axis synchronous reactances are identical. The synchronous reactance of the machine is equal to the sum of magnetizing reactance X_m and leakage reactance X_l:

$$X_s = X_m + X_l. \tag{2.40}$$

The magnetizing reactance can be calculated as

$$X_m = \frac{6\mu_0 l D f K_{w1}^2 N_{ph}^2}{p^2 (K_c l_g + l_m/\mu_r)}, \tag{2.41}$$

where μ_0 is the permeability of the air and the $(K_c l_g + l_m/\mu_r)$ term refers to the effective air gap length in the path of the magnetizing flux. The mechanical air gap clearance, which is the distance between two magnets, is defined by Carter's coefficient to determine the slotting and radial thickness of the PMs. On the other hand, the leakage reactance is a function of the permeance coefficients of the dominant leakage flux paths of the stator such as the tooth-top, slot, and winding overhang leakage flux paths [29]:

$$X_l = 4\pi\mu_0 f \frac{N_{ph}^2 l}{pq} \left(\lambda_{\text{tooth-top}} + \lambda_{\text{slot}} + \lambda_{\text{overhang}} \right). \tag{2.42}$$

The total length of the phase winding is important for per-phase resistance of the stator winding. The total length can be calculated as

$$l_{\text{ph-winding}} = 2(l + l_{\text{end}})N_{ph}, \tag{2.43}$$

where l_{end} is the length of the stator end winding. Therefore, the per-phase resistance of the stator winding can be expressed as

$$R_a = \frac{l_{\text{ph-winding}}}{a\sigma_{\text{Cu}}A_{\text{cond}}}, \tag{2.44}$$

where a is the number of parallel paths, σ_{Cu} is the copper conductivity, and A_{cond} is the cross-sectional area of the conductors [23,29].

The terminal voltage of an isolated PMSG can be written in terms of the induced voltage and the voltage drops due to the synchronous reactance and the resistance of the machine. For a PMSG with negligible saliency, the terminal voltage is obtained by Equation 2.45 [23,29]:

$$V_a = \sqrt{E_f^2 - (I_a X_s \cos\phi + I_a R_a \sin\phi)^2} + I_a X_s \sin\phi - I_a R_a \cos\phi. \tag{2.45}$$

The input mechanical power of the generator is either converted into the electrical power or is lost in the form of resistive and magnetic losses:

$$P_{\text{mech}} = P_E + P_{\text{Cu}} + P_{\text{rot}} + P_{\text{core}}, \tag{2.46}$$

where P_{mech} is the mechanical input power, P_E is the electrical output power, P_{Cu} is the total stator copper loss, P_{rot} is the total rotational losses, and P_{core} is the total core loss of the machine. The efficiency of the generator is

$$\eta = \frac{P_E}{P_{\text{mech}}}. \tag{2.47}$$

The total active power generated by the PM machine can be found as

$$P_E = 3V_a I_a \cos \phi, \tag{2.48}$$

where V_a is the armature terminal voltage and I_a is the armature current. The copper losses are due to the armature resistance and current as [30]

$$P_{Cu} = 3I_a^2 R_a. \tag{2.49}$$

The losses due to the rotation of the machine are the sum of the friction losses in the bearings $P_{friction}$ and windage losses $P_{windage}$ as

$$P_{rot} = P_{friction} + P_{windage}. \tag{2.50}$$

The hysteresis loss densities in the stator teeth (p_{ht}) and yoke (p_{hy}) and the average eddy current loss densities in the stator teeth and yoke ($\bar{p}_{et}, \bar{p}_{ey}$) can be used for the approximation of the total core losses of the machine [30], which can be written as

$$P_{core} = V_{teeth} \left(p_{ht} + \bar{p}_{et}\right) + V_{yoke} \left(p_{hy} + \bar{p}_{ey}\right), \tag{2.51}$$

where the volumes of the stator teeth and yoke are represented by V_{teeth} and V_{yoke}.

Due to the convenience of the controls in the d–q axis quantities, the rotor reference frame equations of the PM synchronous machine can be expressed in dq frame as [31]

$$\frac{di_d}{dt} = \frac{1}{L_d}v_d - \frac{R}{L_d}i_d + \frac{L_q}{L_d}\omega_r i_q, \tag{2.52}$$

$$\frac{di_q}{dt} = \frac{1}{L_q}v_q - \frac{R}{L_q}i_q + \frac{L_d}{L_q}\omega_r i_d - \frac{\lambda \omega_d}{L_q}, \tag{2.53}$$

$$T_e = \frac{3}{2}p\left(\lambda i_q + (L_d - L_q)i_d i_q\right), \tag{2.54}$$

where L_q and L_d are the q- and d-axis inductances, R is the stator resistance, i_q and i_d are the q- and d-axis currents, v_q and v_d are the q- and d-axis voltages, ω_r is the rotor angular speed, λ is the flux amplitude induced in the stator by PMs of the rotor, and T_e is the electromagnetic torque.

The current and torque equations in abc frame can be expressed as

$$\frac{di_a}{dt} = \frac{1}{3L_s}\left[2v_{ab} + v_{bc} - 3R_s i_a + \lambda p \omega_r \left(-2E_{fa} + E_{fb} + E_{fc}\right)\right], \tag{2.55}$$

$$\frac{di_b}{dt} = \frac{1}{3L_s}\left[-v_{ab} + v_{bc} - 3R_s i_b + \lambda p \omega_r \left(E_{fa} - 2E_{fb} + E_{fc}\right)\right], \tag{2.56}$$

$$\frac{di_c}{dt} = -\left[\frac{di_a}{dt} + \frac{di_b}{dt}\right], \tag{2.57}$$

$$T_e = p\lambda(E_{fa}i_a + E_{fb}i_b + E_{fc}i_c), \tag{2.58}$$

where L_s is the stator inductance; R is the stator resistance; i_a, i_b, and i_c are the phase currents; v_{ab} and v_{bc} are the phase-to-phase voltages; and E_{fa}, E_{fb}, and E_{fc} are the induced EMFs. The mechanical part of the PM synchronous machine [31], similar to any electrical machine, can be defined as

$$\frac{d\omega_r}{dt} = \frac{1}{J}(T_m - T_e - F\omega_r), \tag{2.59}$$

$$\frac{d\theta}{dt} = \omega_r, \tag{2.60}$$

where J is the combined inertia of the rotor and wind turbine, F is the combined viscous friction of the rotor and turbine, T_m is the mechanical shaft torque, and θ is the rotor angular position.

Figure 2.24 shows a WPS where a PMSG is connected to a full bridge rectifier followed by a boost converter. In this case, the boost converter controls the electromagnetic torque. The supply side converter regulates the DC link voltage and controls the input power factor. One drawback of this configuration is the use of the diode rectifier that increases the current amplitude and losses. Also, the commutation of the diodes leads to a voltage drop of about 5–10%. The grid side converter (GSC) can be used to control the active and reactive power being fed to the grid. The automatic voltage regulator (AVR) collects the information on the turbine's speed, DC link voltage, current, and grid side voltage, and it calculates the pulse width modulation (PWM) pattern (control scheme) for the converter. This configuration has been considered for small size (<50 kW) WPSs [32].

Most of the DD generators are electrically excited synchronous generators. Manufacturers such as Enercon [13] and Lagerwey build synchronous generators. However, manufacturers such as Zephyros [33], Jeumont, and Vensys work on PMSGs. PMSGs are more efficient and do not require any excitation. Although Enercon and Lagerwey started developing DD generators in the early 1990s, the PMs were too expensive. Due to the price drop of magnets over the past 10 years, many wind turbine manufacturers started developing DD wind turbines with PMSMs [34].

Figure 2.25 shows a PMSG where the PWM rectifier is placed between the generator and the DC link, and PWM inverter is connected to the utility. In this case, the back-to-back converter can be used as the interface between the grid and the stator windings of the PMSG [35]. The turbine can be operated at its maximum efficiency and the variable speed operation of the PMSG can be controlled by using a power converter that is able to handle the maximum power flow. The stator terminal voltage can be controlled in

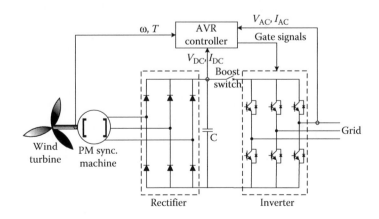

FIGURE 2.24 PMSG with a rectifier/inverter.

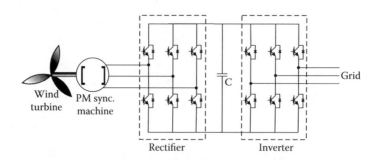

FIGURE 2.25 PMSG with a back-to-back inverter.

several ways [36]. In this system, utilizing the field orientation control (FOC) allows the generator to operate near its optimal working point in order to minimize the losses in the generator and power electronic circuit. However, the performance depends on the knowledge of the generator parameter that varies with temperature and frequency. The main drawbacks are the cost of the PMs that increases the price of the machine, and the demagnetization of the PMs. In addition, it is not possible to control the power factor of the machine [14].

Alternatively, a diode bridge for the rectification stage followed by a DC/DC converter, and a voltage source inverter (VSI) can be used as the power electronic interface between the generator and the grid [37], as shown in Figure 2.26.

The main advantages of PM excitation in comparison with electrical excitation are lower losses (no excitation losses) and lower weight (roughly a factor of 2 in active generator material), which results in lower cost. The disadvantage is that the excitation is not controllable. Usually, PM machines are designed as radial flux machines. Therefore, magnets are magnetized in a radial axis, but there are some examples of PM machines of different designs such as axial flux and transverse flux generators [38].

Axial flux generators (where flux is positioned along the axial axis) are smaller, but heavier and more expensive than radial flux machines [15]. This is mainly due to the fact that in axial flux machines, the force intensity is not optimal for all radii, and the radius where the force works is not maximum everywhere.

The use of *transverse flux generators* (two opposite PMs of different polarity in the transverse direction from one pole pair [39]) in wind turbine applications has become very popular, due to the high force intensities of this machine type [32]. The main issue with these machines is that due to the existence of a large air gap, this high force density weakens. The advantage of the transverse flux generators is the simple stator winding geometry, which offers the possibility of applying high-voltage insulation. The disadvantages are

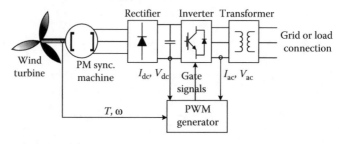

FIGURE 2.26 Block diagram of the simulated wind-turbine-driven PMSG.

TABLE 2.5

Zephyros Wind Turbine Technical Specifications

Rotor diameter	70.65 m
Rotor speed	Variable, nominal 23.5 rpm
Nominal power	2.0 MW
Transmission	DD generator, single main bearing
Rated wind speed	13 m/s
Cut-in/cut-out wind speed	3–25 m/s
Survival wind speed	70 m/s
Rotor speed control	Blade pitch
Wind class	I and S, according to IEC 61400-1
Generator mass	49 tons
Rotor mass	36 tons
Nacelle mass	12 tons

the very low power factor and the complex construction, which may result in mechanical problems and audible noise.

Zephyros is an example of a commercially available wind turbine with PMSG. The technical specifications of a Zephyros wind turbine are presented in Table 2.5 [33].

The wind speed versus output power curve of the Zephyros wind turbine is given in Figure 2.27 [33].

The wind turbine produces mechanical torque that drives the PMSG. For this application, the pitch angle controller is not used and the pitch angle is kept constant at 10°. The wind speed is kept constant at 10 m/s, as an input to the wind turbine. The PMSG is a 300 V_{dc}, 2300 rpm machine with a rated base torque of 10 N m and a maximum torque of 14.2 N m. The electrical parameters are as follows:

Stator resistance: $R_s = 0.4578 \, \Omega$.
dq-axis inductances: $L_d = 3.34 \, mH$ and $L_q = 3.34 \, mH$.
Flux induced by magnets: $\lambda = 0.171 \, Wb$.

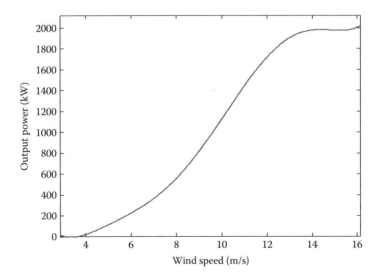

FIGURE 2.27 Wind speed versus output power curve of the Zephyros wind turbine.

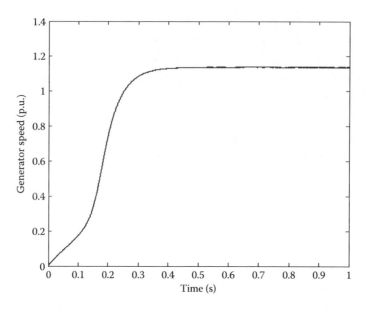

FIGURE 2.28   Rotor speed of the PMSG.

The mechanical parameters of the generator are as follows:

Inertia: $J = 0.001469 \, \text{kg m}^2$
Friction factor: $F = 0.0003035 \, \text{N ms}$
Number of pole pairs: $p = 4$

The rotor speed of the generator is recorded for one second, as shown in Figure 2.28. According to Figure 2.28, it is seen that for the given conditions, the rotor speed is above the synchronous speed.

The mechanical torque produced by the wind turbine, which is the input torque for the PMSG, is represented in Figure 2.29. It should be noted that the transient response of the torque for the first fractions of the simulation is due to the turbine dynamics and the stall condition of the generator at time zero. The speed of the generator increases as the torque increases and reaches the steady-state condition as the torque settles.

The terminal voltage of the PMSG is rectified as shown in Figure 2.30. The DC bus voltage measured at the terminals of the DC bus capacitor is shown in Figure 2.30. The DC bus voltage also reaches its steady-state value after the transient conditions and it is kept constant during the rest of the operation.

The voltage measured across the terminals of a switch in the inverter is shown in Figure 2.31, which is the PWM chopped DC bus voltage.

The output of the DC/AC inverter and line-to-line voltage after the transformer is shown in Figure 2.32. This is a PWM sinusoidal voltage that can supply AC loads.

2.7.3   Induction Machines

Three-phase induction machines are generally used in motor applications. However, they can also be effectively used as generators in electrical power systems [33]. The main issue with induction machines used as electric power generators is the need for an external reactive power source that will excite the induction machine, which is certainly not required

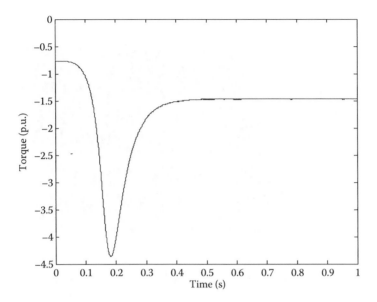

FIGURE 2.29 Torque of the direct-driven PMSG.

for synchronous machines in similar applications. If the induction machine is connected to the grid, the required reactive power can be provided by the power system [40]. The induction machine may be used in cogeneration with other synchronous generators or the excitation can be supplied from capacitor banks (only for stand-alone self-excited generators application) [41–46].

The self-excitation phenomenon in an induction machine can be easily explained using the magnetic saturation in the rotor. If the induction machine is driven by a prime mover,

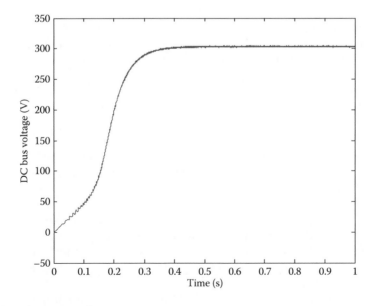

FIGURE 2.30 Rectifier output voltage.

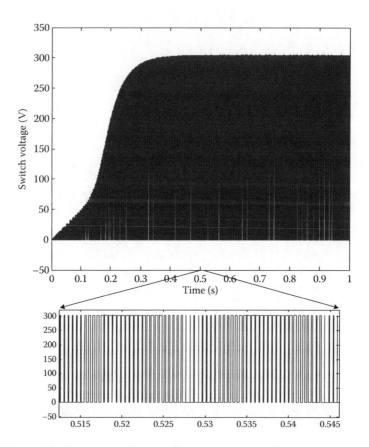

FIGURE 2.31 Voltage of the first switch of the inverter.

the residual magnetism in the rotor generates a small voltage, which forms a capacitive current, which in addition provides feedback that causes a further increase in the voltage. This is how stand-alone induction generators produce electricity. For stand-alone induction generator applications, the reactive power required for excitation can be supplied using static Volt-Ampere-reactive (VAr) compensators [47,48] or static compensators (STATCOMs) [49].

Brushless rotor construction does not need a separate source for excitation; hence its cost can be relatively low. Induction generators are often used in wind turbine applications since they require no maintenance and they offer self-protection against severe overloads, short circuits, and self-excitation [42–46]. The only drawback of these types of generators can be their inherent generated voltage and frequency regulation under varied loads [50].

The electrical parts of an induction machine can be modeled by a fourth-order state-space equivalent circuit and the mechanical part of the machine can be represented by a second-order system [51,52]. The equivalent electrical circuits for the *dq* reference frame are shown in Figure 2.33a and b. It should be noted that all electrical variables are referred to the stator in these equivalent circuits.

In Figure 2.33, the subscripts *d*, *q*, r, s, l, and m refer to the *d*-axis parameter, *q*-axis parameter, rotor parameter, stator parameter, leakage inductance, and magnetizing inductance.

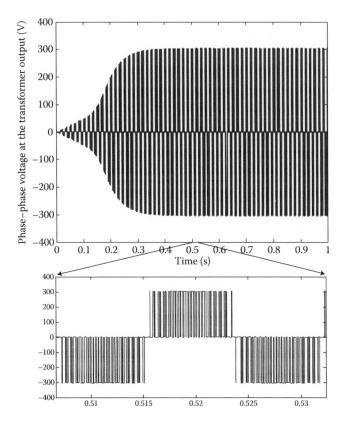

FIGURE 2.32 Phase-to-phase voltage at the transformer output.

The electrical equations can be expressed as [51–53]

$$V_{qs} = R_s i_{qs} + \frac{d\varphi_{qs}}{dt} + \omega\varphi_{ds}, \tag{2.61}$$

$$V_{ds} = R_s i_{ds} + \frac{d\varphi_{ds}}{dt} - \omega\varphi_{qs}, \tag{2.62}$$

$$V'_{qr} = R'_r i'_{qr} + \frac{d\varphi'_{qr}}{dt} + (\omega - \omega_r)\varphi'_{dr}, \tag{2.63}$$

$$V'_{dr} = R'_r i'_{dr} + \frac{d\varphi'_{dr}}{dt} - (\omega - \omega_r)\varphi'_{qr}, \tag{2.64}$$

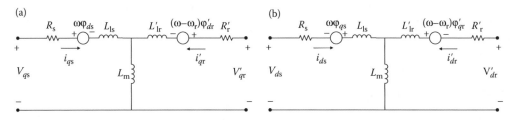

FIGURE 2.33 Induction machine electrical equivalent circuits.

where V_{qs} and i_{qs} are the q-axis stator voltage and current, V'_{qr} and i'_{qr} are the q-axis rotor voltage and current, V_{ds} and i_{ds} are the d-axis stator voltage and current, V'_{dr} and i'_{dr} are the d-axis rotor voltage and current, R_s and L_{ls} are the stator resistance and leakage inductance, R'_r and L'_{lr} are the rotor resistance and leakage inductance, L_m is the magnetizing inductance, ω is the rotor angular speed, ω_r is the electrical angular speed, φ_{ds} and φ_{qs} are the stator q- and d-axis fluxes, and φ'_{qr} and φ'_{dr} are the rotor q- and d-axis fluxes.

The fluxes and equivalent stator and rotor inductances can be represented as [51–53]

$$\varphi_{qs} = L_s i_{qs} + L_m i'_{qr}, \tag{2.65}$$

$$\varphi_{ds} = L_s i_{ds} + L_m i'_{dr}, \tag{2.66}$$

$$\varphi'_{qr} = L'_r i'_{qr} + L_m i_{qs}, \tag{2.67}$$

$$\varphi'_{dr} = L'_r i'_{dr} + L_m i_{ds}, \tag{2.68}$$

$$L_s = L_{ls} + L_m, \tag{2.69}$$

$$L'_r = L'_{lr} + L_m. \tag{2.70}$$

The torque of the machine is related to the q- and d-axis currents and fluxes of the machine as

$$T_e = \frac{3}{2} p(\varphi_{ds} i_{qs} - \varphi_{dq} i_{ds}), \tag{2.71}$$

where T_e is the generator torque and p is the number of pole pairs.

The mechanical system equations are

$$\frac{d\omega_m}{dt} = \frac{1}{2J}(T_m - T_e - F\omega_m), \tag{2.72}$$

$$\frac{d\theta_m}{dt} = \omega_m, \tag{2.73}$$

where T_m is the input mechanical torque, F is the friction coefficient, J is the inertia constant, and θ_m is the rotor angular position.

2.7.3.1 *Conventional Control Scheme*

The conventional scheme for frequency control of an induction generator consists of a speed governor, which regulates the prime mover, and switched capacitor banks, to provide reactive power to the load and excitation of the induction generator, as shown in Figure 2.34. The input power for the induction generator has to match the load power demand. The speed of the wind is not controllable; therefore, the induction generator should be controlled according to the load variations [54]. Increasing the speed of the shaft, in the case of any increase in the load, is very difficult. Therefore, the frequency control technique has a very poor performance. Another issue is that voltage regulation requires precise control of the reactive power source.

2.7.3.2 *Voltage and Frequency Control with Load Regulation*

By employing a variable impedance controller, as shown in Figure 2.35, the turbine speed governor can be eliminated and the efficiency of the system can be increased [55]. Through

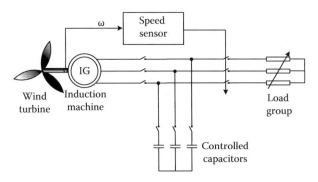

FIGURE 2.34 Conventional control scheme for the induction generator.

this method, voltage and frequency control is achieved by using an impedance controller through a bridge rectifier, chopper, and a DC side resistor R_{dc}. A large AC capacitor bank is used to meet the power demand at the desired power factor.

The main purpose of an impedance controller is to keep the total real and reactive powers at a constant level. In the case of any load changes, the impedance controller will absorb the active or reactive power, which has not been used by the load. The excess power will be redirected to R_{dc} and will be consumed on this resistance. Thus, the total reactance of the system is adjusted as a function of the duty cycle of the chopper switch as shown in Figure 2.35. On the other hand, the excess power, which is not absorbed by the load, will be released through the resistor. Therefore, the prime mover requires no regulation and can always be operated at the required power, voltage, and frequency [56].

The impedance controller can regulate the voltage and frequency of the induction generator more efficiently than the conventional control strategy. However, the excess energy is consumed in the R_{dc} resistance, without effectively being used. Using the excess energy in storage devices or injecting it to the grid would be a more efficient solution for load regulation. The main concern is the need of large AC capacitors that provide the required reactive power for the excitation process.

2.7.3.3 Improved Voltage and Frequency Control with a VSI

Utilizing a VSI instead of a bridge rectifier, the control range can be extended and the excitation capacitors can be eliminated. As shown in Figure 2.36, the AC side of the VSI is

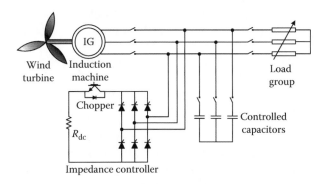

FIGURE 2.35 Induction generator voltage and frequency control with load regulation.

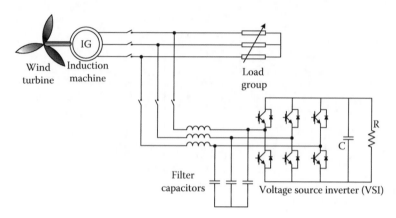

FIGURE 2.36 Induction generator controlled by VSI.

connected in parallel to the load using a small inductor–capacitor filter, whereas the DC side of the inverter consists of an electrolytic DC capacitor and a resistor [57].

The use of the VSI improves the control range since it not only provides the reactive power required for the excitation of the induction generator at full load, but also compensates the inductive-resistive loads in the system.

The drawback with using the VSI is the startup process, which can be quite complicated. This control strategy requires additional DC voltage in order to provide reactive power for the startup of the induction generator, because the existing filter capacitor cannot support enough reactive power for the self-excitation of the induction generator. This option may not be feasible for some applications, where the startup power source is not available.

2.7.3.4 Advanced Voltage and Frequency Control Using VSI

The new control strategy consists of an excitation capacitor (instead of filter capacitors) to provide the required reactive power for the startup condition and eliminate the precharging requirement of the DC capacitor. It is important to overdesign this capacitor, because they must be larger than the filter ones, but smaller than the capacitors used for the conventional control strategy. The structure of the induction machine controlled by a VSI with exciter capacitors is presented in Figure 2.37.

The main advantage of this topology is that the whole system is more reliable, since there is no need for a startup power source [57]. In addition, excitation capacitors are much smaller than the ones used for the conventional control, which makes this control flexible enough for various low-cost applications of wind turbines.

2.7.3.5 Back-to-Back Connected PWM VSI

The common structure of a squirrel cage induction generator with back-to-back converters is shown in Figure 2.38. In this structure, the stator winding is connected to the utility grid through a four-quadrant power converter. Two PWM VSIs are connected back-to-back through a DC link. The stator side converter regulates the electromagnetic torque and supplies reactive power, while the GSC regulates the real and reactive power delivered from the

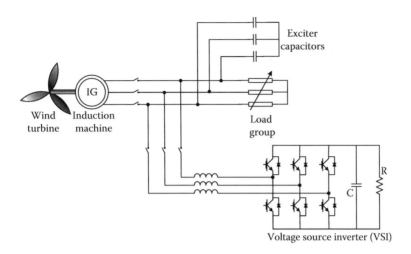

FIGURE 2.37 Induction generator controlled by VSI with exciter capacitors.

system to the utility and regulates the DC link. This topology has several practical advantages, and one of them is the possibility of a fast transient response for speed variations. In addition, the inverter can operate as a VAR/harmonic compensator [58].

On the other hand, the main drawback is the complex control system. Usually an FOC is used to control this topology when its performance relies on the generator parameters, which vary with temperature and frequency. Hence, in order to supply the magnetizing power requirements, that is, to magnetize the machine, the stator side converter must be oversized 30–50% with respect to the rated power.

2.7.3.6 DFIG

Figure 2.39 presents a topology that consists of a DFIG with AC/DC and DC/AC converters, as a four-quadrant AC/AC converter using isolated gate bipolar transistors (IGBTs) connected to the rotor windings. In the DFIG topology, the induction generator is not a squirrel cage machine and the rotor windings are not short circuited. Instead, the rotor windings are used as the secondary terminals of the generator to provide the capability of controlling the machine power, torque, speed, and reactive power. To control the active and reactive power flow of the DFIG topology, the rotor side converter (RSC) and GSC should be controlled separately [59–62].

Wound rotor induction machines can be supplied from both the rotor and stator sides. The speed and the torque of the wound rotor induction machine can be controlled by

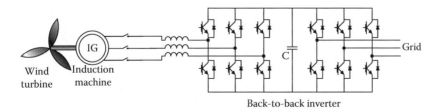

FIGURE 2.38 Induction machine controlled by a back-to-back inverter.

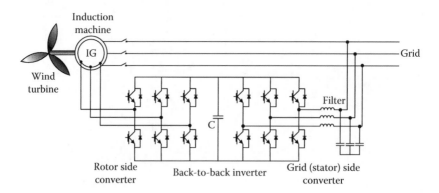

FIGURE 2.39 Doubly fed induction machine topology.

regulating voltages from both the rotor and the stator sides of the machine. The DFIG can be considered as a synchronous/asynchronous hybrid machine. In the DFIG, like the synchronous generator, the real power depends on the rotor voltage magnitude and angle. In addition, like the induction machine, the slip is also a function of the real power [63].

DFIG topology offers several advantages in comparison to systems using direct-in-line converters [64,65]. These benefits are as follows:

1. The main power is transferred through the stator windings of the generator that are directly connected to the grid. Around 65–75% of the total power is transmitted through the stator windings and the remaining 25% of the total power is transmitted using the rotor windings, that is, through the converters. Since the inverter rating is 25% of the total system power, the inverter cost and size are considerably reduced.

2. While the generator losses are the same in both topologies (direct-in-line and DFIG), the inverter losses can be reduced from around 3% to 0.75%, since the inverter transfers only 25% of the total power. Therefore, an efficiency improvement of approximately 2–3% can be obtained.

3. DFIG topology offers a decoupled control of generator active and reactive powers [66,67]. The cost and size of the inverter and EMI filters can be reduced since the inverter size is reduced. In addition, the inverter harmonics are lowered since the inverter is not connected to the main stator windings.

In the rotor circuit, two voltage-fed PWM converters are connected back-to-back, while the stator windings are directly connected to the AC grid side, as shown in Figure 2.40. The direction and magnitude of the power between the rotor windings and the stator windings can be controlled by adjusting the switching of the PWM signals of the inverters [68–70].

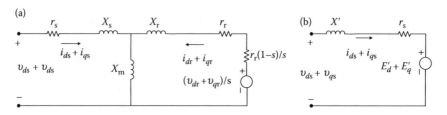

FIGURE 2.40 (a) DFIG steady state and (b) dynamic equivalent electrical circuits.

This is very similar to connecting a controllable voltage source to the rotor circuit [71]. This can also be considered as a conventional induction generator without a zero rotor voltage.

In order to benefit from the variable speed operation, the optimum operating point of the torque–speed curve should be tracked precisely [72]. By controlling the torque of the machine, the speed can be adjusted. Thus, by using the instantaneous rotor speed value and controlling the rotor current i_{ry} in the stator flux-oriented reference frame, the desired active power can be obtained. Operating at the desired active power results in the desired speed and torque [60]. On the other hand, the GSC is controlled to keep the DC link voltage fixed and independent of the direction of rotor power flow. The decoupled control of the active and reactive power flows between the rotor and grid is provided by vector-oriented control of supply voltage.

The per unit electrical dq frame equations of the DFIG, in phasor form, when the stator transients are neglected [71,73], can be expressed as

Stator voltage:

$$v_{qs} = r_s i_{qs} - \omega_s \varphi_{ds},$$ (2.74)

$$v_{ds} = r_s i_{ds} + \omega_s \varphi_{qs}.$$ (2.75)

Rotor voltage:

$$v_{qr} = r_r i_{qr} + s\omega_s \varphi_{dr} + \frac{d\varphi_{qr}}{dt},$$ (2.76)

$$v_{dr} = r_r i_{dr} - s\omega_s \varphi_{qr} + \frac{d\varphi_{dr}}{dt}.$$ (2.77)

Flux linkage:

$$\varphi_{ds} = -L_{ss} i_{qs} + L_m i_{qr},$$ (2.78)

$$\varphi_{ds} = -L_{ss} i_{ds} + L_m i_{dr},$$ (2.79)

$$\varphi_{qr} = -L_m i_{qs} + L_{rr} i_{qr},$$ (2.80)

$$\varphi_{dr} = -L_m i_{ds} + L_{rr} i_{dr}.$$ (2.81)

Electromagnetic torque:

$$T_{em} = \varphi_{qr} i_{dr} - \varphi_{dr} i_{qr}.$$ (2.82)

The electrical equivalent circuits for the steady-state and dynamic conditions are given in Figure 2.40a and b.

Since the rotor is short circuited in a regular squirrel cage induction machine, the rotor voltages given by Equations 2.76 and 2.77 are zero. In order to eliminate the rotor currents, voltage equations can be rewritten in terms of transient reactances as

$$v_{qs} = r_s i_{qs} + X' i_{ds} + E'_q,$$ (2.83)

$$v_{ds} = r_s i_{ds} - X' i_{qs} + E'_d,$$ (2.84)

where E'_q and E'_d are the voltage behind the transient impedance for the q- and d-axes, and X' is the transient reactance. These transient voltages and transient reactance can be expressed by

$$E'_q = \frac{-\omega_s L_m \varphi_{dr}}{L_{rr}}, \tag{2.85}$$

$$E'_d = \frac{\omega_s L_m \varphi_{qr}}{L_{rr}}, \tag{2.86}$$

$$X' = \omega_s \left(\frac{L_{ss} - L_m^2}{L_{rr}} \right). \tag{2.87}$$

The rotor flux linkages can be represented in terms of E'_q and E'_d [74,75]; hence the rotor circuit dynamics can be obtained from

$$\frac{dE'_q}{dt} = -s\omega_s E'_d - \omega_s v'_{dr} - \frac{1}{T'_0} \left(E'_q - (X - X')i_{ds} \right), \tag{2.88}$$

$$\frac{dE'_d}{dt} = s\omega_s E'_q + \omega_s v'_{qr} - \frac{1}{T'_0} \left(E'_d + (X - X')i_{qs} \right), \tag{2.89}$$

where T'_0 represents the rotor circuit time constant. The transient rotor q- and d-axis voltages are

$$v'_{qr} = \frac{v_{qr} X_m}{(X_m + X_r)}, \tag{2.90}$$

$$v'_{dr} = \frac{v_{dr} X_m}{(X_m + X_r)}. \tag{2.91}$$

The electromagnetic torque can be expressed independent of the rotor currents i_{dr} and i_{qr} for unity stator angular speed:

$$T_{em} = E'_d i_{ds} + E'_q i_{qs}. \tag{2.92}$$

By substituting Equations 2.88 and 2.89 into Equation 2.92, the electromagnetic torque would be

$$T_{em} = v_{ds} i_{ds} + v_{qs} i_{qs} - r_s \left(i_{ds}^2 + i_{qs}^2 \right). \tag{2.93}$$

Since the stator resistance losses are small enough, they can be neglected; therefore the electromagnetic power or torque per unit would be

$$P_s = T_{em} = v_{ds} i_{ds} + v_{qs} i_{qs}. \tag{2.94}$$

The reactive power injected/absorbed to/from grid can be calculated by

$$Q_s = v_{qs} i_{ds} + v_{ds} i_{qs}. \tag{2.95}$$

The power absorbed/injected from/to the grid would be

$$P_r = v_{dr}i_{dr} + v_{qr}i_{qr}, \tag{2.96}$$

$$Q_r = v_{qr}i_{dr} + v_{dr}i_{qr}. \tag{2.97}$$

Assuming no loss power electronic converters are used with the DFIG, the active power supplied to the grid (P_e) is equal to the sum of the stator power (P_s) and the rotor power (P_r). Similarly, the total reactive power (Q_e) is the summation of the stator reactive power (Q_s) and the reactive power of the GSC (Q_g).

The rotor motion equation of the DFIG can be represented as

$$2H_r \frac{d\Delta\Omega_r}{dt} = T_{em} - T_m, \tag{2.98}$$

where $\Delta\Omega_r$ is the angular rotor speed deviation and H_r is the generator rotor inertia. In all of the topologies using gearboxes, the rotor and wind turbine interactions differ from a DD system [73]. The effect of the turbine–generator mill coupling can be described by

$$2H_m \frac{d\Delta\Omega_m}{dt} = T_m - T_\omega, \tag{2.99}$$

where H_m is the wind turbine shaft inertia, $\Delta\Omega_m$ is the wind turbine angular speed deviation, and T_ω is the wind turbine prime torque from the wind. The mechanical torque T_m that drives the generator shaft can be described as the twist positions between the turbine shaft and generator rotor shaft and can be obtained from the following equation:

$$T_m = K_s(\theta_r - \theta_m), \tag{2.100}$$

where θ_r is the rotor angular position, θ_m is the wind turbine angular position, and K_s is the stiffness coefficient of the shaft.

Figure 2.41 shows the principle of controlling power flow (current) by controlling the rotor voltage of a doubly fed machine. Using a PWM inverter in a rotor circuit, it is possible to control the magnitude and phase of the rotor voltage, which controls the slip of the machine. Therefore, the power flow can be controlled by manipulating the slip [76].

The power electronic interface of the conventional induction machine is usually designed for rating the power of the machine. However, only during high wind gusts, maximum rated power flows through the system (2–5% of the time). Therefore, the power electronic interfaces of these machines are usually oversized.

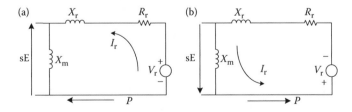

FIGURE 2.41 Power transfer of DFIG by controlling the slip: (a) power transfer to DFIG and (b) power transfer from DFIG.

Nevertheless, the oversizing problem can be solved by using a DFIG. Since the speed range of the turbine is still wide, a power converter, which is rated for much lower powers, can be placed only on the rotor side and the stator can be connected directly to the grid. The most important advantage of the DFIG is that, since the power going through the rotor is usually around 25–30% of the power going through the stator, the power electronic interface can be designed for only 25–30% of the total power.

Using the variable speed capability of the turbine, it is possible to obtain optimal energy gain and achieve a speed range from roughly 60% to 110% of the rated speeds [77]. By selecting the gearbox ratio so that the synchronous speed of the generator just falls in the middle of the speed range (at 85% of rated speed), it is possible to use the lowest converter power rating. A converter rating of not more than 35% of the rated turbine power would be sufficient.

The rotor converter can control the slip of the induction generator to obtain a wide speed range (Figure 2.42). If the input wind power surpasses the stability limit, the generator shifts to subsynchronous mode.

On the other hand, if the input wind power decreases, it is necessary to feed additional electric power over the slip-rings into the rotor, in order to prevent the transition to motor-driven mode, in which the generator operates subsynchronously.

Figure 2.43 shows the torque–speed characteristic of the DFIG, where $\omega < \omega_s$ corresponds to the operation mode shown in Figure 2.43a and $\omega > \omega_s$ corresponds to the operation mode shown in Figure 2.43b.

The total power of a doubly fed induction machine can be calculated by considering all the losses, and the absorbed/supplied power from rotor side,

$$P_e = P_m - P_g = sP_g = P_r + P_{\chi r}, \tag{2.101}$$

where P_m is the total (mechanical) power from the wind turbine's shaft, P_g is the power situated in the air gap, P_r is the rotor power, and $P_{\chi r}$ is the power of rotor losses, as shown in Figure 2.44.

Hence, the power factor of the machine can be controlled by controlling the magnitude and phase angle of rotor current. The rotor current is

$$I_r = \frac{V\angle 0° - (V_r/s)\angle\theta_r}{(R_s + R_r/s) + j(X_s + X_r)}. \tag{2.102}$$

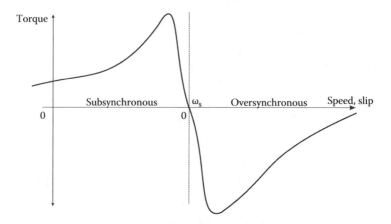

FIGURE 2.42 Torque–speed characteristic of the DFIG.

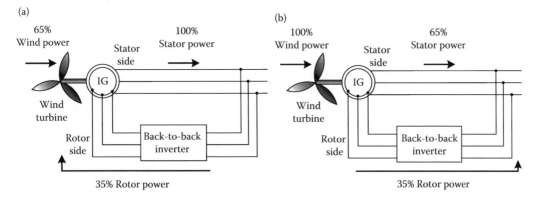

FIGURE 2.43 Operation modes of the DFIG: (a) subsynchronous operation $\omega < \omega_s$ and (b) oversynchronous operation $\omega > \omega_s$.

Therefore, the power factor can be adjusted by controlling the angle (using a PWM inverter) through

$$P_e = sP_g = 3\left[I_r^2 R_r + V_r I_r \cos \varphi_r\right], \tag{2.103}$$

where φ_r represents the angle between the rotor voltage (V_r) and the rotor current (I_r). From Equation 2.103, it can be observed that if the phase angle of the output current is controlled, the power factor can also be controlled.

The DFIG allows the harvesting of energy for a wide range of turbine torques and speeds. This is achievable by independently controlling the frequency of the injected rotor voltage, while the stator frequency is kept constant.

2.7.3.7 Voltage and Frequency Control Using Energy Storage Devices

By utilizing the induction generator as a voltage source connected to a variable load with additional battery storage bank, the voltage and frequency of the induction generator can be controlled in the case of load variations. The main purpose of batteries is to assist the

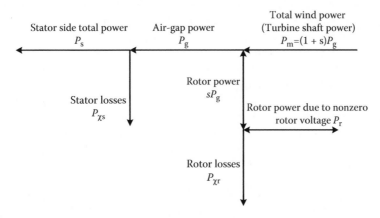

FIGURE 2.44 Total power and power flow of DFIG.

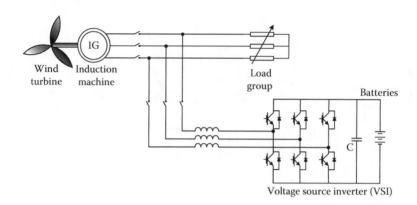

FIGURE 2.45　Voltage and frequency control using energy storage.

induction generator in meeting the load demand. When the load current is less than the generator current, the extra current is used to charge the battery energy storage. On the other hand, when the load current is larger than the generator current, the current is supplied from the battery to the load. With this strategy, the voltage and frequency of the generator can be manipulated for various load conditions.

Having additional energy storage decreases the system inertia, improves the behavior of the system in the case of disturbances, compensates transients, and therefore improves the overall system efficiency [78]. However, it adds to the initial cost of the system and requires periodical maintenance depending on the storage devices.

Therefore, the voltage and frequency management discussed in Section 2.6.3.4 can be improved by using batteries as the controllable load of the VSI as presented in Figure 2.45. In this way, the load regulation is achieved by monitoring the power flow to the batteries. A bidirectional inverter/converter can be used for the power flow from/to the batteries. As an alternative, the battery voltage can be converted to AC voltage with another individual inverter to provide power to the AC loads.

Another alternative energy storage source is the superconducting magnetic energy storage (SMES). The SMES can store electric energy in magnetic form, which can be used as a storage device in wind power applications [79]. In SMES systems, the stored energy can be charged or discharged very fast with a very high power density. A secondary battery pack or flywheels generate an increase in the energy density of the overall storage system of the wind turbine. Using an SMES and batteries as shown in Figure 2.46 leads to higher energy and power densities, without employing any rotating parts.

Storage systems can be connected in various forms to the wind turbine systems [80–84]. Generally, a bidirectional DC/DC converter is required for the integration of the storage system in the DFIG system [85]. In this topology, one of the converters regulates the storage power, whereas the other is responsible for DC bus voltage control. The bidirectional energy storage topology for DFIGs in wind applications is shown in Figure 2.47.

Alternately, in order to reduce the switching transients and losses, multilevel converters can be used with lower switching frequencies [86,87]. A multilevel converter, which can substitute the one shown in Figure 2.47, is presented in Figure 2.48.

Power management of the energy storage devices in wind applications is achievable by utilizing single- or multistage converters [88]. A multiple-level energy storage system is presented in Figure 2.49, which is useful for the long- and short-term storage systems. Fluctuations in the generator power of each wind turbine and the projected power value

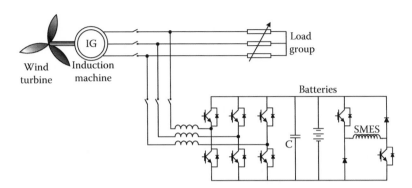

FIGURE 2.46 Energy storage with SMES and batteries for wind applications.

should be used to determine a reference power value for the individual storage levels. Instant fluctuations can be used for short-term storage or can be compensated for using short-term storage. The difference from the predicted wind power can be compensated for using long-term storage. The sum of the powers should be equal to the total reference output by extracting and attributing the various frequency components to the appropriate storage levels [78]. Storage devices can function as ultracapacitors for short-term storage and batteries for long-term storage levels.

The produced excess energy of a wind generator can also be converted to kinetic energy and stored in this form by using a flywheel. In this approach, the flywheel can be driven by another induction generator and this induction generator can be controlled by a cascaded bidirectional rectifier/inverter topology [89]. This energy can be converted to electrical energy for future use if the power production of the wind generator is not sufficient. The flywheel energy storage systems can be used as power regulators over short periods of time for electrical power quality improvement and to alleviate the voltage and frequency fluctuations. Therefore, the storage system's lifetime and efficiency can be increased [90–93]. These systems improve the power quality of the energy produced by wind generators and, therefore, they can be connected to the grid with less fluctuations and harmonics [94].

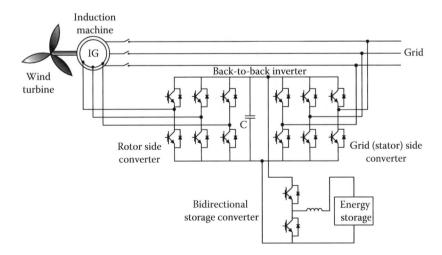

FIGURE 2.47 Energy storage with bidirectional converter in DFIG systems.

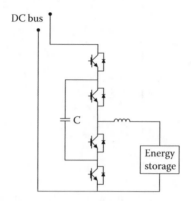

FIGURE 2.48 Multilevel converter for energy storage.

The block diagram of a DFIG is shown in Figure 2.50. The wind turbine produces mechanical torque that drives the induction generator.

In Figure 2.50, RSC stands for the rotor side converter and GSC stands for the grid side converter. The wind turbine produces the mechanical torque T_m that drives the induction generator with the rotor speed of ω_r. β represents the pitch angle, $\omega_{reference}$ represents the reference rotor speed (1 p.u.), V_{dc} represents the DC bus voltage, $I_{abc\_(Stator\ bus)}$ stand for the phase currents measured from the stator windings, $I_{abc\_(RSC\ bus)}$ is the phase currents measured from the rotor side, $I_{abc\_(GSC\ bus)}$ is the phase current of the GSC, Q_{ref} is the reference reactive power, $Q_{(Grid\ bus)}$ is the measured reactive power that is injected to the grid, $V_{abc\_(Grid\ bus)}$ is the phase voltages measured from the grid bus, $I_{abc\_(GSC\ bus)}$ is the GSC's output phase current, and I_{q_ref} is the reference q-axis current generated by GSC.

The pitch angle controller, the RSC controller, and the GSC controller for this application have subblocks, which are described as follows [95,96].

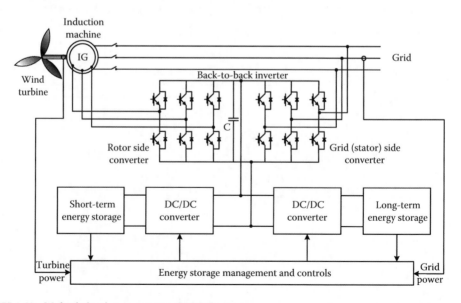

FIGURE 2.49 Multiple-level energy storage in DFIG system.

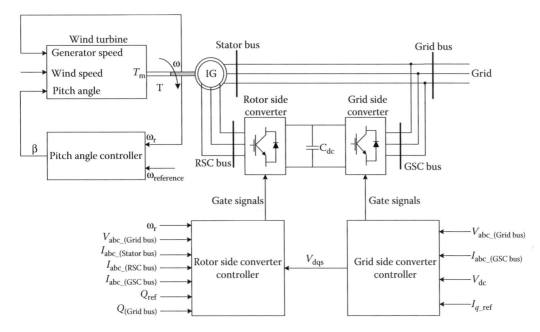

FIGURE 2.50 Block diagram of a DFIG topology.

2.7.3.7.1 The Pitch Angle Controller

For the pitch angle controller, the actual rotor speed is subtracted from the reference rotor speed and the error is processed by a PI controller. The output of the PI controller might be higher than the maximum pitch angle value, depending on the speed difference. Therefore, the output of the PI controller is limited using a saturation block. The pitch angle controller block diagram is shown in Figure 2.51. This controller helps reduce the rotor speed to a reasonable value in high wind speed conditions.

2.7.3.7.2 The RSC Controller

The RSC is used to control the wind turbine output power. This converter can also be used to regulate the voltage (or reactive power) measured at the grid terminals. The power is regulated to track the predefined power–speed characteristic of the wind turbine, and the mechanical power curve can be obtained for different wind speeds. The actual speed of the turbine (ω_r) and power control loop are measured and the corresponding mechanical power is used as the reference power [69,95,96]. The block diagram of the proposed RSC controller is depicted in Figure 2.52.

The power losses are added to the actual electrical output power, which is measured at the grid connection bus of the DFIG topology. The resultant power is then compared with the reference power that corresponds to the operating speed obtained from the tracking characteristic or the speed–power curve. The power error is processed using a PI controller

FIGURE 2.51 Block diagram of the pitch angle controller.

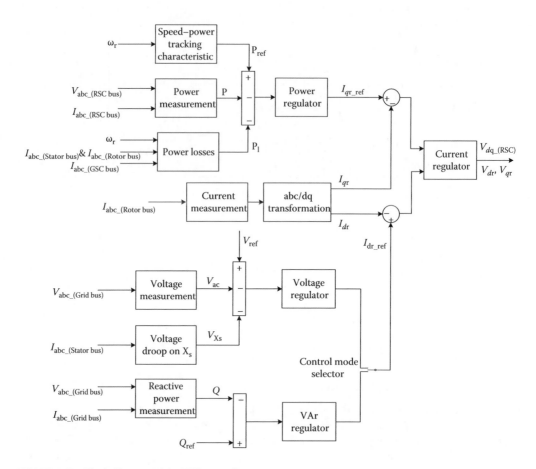

FIGURE 2.52 Block diagram of the RSC controller.

to track the reference power. The reference rotor current I_{qr_ref} is produced by the PI regulator, which must be injected to the rotor side by the RSC. The electromagnetic torque T_{em} is produced by this current component (according to Equation 2.94). A current regulator reduces the error between the actual and the reference q-axis current I_{qr} to zero. On the other hand, the d-axis component of the current is related to the output voltage of the RSC or the reactive power produced/absorbed by this converter (Equation 2.97). Either the voltage regulation or the reactive power regulation can be used to determine the reference d-axis current I_{dr_ref}. The current regulator also uses the error between this reference current and the measured d-axis current. The output of the current regulator is the voltage that must be produced by the RSC.

2.7.3.7.3   The GSC Controller

The GSC is used to regulate the DC bus capacitor voltage. The energy stored or released by the GSC to/from the DC link capacitor allows the generation or absorption of the reactive power. The block diagram of the GSC controller is shown in Figure 2.53.

    The grid phase currents (converted to dq-axis currents) and the DC bus voltages are measured in the GSC controller. The DC bus voltage is regulated by a control loop where the output is the reference d-axis current for the current regulator, $I_{d\_ref(GSC\ bus)}$. It should

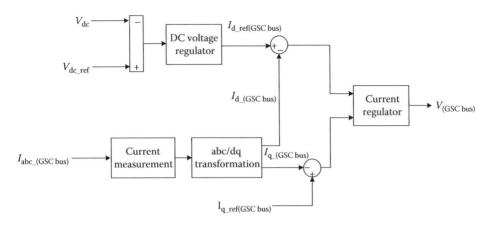

FIGURE 2.53 Block diagram of the GSC controller.

be noted that $I_{d(\text{GSC bus})}$ is the current that is in the same phase with the grid voltage, which controls the active power. The current regulator in the inner loop controls the magnitude and phase of the output voltage of the GSC ($V_{(\text{GSC bus})}$).

The active power flow that is measured from the grid bus shown in Figure 2.51 is obtained as shown in Figure 2.54.

The reactive power flow that is also measured from the grid bus of Figure 2.50 is shown in Figure 2.55. The reference reactive power is set to zero and the actual reactive power flow tracks its reference with a small amount of oscillations as shown in Figure 2.55. The reactive power is controlled by the RSC, which was described earlier. Since the reactive power regulation is selected as in Figure 2.52, the grid bus voltage is also regulated. If the grid bus voltage regulation was preferred in the control loop, the necessary reactive power would be injected/absorbed to/from the grid.

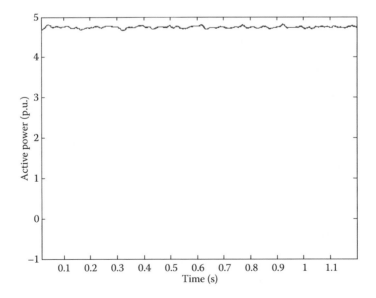

FIGURE 2.54 Active power output to the grid side.

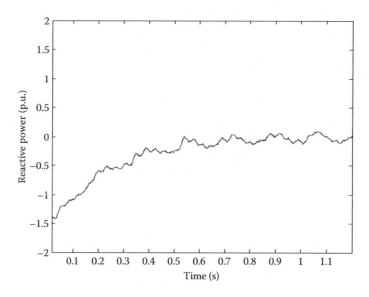

FIGURE 2.55   Reactive power output to the grid side.

The DC bus voltage that is the DC bus capacitor voltage of the back-to-back converter is shown in Figure 2.56. It can be observed that the DC bus voltage tracks the reference DC bus voltage that is set to 1200 V.

Finally, the grid side voltage and current are shown in Figures 2.57 and 2.58, respectively. Due to the power flow through the back-to-back converter, the voltage and current have some harmonics; however, the harmonic distortion is not significant.

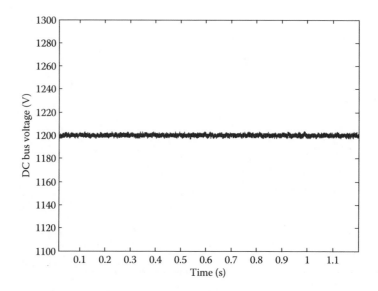

FIGURE 2.56   DC bus voltage between the converters.

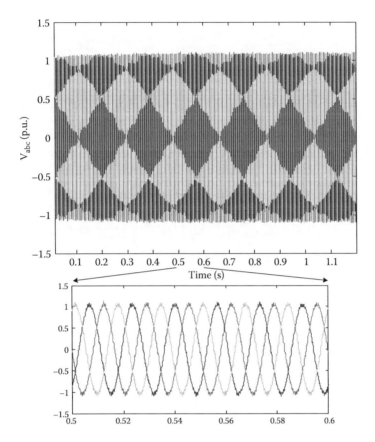

FIGURE 2.57 **(See color insert following page 80.)** Three-phase voltage measured from the grid side bus.

2.8 Synchronous Generators

Synchronous generators are commonly used for variable speed wind turbine applications, due to their low rotational synchronous speeds that produce voltage at grid frequency.

Based on the dynamics of the stator, field, and damper windings, a synchronous machine model can be described as the equivalent q- and d-axis electrical circuits shown in Figure 2.59. The model is represented in dq rotor reference frame. In this model, all the rotor parameters and electrical quantities are transferred to the stator side [97,98].

In the figure, d and q subscripts refer to the d- and q-axis quantities, r and s subscripts refer to the rotor and stator quantities, l and m refer to the leakage and magnetizing inductances, and f and k refer to the field and damper winding quantities.

The voltage equations for the equivalent circuits can be represented as

$$V_d = R_s i_d + \frac{d\varphi_d}{dt} - \omega_R \varphi_q, \tag{2.104}$$

$$V_q = R_s i_q + \frac{d\varphi_q}{dt} + \omega_R \varphi_d, \tag{2.105}$$

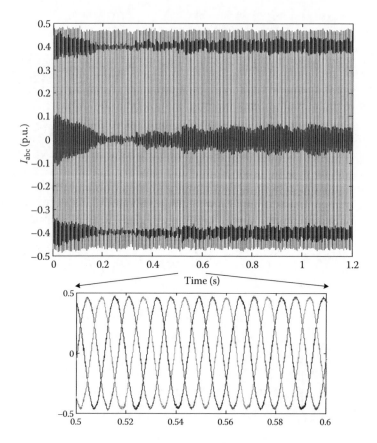

FIGURE 2.58   Three-phase current measured from the grid side bus.

$$V'_{fd} = R'_{fd}i'_{fd} + \frac{d\varphi'_{fd}}{dt}, \tag{2.106}$$

$$V'_{kd} = R'_{kd}i'_{kd} + \frac{d\varphi'_{kd}}{dt}, \tag{2.107}$$

$$V'_{kq1} = R'_{kq1}i'_{kq1} + \frac{d\varphi'_{kq1}}{dt}, \tag{2.108}$$

$$V'_{kq2} = R'_{kq2}i'_{kq2} + \frac{d\varphi'_{kq2}}{dt}, \tag{2.109}$$

where the fluxes can be described as

$$\varphi_d = L_d i_d + L_{md}\left(i'_{fd} + i'_{kd}\right), \tag{2.110}$$

$$\varphi_q = L_q i_q + L_{mq}i_{kq}', \tag{2.111}$$

$$\varphi_{fd}' = L_{fd}'i_{fd}' + L_{md}(i_d + i_{kd}'), \tag{2.112}$$

$$\varphi_{kd}' = L_{kd}'i_{kd}' + L_{md}\left(i_d + i_{fd}'\right), \tag{2.113}$$

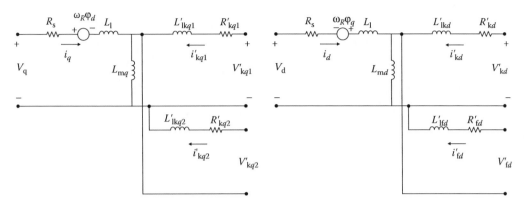

FIGURE 2.59 Synchronous machine *q*- and *d*-axis equivalent electrical circuits.

$$\varphi_{kq1}' = L_{kq1}'i_{kq1}' + L_{mq}i_q, \tag{2.114}$$

$$\varphi_{kq2}' = L_{kq2}'i_{kq2}' + L_{mq}i_q. \tag{2.115}$$

The mechanical system of the synchronous machine can be described by

$$\Delta\omega(t) = \frac{1}{2J}\int_0^t (T_m - T_e)dt - K_d\Delta\omega(t), \tag{2.116}$$

where $\Delta\omega(t)$ is the speed variation of the machine, J is the inertia constant, T_m is the mechanical input torque, T_e is the induced electromagnetic torque, and K_d is the damping factor representing the effect of damper windings.

Synchronous generators can be an appropriate selection for variable speed operation of wind turbines [99,100]. Since a pitch control mechanism increases the cost of the turbine and causes stress on the turbine and generator, the flexibility of operation speed must be achieved through other means [101]. Synchronous generators in variable speed operation will generate variable voltage and variable frequency power. The output voltage of the synchronous generator can be controlled by using an AVR for the excitation voltage of the field voltage. However, induction generators require controlled capacitors for voltage control. In addition, their operating speed should be over the synchronous speed in order to operate in generating mode [102].

More efficient than synchronous generators are multipole synchronous generators since the gear can be eliminated and the turbine and generator can operate in DD mode [103,104]. However, synchronous generators without multipoles require gearboxes in order to produce the required frequency for grid connection. On the other hand, either a DC voltage source or an AC/DC converter is required for synchronous generators in wind applications in order to produce the required excitation voltage for the field windings. The synchronous generator connection with the wind turbine is shown in Figure 2.60.

If multipole synchronous generators are not used, the turbine–generator coupling can be provided using a gearbox as shown in Figure 2.61. An IGBT bridge can be used as an alternative to the diode bridge for the stator winding rectifier. In this case, the stator side converter is used for active power control, while the GSC is used for reactive power control or DC bus voltage regulation. This option can also be applied to the topology shown in Figure 2.60.

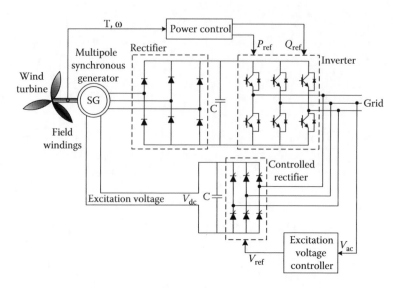

FIGURE 2.60　Multipole synchronous generator for wind turbine applications.

In both topologies shown in Figures 2.60 and 2.61, instead of using a controlled rectifier, a diode bridge followed by a DC/DC converter can be used to control the excitation voltage by the DC/DC converter.

As an alternative to the controlled rectifier, the DC bus voltage or current can be regulated using a DC/DC converter at the DC link [105,106]. This topology is shown in Figure 2.62.

In this topology, the DC/DC converter is controlled by a PWM generator and the duty cycle is determined by a controller that regulates the DC link current [105]. The DC link current controller is shown in Figure 2.63, where I_{dc}^* is the reference DC link current, I_{dc} is the measured DC link current, V_{in} is the input voltage of the converter, and V_{dc} is the output voltage of the converter. Here the reference DC link current is determined based on

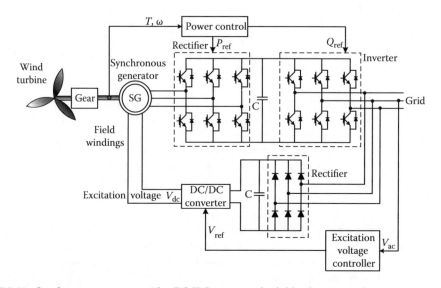

FIGURE 2.61　Synchronous generator with a DC/DC converter for field voltage control.

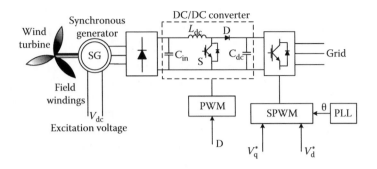

FIGURE 2.62 Synchronous generator with DC link regulation.

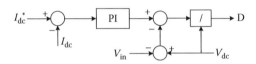

FIGURE 2.63 Current loop for the DC/DC converter duty cycle.

the current–torque relationship of the generator in order to capture maximum power from the wind turbine.

On the other hand, the grid side inverter is controlled by a sinusoidal pulse width modulation (SPWM) generator using the reference d- and q-axis reference voltages, V_d^* and V_q^*. The d-axis reference voltage is determined using the d-axis current controller as shown in Figure 2.64. The reference power factor determines the reference d-axis current, i_d^*, through the power factor controller, then the measured d-axis current i_d is compared and the error is reduced to zero by a PI controller in order to determine the reference d-axis voltage, V_d^*.

The q-axis reference voltage is determined by the difference between the reference and the measured q-axis currents, i_q and i_q^*. The reference q-axis current is the output of the voltage controller, which uses the difference reference DC voltage, V_{dc}^*, and measured DC voltage, V_{dc}, as shown in Figure 2.65.

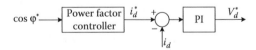

FIGURE 2.64 The d-axis voltage controller.

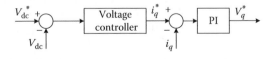

FIGURE 2.65 The q-axis voltage controller.

2.9   Wind Harvesting Research and Development

Due to the increased popularity of installed wind power plants, especially in Europe, wind turbines are becoming a significant and important part of the global energy systems. Wind-farm installation in Europe grew by 38% in 2007, whereas it grew by "only" 19% in 2006, bringing the total capacity to about 67 GW. The only issue is the transmission of the produced power, since the transmission capacity of power lines is limited [107].

Significant funds are invested in various research and development projects of wind energy harvesting. The research and development efforts in wind energy harvesting can be categorized into three different areas: (a) developments in control systems, (b) developments in electrical machine design, and (c) developments in distribution and grid-connected topologies.

2.9.1   Developments in Control Systems

Wind turbines may operate at fixed or variable speeds. Those operating at FS have a gearbox in between the turbine's rotor and the shaft of the electric machine, and therefore, they are connected directly to the grid as long as the wind speed is within operation limits. A squirrel cage induction generator is usually used when power is limited by the stall control principle. That means, if the wind speed increases, the coefficient is decreased inherently; hence the overall produced power stays at the rated power of the turbine.

Controlling these kinds of machines can be done by various methods. One method is to have a stator that can operate with two or more different number of poles [108–110]. Thus, the machine can work with different mechanical speeds with respect to the different number of poles used; therefore, more energy can be produced. Another example of these control methods is to engage an electronically variable rotor resistance, so that the mechanical loads can be reduced, and the turbine can work with variable wind speeds [111–113]. In addition, back-to-back converters or doubly fed topologies can be used for the control of variable speed wind turbines [49,60,63–67,69,71,73–75,95,114,115].

Other developments in DFIG turbines are on overall system monitoring (Supervisory Control and Data Acquisition, SCADA) systems. SCADA systems are becoming very complex because they control and monitor the entire wind turbine system. There are several other electric machine types, which can be employed with wind turbines as auxiliary devices and controls. Stepper motors are preferred to pitch control, BLDCs can be used for yaw control, and there are many research activities on different control methods for the pitch angle of the blades. This is mainly due to stability reasons, since by increasing the size of the turbines, the blades become longer and longer, so during higher wind speeds, there is a difference between the forces over the blades on the bottom and on the top of the blades [116].

Recent research on DD turbines is focused on issues related to the control of unconventionally big machines with lots of pairs of poles, that is, for small speeds around 15–20 rpm [23,35,36,117,118]. These are machines that produce high torque (due to the small speeds); however, for high wind speed conditions, the pitch control is used in order to regulate the speed and power. Table 2.6 summarizes the advantages and drawbacks of all the various types of wind turbines.

Power quality is another issue in the research and development of wind turbines. The power quality of variable speed turbines is better than that of the FS ones. It is possible

TABLE 2.6

Comparison of Three Wind Turbine Concepts

Description	FS	DFIG	DD
Cost, size, and weight	+	+/−	−
Blades audible noise	−	−	+
Suitability for 50 or 60 Hz grid frequency	−	+	+
Energy yield			
Variable speed	−	+	+
Gearbox	−	−	+
Generator	+	+	−
Converter	+	+/−	−
Power quality			
Flicker effect	−	+	+
Grid voltage and frequency control possibility	−	+	+
Harmonics	+	−	−
Reliability and maintenance			
Brushes	+	−	− (+ if PM)
Gearbox	−	−	+
Mechanical loads	−	+	+
Complexity	+	−	−
Grid faults			
Fault currents	+	+	+/−
Restoring voltage	−	+	+

Source: Modified from M.R. Dubois, H. Polinder, and J.A. Ferreira, *Proceedings of the Conference on Wind Energy in Cold Climates*, Matane, Canada, 2001.

Notes: +, strength; −, weakness.

to observe this from any power curve, but the drawback lies in the power electronics harmonics, which are produced by the converters in the variable speed turbines.

During the FS operation of induction machines, high voltage drops may occur due to high currents during grid faults. During the recovering stage (getting back to the prefault condition), induction machines require plenty of reactive power, which results in high torque and may damage the gearbox or other stressed mechanical components. DFIG has extremely high currents during faults, and needs to be disconnected in milliseconds, in order to protect the power converter. Due to the connection to the grid through the inverter, wind turbines can be easily disconnected from the grid anytime. Moreover, after grid failures, wind turbines can help conventional power stations recover the normal prefault operating conditions [119].

Another popular research and development topic is on the frequency conversion of variable speed turbines [120–124]. Since the input voltage and input frequency are variable, this variable AC magnitude and frequency must be converted to constant frequency AC. A lot of research was done to cause this conversion to be implemented using back-to-back connected voltage source converters [125–128].

Not only the generator side voltage source converter can provide the voltage step-up function (needed when the input voltage is low, such as at the low speed of a DD generator), but also the converter stages can provide reactive power to the induction machine. On the generator side this may be needed for magnetization or control of DFIG, and on the grid side it may be used to support voltage control of the grid [129] by supplying or absorbing reactive power.

2.9.2   Developments in Machine Design

It is not possible to use conventional machines in wind turbine applications. Therefore different designs of induction, doubly fed, and synchronous PM machines are used in generation mode. Most of the mentioned machines need a gearbox that converts the mechanical speed of the rotor of the turbine (50–300 rpm) to mechanical speed of the rotor of the generator (750, 1000, or 1500 rpm). Still, the main problem is the gearbox because of its mechanical losses, efficiency, and maintenance.

Hence, the DD machine train has been used in the past few years, in which the rotor is directly connected to the electrical machine, and there is no gearbox. The DD machines are usually PM machines with multiple pairs of poles and with nominal speeds within the range of the turbine's rotor speeds. Moreover, they are more compact; therefore, their installation is easier. One of the issues related to the cost of DD turbines is transportation. The 1.5 or 2 MW generator has an outer diameter of about 4 m, so that it can be transported by regular means. The 4.5 MW generators with a diameter on the order of 10 m are made in segments so that they can be transported separately [130].

Another advantage of DD train generators is that they can be designed for high voltages. This does not improve the efficiency, but for high powers in the order of MW, when the voltage level is changed from 400 or 750 V to 5 kV, the copper losses are reduced.

DD generators can be built with lower diameters; therefore, they do not pose transportation since they can be assembled on-site. The on-site assembly allows for a substantial increase in the final height of the wind turbine.

It is possible to increase the power density of a given machine by increasing its rotational speed. However, it is not possible to compare, based on power density, machines with different rotational speeds. They could be compared by their torque density, since it is independent of rotational speed. This is valid only up to a certain speed, which is largely above the typical speed found in wind turbines. Torque density is defined as

$$T_d = \frac{T}{((d_0^2 \pi)/4)L_a},\qquad(2.117)$$

where T is the machine nominal torque (kN m), T_d is the machine torque density (kN m/m³), d_0 is the stator outer diameter (active outer diameter only), and L_a is the machine total axial length (active length only including stator end windings) [117].

2.9.3   Developments in Distribution and Grid-Connected Topologies

The International Energy Agency has established a research team to analyze 19 national wind and grid studies. The first draft of the analysis showed that for every 100 MW of wind power, another 100 MW of fossil, nuclear, or hydroelectric is needed as a back-up. Moreover, sudden high wind speeds cause large and unanticipated power flows that saturate grids, not only in the particular location, but in the neighboring grids as well [131].

This phenomenon is the most important problem for grid operation, in which schedules are generally used for electrical energy transmission. However, unanticipated power flows could overload lines all around Northern and Western Europe, where the installed wind power is between 15% and 20% of the system [131]. In several occasions, wind farms have been shut down due to high wind speed. These cases may happen during night when consumption is minimal. Furthermore, in 2007, Denmark paid Sweden 20 cents/kWh just to relieve the generated power due to sudden high winds, to avoid disconnecting the transmission lines.

One solution for such problems is to build hybrid systems of wind turbines and solar panels, or diesel generators with long- and short-term energy storage devices. Therefore, during the daytime, solar panels can be used as the prime energy generator, and the diesel generator can be turned on if additional power is needed (during peak time). On the other hand, during the night, wind might be strong enough to use solely the wind turbines. This concept needs energy storage units (batteries, ultracapacitors, or flywheels) and has an increased system cost; however it is highly reliable. In conclusion, hybrid topologies for renewable energy technologies offer sustained power production and satisfy sustained load demands.

2.10 Summary

Due to the environmental concerns and diminishing fossil fuels, wind energy harvesting has been one of the most attractive solutions for future energy problems. The WECS are becoming more cost-effective as the technology is cumulatively improving. Therefore, this clean source of energy will be wider and even more economical in the near future competing existing energy sources and other renewable sources of energy. In this chapter, first theoretical background of wind energy is discussed along with wind energy terminology. Different types of wind turbines and electrical machines are reviewed throughout this chapter along with various power electronic interfaces.

References

1. H. Holttinen, P. Meibom, A. Orths, F. Van Hulle, C. Ensslin, L. Hofmann, J. McCann et al., "Design and operation of power systems with large amounts of wind power, first results of IEA collaboration," *Global Wind Power Conference*, Adelaide, Australia, September 18–21, 2006.
2. World Wind Energy Association, Press Release, Bonn, Germany, February 21, 2008.
3. Danish Wind Power Association, available at: http://www.windpower.org/en
4. Energy Information Administration, Official Energy Statistics from the U.S. Government, available at: http://www.eia.doe.gov
5. G.M. Masters, *Renewable and Efficient Electric Power Systems*, New York: Wiley, 2004.
6. NREL, National Renewable Energy Laboratory, Renewable Resource Data Center, available at: http://www.nrel.gov/rredc/

7. T. Price, "James Blyth—Britain's first modern win power pioneer," *Wind Engineering*, 29, 191–200, 2003.

8. "History of Wind Energy," *Encyclopaedia of Energy*, Vol. 6, p. 420, March 2004.

9. American Wind Energy Association, available at: http://www.awea.org

10. S. Heier, *Grid Integration of Wind Energy Conversion Systems*, New York: Wiley, 1998.

11. Wind Turbines, News and Information about Wind Generator Technologies and Innovations, Alternative Energy News, available online: http://www.alternative-energy-news.info/technology/wind-power/wind-turbines/

12. M. Bergey, T. Bartholf, D. Blottersdorf, A. Caldwell, D. Calley, C. Hansen, B. Markee, V. Nelson, G. Norton, and R. Sherwin, Technical Report: The U.S. Small Wind Turbine Industry Roadmap, A 20-year industry plan for small wind turbine technology, American Wind Energy Association, Small Wind Turbine Committee.

13. Enercon German Wind Turbine Manufacturer, available at: http://www.enercon.com

14. J.R. Hendershot Jr. and T.J.E. Miller, *Design of Brushless Permanent-Magnet Motors*, England: Oxford Magna Physics Publications, 1994.

15. H.-W. Lee, "Advanced control for power density maximization of the brushless DC generator," PhD dissertation, University of Arlington, Texas, 2003.

16. H.-W. Lee, T.-H. Kim, and M. Ehsani, "Maximum power throughput in the multiphase brushless DC generator," *IEE Proceedings on Electric Power Applications*, 152, 501–508, 2005.

17. H.-W. Lee, T.-H. Kim, and M. Ehsani, "Practical control for improving power density and efficiency of the BLDC generator," *IEEE Transactions on Power Electronics*, 20, 192–199, 2005.

18. M.J. Khan and M.T. Iqbal, "Simplified modeling of rectifier-coupled brushless DC generators," *International Conference on Electrical and Computer Engineering*, pp. 349–352, December 2006.

19. H.-W. Lee, T.-H. Kim, and M. Ehsani, "State of the art and future trends in position sensorless brushless DC motor/generator drives," *Proceedings of the IEEE 31st Annual Conference of Industrial Electronics Society*, November 2005.

20. J.R. Hendershot Jr and T.J.E. Miller, *Design of Brushless Permanent-Magnet Motors*, Oxford, UK: Oxford Magna Physics Publications, 1994.

21. R. Krishnan and G.H. Rim, "Modeling, simulation, and analysis of variable-speed constant frequency power conversion scheme with a permanent magnet brushless DC generator," *IEEE Transactions on Industrial Electronics*, 37, 291–296, 1990.

22. W.D. Jones, "I've got the power: Backyard wind turbines turn energy consumers into suppliers," *IEEE Spectrum*, 43 (10), 2006.

23. M.A. Khan, P. Pillay, and M. Malengret, "Impact of direct-drive WEC systems on the design of a small PM wind generator," *IEEE Bologna Power Technology Conference*, Vol. 2, pp. 7, June 2003.

24. L. Soderlund, J.-T. Eriksson, J. Salonen, H. Vihriala, and R. Perala, "A permanent-magnet generator for wind power applications," *IEEE Transactions on Magnetics*, 32, 2389–2392, 1996.

25. T.J.E. Miller, *Brushless Permanent-Magnet and Reluctance Motor Drives*, New York: Oxford University Press, 1989.

26. J.F. Giecras and M. Wing, *Permanent Magnet Motor Technology, Design and Applications*, New York: Marcel Dekker, 1997.

27. S. Huang, J. Luo, and T.A. Lipo, "Analysis and evaluation of the transverse flux circumferential current machine," *IEEE Industry Applications Society Annual Meeting*, New Orleans, pp. 378–384, October 5–9, 1997.

28. K. Shaarbafi, J. Faiz, M.B.B. Sharifian, and M.R. Feizi, "Slot fringing effect on the magnetic characteristics of the electrical machines," *IEEE International Conference on Electronics, Circuits, and Systems*, Vol. 2, pp. 778–781, December 2003.

29. C.C. Mi, G.R. Slemon, and R. Bonert, "Minimization of iron losses of permanent magnet synchronous machines," *IEEE Transactions on Energy Conversion*, 20, 121–127, 2005.

30. G.R. Slemon and X. Liu, "Core losses in permanent magnet motors," *IEEE Transactions on Magnetics*, 26, 1653–1655, 1990.

31. D. Grenier, L.-A. Dessaint, O. Akhrif, Y. Bonnassieux, and B. LePioufle, "Experimental non-linear torque control of a permanent magnet synchronous motor using saliency," *IEEE Transactions on Industrial Electronics*, 44 , 680–687, 1997.

32. Tog Inge Reigstad: "Direct driven permanent magnet synchronous generators with diode rectifiers for use in offshore wind turbines," PhD thesis, Norwegian University of Science and Technology Department of Electrical Power Engineering, June 2007.

33. Zephyros Permanent Magnet Direct Drive Wind Turbines, available at: http://www.peeraer.com/zephyros/

34. H. Polinder, S.W.H. de Haan, M.R. Dubois, and Johannes G. Slootweg, "Basic operation principles and electrical conversion systems of wind turbines," *EPE Journal, European Power Electronics and Drives Association*, 15, 43–50, 2005.

35. H. Li and Z. Chen, "Optimal direct-drive permanent magnet wind generator systems for different rated wind speeds," *European Conference on Power Electronics and Applications*, pp. 1–10, September 2007.

36. A. Grauers, "Design of direct-driven permanent-magnet generators for wind turbines," PhD dissertation, Chalmers University of Technology, Goteburg, 1996.

37. J.A. Baroudi, V. Dinahvi, and A.M. Knight, "A review of power converter topologies for wind generators," *Renewable Energy*, 32, 2369–2385, 2007.

38. Y. Chen, P. Pillay, and A. Khan, "PM wind generator topologies," *IEEE Transactions on Industry Applications*, 41 (6), 1619–1626, 2005.

39. J.F. Gieras, "Performance characteristics of a transverse flux generator," *IEEE International Conference on Electric Machines and Drives*, pp. 1293–1299 May 2005.

40. Z. Miao and L. Fan, "The art of modeling and simulation of induction generator in wind generation applications using high-order model," *Simulation Modeling Practice and Theory*, 16, 1239–1253, 2008.

41. J. Marques, H. Pinheiro, H.A. Gründling, J.R. Pinherio, and H.L. Hey, "A survey on variable-speed wind turbine system," *Cientifico Greater Forum of Brazilian Electronics of Power, COBEP'03*, Cortaleza, Vol. 1, pp. 732–738, 2003.

42. M.M. Neam, F.F.M. El-Sousy, M.A. Ghazy, and M.A. Abo-Adma, "The dynamic performance of an isolated self-excited induction generator driven by a variable-speed wind turbine," *International Conference on Clean Electric Power*, pp. 536–543, May 2007.

43. G.S. Kumar and A. Kishore, "Dynamic analysis and control of output voltage of a wind turbine driven isolated induction generator," *IEEE International Conference on Industrial Technology*, pp. 494–499, December 2006.

44. M. Orabi, M.Z. Youssef, and P.K. Jain, "Investigation of self-excited induction generators for wind turbine applications," *Canadian Conference on Electrical and Computer Engineering*, Vol. 4, pp. 1853–1856, May 2004.

45. D. Seyoum, M.F. Rahman, and C. Grantham, "Inverter supplied voltage control system for an isolated induction generator driven by a wind turbine," *Industry Applications Conference (38th IAS Annual Meeting)*, Vol. 1, pp. 568–575, October 2003.

46. E. Muljadi, J. Sallan, M. Sanz, and C.P. Butterfield, "Investigation of self-excited induction generators for wind turbine applications," *IEEE Industry Applications Conference (34th IAS Annual Meeting)*, Vol. 1, pp. 509–515, October 1999.

47. T. Ahmed, O. Noro, K. Matsuo, Y. Shindo, and M. Nakaoka, "Wind turbine coupled three-phase self-excited induction generator voltage regulation scheme with static VAR compensator controlled by PI controller," *International Conference on Electrical Machines and Systems*, Vol. 1, pp. 293–296, November 2003.

48. T. Ahmed, O. Noro, E. Hiraki, and M. Nakaoka, "Terminal voltage regulation characteristics by static Var compensator for a three-phase self-excited induction generator," *IEEE Transactions on Industry Applications*, 40, 978–988, 2004.

49. W. Qiao, G.K. Veneyagamoorthy, and R.G. Harley, "Real-time implementation of a STATCOM on a wind farm equipped with doubly fed induction generators," *IEEE Industry Applications Conference (41st IAS Annual Meeting)*, Vol. 2, pp. 1073–1080, October 2006.

50. J.M. Elder, J.T. Boys, and J.L. Woodward, "The process of self-excitation in induction generators," *IEEE Proceedings, Part B*, 130, 103–108, 1983.

51. P.C. Krause, O. Wasynczuk, and S.D. Sudhoff, *Analysis of Electric Machinery*, New York: IEEE Press, 2002.

52. N. Mohan, T.M. Undeland, and W.P. Robbins, *Power Electronics: Converters, Applications, and Design*, New York: Wiley, 1995.

53. I. Boldea and S.A. Nasar, *The Induction Machine Handbook*, Boca Raton, FL: CRC Press, November 2001.

54. R. Gagnon, B. Saulnier, G. Sybille, and P. Giroux, "Modeling of a generic high-penetration no-storage wind-diesel system using Matlab/Power System Blockset," *Global Windpower Conference*, Paris, France, April 2002.

55. H. Weh and H. May, "Achievable force densities for permanent magnet machines in new configurations," *International Conference on Electrical Machines*, pp. 1107–1111, 1986.

56. G.D. Hoops, "Terminal impedance control of a capacitor excited induction generator," PhD thesis, Department of Electrical and Computer Engineering, University of Toronto, 1988.

57. N.H. Malik and A.H. Al-Bahrani, "Influence of the terminal capacitor on the performance characteristics of a self excited induction generator," *IEE Proceedings*, Vol. 137, Part C, pp. 168–173, March 1990.

58. C. Ma and R. Cheung, *Advanced Voltage and Frequency Control for the Stand-Alone Self-Excited Induction Generator*, Canada: Electrical & Computer Engineering Department of Ryerson University.

59. M.T. Abolhassani, H.A. Toloyat, and P. Enjeti, "Stator flux-oriented control of an integrated alternator/active filter for wind," *Proceedings of the IEEE International Electric Machines and Drives Conference*, Vol. 1, pp. 461–467, June 2003.

60. Eel-Hwan, S.-B. Oh, Y.-H. Kim, and C.-S. Kim, "Power control of a doubly fed induction machine without rotational transducers," *Proceedings of the Power Electronics and Motion Control Conference*, Vol. 2, pp. 951–955, August 2000.

61. H. Azaza and A. Masmoudi, "On the dynamics and steady state performance of a vector controlled DFM drive systems," *IEEE International Conference on Man and Cybernetics*, Vol. 6, p. 6, October 2002.

62. A. Tapia, G. Tapia, J.X. Ostolaza, and J.R. Saenz, "Modeling and control of a wind turbine driven DFIG," *IEEE Transactions on Energy Conversion*, 18, 194–204, 2003.

63. L. Jiao, B.-Teck Ooi, G. Joos, and F. Zhou, "Doubly-fed induction generator (DFIG) as a hybrid of asynchronous and synchronous machines," *Electric Power Systems Research*, 76, 33–37, 2005.

64. S. Muller, M. Diecke, and Rik W. De Doncker, "Doubly fed induction generator systems for wind turbines," *IEEE Industry Applications Magazine*, 8, 26–33, 2002.

65. R. Pena, J.C. Clare, and G.M. Asher, "A doubly fed induction generator using back-to-back PWM converters supplying an isolated load from a variable speed wind turbine," *IEEE Proceedings on Electric Power Applications*, 143, 380–387, 1996.

66. L. Xu and Y. Tang, "Stator field oriented control of doubly-excited induction machine in wind power generation system," *Proceedings of the 25th Mid West Symposium on Circuit and Systems*, pp. 1449–1466, August 1992.

67. L. Xu and W. Cheng, "Torque and reactive power control of a doubly fed induction machine by position sensorless scheme," *IEEE Transactions on Industrial Applications*, 31, 636–642, 1995.

68. S. Doradla, S. Chakrovorty, and K. Hole, "A new slip power recovery scheme with improved supply power factor," *IEEE Transactions on Power Electronics*, 3, 200–207, 1988.

69. R. Pena, J. Clare, and G. Asher, "Doubly fed induction generator using back-to-back pwm converters and its application to variable-speed wind-energy conversion," *IEE Proceedings on Electric Power Applications*, 143, 231–241, 1996.

70. Y. Tang and L. Xu, "A flexible active and reactive power control strategy for a variable speed constant frequency generating systems," *IEEE Transactions on Power Electronics*, 10, 472–478, 1995.

71. A Feijo, J. Cidrs, and C. Carrillo, "Third order model for the doubly-fed induction machine," *Electric Power System Research*, 56, 121–127, 2000.
72. B.H. Chowdhury and S. Chellapilla, "Double-fed induction generator control for variable speed wind power generation," *Electric Power Systems Research*, 76, 786–800, 2006.
73. Y. Lei, A. Mullane, G. Lightbody, and R. Yacamini, "Modeling of the wind turbine with a doubly fed induction generator for grid integration studies," *IEEE Transactions on Energy Conversion*, 21, 257–264, 2006.
74. F. Wu, X.-P. Zhang, K. Godfrey, and P. Ju, "Modeling and control of wind turbine with doubly fed induction generator," *Power Systems Conference and Exposition*, pp. 1404–1409, October–November 2006.
75. J.B. Ekanayake, L. Holdsworth, X. Wu, and N. Jenkins, "Dynamic modeling of doubly fed induction generator wind turbines," *IEEE Transactions on Power Systems*, 18, 803–809, 2003.
76. W. Hofmann and A. Thieme, "Control of a double-fed induction generator for wind-power plants," *Power Quality Proceedings, PCIM*, Nurnberg, pp. 275–282, 1998.
77. I. Cadirici and M. Ermis, "Double-output induction generator operating at subsynchronous and supersynchronous speeds: Steady-state performance optimisation and wind-energy recovery," *IEE Proceedings-B on Electric Power Applications*, 139, 429–441, 1992.
78. C. Abbey and G. Joos, "Energy storage and management in wind turbine generator systems," *Proceedings of the 12th International Power Electronics and Motion Control Conference*, pp. 2051–2056, August 2006.
79. T. Ise, M. Kita, and A. Taguchi, "A hybrid energy storage with a SMES and secondary battery," *IEEE Transactions on Applied Superconductivity*, 15, 1915–1918, 2005.
80. J.H.R. Enslin, J. Knijp, C.P.J. Jansen, and P. Bauer, "Integrated approach to network stability and wind energy technology for on-shore and offshore applications," *Power Quality Conference*, pp. 185–192, May 2003.
81. L. Ran, J.R. Bumby, and P.J. Tavner, "Use of turbine inertia for power smoothing of wind turbines with a DFIG," *Proceedings of the 11th International Conference on Harmonics and Quality Power*, pp. 106–111, September 2004.
82. K. Strunz and E.K. Brock, "Hybrid plant of renewable stochastic source and multilevel storage for emission-free deterministic power generation," *Quality and Security Electric Power Delivery Systems CIGRE/IEEE PES International Symposium*, pp. 214–218, October 8–10, 2003.
83. J.P. Barton and D.G. Infield, "Energy storage and its use with intermittent renewable energy," *IEEE Transactions on Energy Conversion*, 19, 441–448, 2004.
84. R. Cardenas, R. Pena, G. Asher, and J. Clare, "Power smoothing in wind generation systems using a sensorless vector controlled induction machine driving a flywheel," *IEEE Transactions on Energy Conversion*, 19, 206–216, 2004.
85. C. Abbey and G. Joos, "Supercapacitor energy storage for wind energy applications," *IEEE Transactions on Industry Applications*, 43, 769–776, 2007.
86. C. Abbey and G. Joos, "Short-term energy storage for wind energy applications," *Industry Applications Conference*, Vol. 3, pp. 2035–2042, October 2005.
87. F. Zhang, F.Z. Peng, and Z. Qian, "Study of the multilevel converters in dc-dc applications," *IEEE 35th Annual Power Electronics Specialists Conference*, Vol. 2, pp. 1702–1706, June 2004.
88. Y. Hu, J. Tatler, and Z. Chen, "A bi-directional DC/DC power electronic converter for an energy storage device in an autonomous power system," *Proceedings of the Fourth Power Electronics and Motion Control Conference*, Vol. 1, pp. 171–176, August 2004.
89. K. Ghedamsi, D. Aouzellag, and E.M. Berkouk, "Control of wind generator associated to a flywheel energy storage system," *Renewable Energy*, 33, pp. 2145–2156, 2008.
90. G. Cimuca, M.M. Radulescu, C. Saudemont, and B. Robyns, "Performance analysis of an induction machine-based flywheel energy storage system associated to a variable-speed wind generator," *Proceedings of the Ninth International Conference on Optimization of Electrical and Electronic Equipments*, Vol. 2, Brasov, Romania, pp. 319–326, May 2004.

91. F. Hardan, J.A.M. Blejis, R. Jones, and P. Bromley, "Bi-directional power control for flywheel energy storage system with vector-controlled induction machine drive," *Proceedings of the IEEE Seventh International Conference on Power Electronics and Variable Speed Drives*, pp. 477–482, September 1998.

92. L. Leclercq, B. Robyns, and J.M. Grave, "Control based on fuzzy logic of a flywheel energy storage system associated with wind and diesel generators," *Mathematics and Computers in Simulation*, 63, 271–280, 2003.

93. R. Cadenas, R. Pena, G. Asher, J. Clare, and R.B.-Gimenez, "Control strategies for enhanced power smoothing in wind energy systems using a flywheel driven by a vector-controlled induction machine," *IEEE Transactions on Industrial Electronics*, 48, 625–635, 2001.

94. J.P. Barton and D.G. Infield, "Energy storage and its use with intermittent renewable energy," *IEEE Transactions on Energy Conversion*, 19, 441–448, 2004.

95. V. Akhmatov, "Variable-speed wind turbines with doubly-fed induction generators, Part I: Modelling in dynamic simulation tools," *Wind Engineering*, 26, 85–108, 2002.

96. N.W. Miller, J.J. Sanchez-Gasca, W.W. Price, and R.W. Delmerico, "Dynamic modeling of GE 1.5 and 3.6 MW wind turbine-generators for stability simulations," *GE Power Systems Energy Consulting, IEEE WTG Modeling Panel*, 3, 1977–1983.

97. P.C. Krause, *Analysis of Electric Machinery*, New York: McGraw-Hill, 1986.

98. I. Kamwa, M. Pilote, P. Viarogue, B. Mpanda-Mabwe, M. Crappe, and R. Mahfoudi, "Experience with computer-aided graphical analysis of sudden-short circuit oscillograms of large synchronous machines," *IEEE Transactions on Energy Conversion*, 10, 407–414, 1995.

99. J.A. Sanchez, C. Veganzones, S. Martinez, F. Blazquez, N. Herrero, and J.R. Wilhelmi, "Dynamic model of wind energy conversion systems with variable speed synchronous generator and full-size power converter for large-scale power system stability studies," *Renewable Energy*, 33, 1186–1198, 2008.

100. A. Bouscayrol, P. Delarue, and X. Guillaud, "Power strategies for maximum control structure of a wind energy conversion system with a synchronous machine," *Renewable Energy*, 30, 2273–2288, 2005.

101. G. Raina and O.P. Malik, "Variable speed wind energy conversion using synchronous machine," *IEEE Transactions on Aerospace and Electronic Systems*, 21, 100–105, 1985.

102. G. Raina and O.P. Malik, "Wind energy conversion using a self-excited induction generator," *IEEE Transactions on Power Apparatus and Systems*, 102, 3933–3936, 1983.

103. B. Borowy and Z. Salameh, "Dynamic response of a stand alone wind energy conversion system with battery energy storage to a wind gust," *IEEE Transactions on Energy Conversion*, 12, 73–78, 1997.

104. Z. Chen and E. Spooner, "Grid power quality with variable speed wind turbine," *IEEE Transactions on Energy Conversion*, 16, 148–154, 2001.

105. S.-H. Song, S.-I. Kang, and N.-K. Hahm, "Implementation and control of grid connected AC-DC-AC power converter for variable speed wind energy conversion system," *18th Annual IEEE Applied Power Electronics Conference and Exposition*, Vol. 1, pp. 154–158, February 2003.

106. J.A. Baroudi, V. Dinavahi, and A.M. Knight, "A review of power converter topologies for wind generators," *IEEE International Conference on Electric Machines and Drives*, May, pp. 458–465, 2005.

107. P. Fairly, "Can wind energy continue double-digit growth?" *IEEE Spectrum*, 45, 16, 2008.

108. K. Yukita, S. Washizu, K. Taniguchi, M. Oshima, N. Hayashi, K Ichiyanagi, and Y. Goto, "A pole change generator for micro windmill with maximum power point tracking system," *Proceedings of the IEEE 22nd International Symposium on Intelligent Control*, Singapore, pp. 640–645, October 2007.

109. J.H.H. Alwash, K.S. Ismail, and J.F. Eastham, "A novel 16/6 phase modulated winding," *IEEE Transactions on Energy Conversion*, 15, 188–190, 2000.

110. J.W. Kelly, E.G. Strangas, and J.M. Miller, "Control of a continuously operated pole-changing induction machine," *Proceedings of the IEEE International Electric Machines and Drives Conference*, col. 1, pp. 211–217, June 2003.

111. R. Gagnon, B. Saulnier, G. Sybille, and P. Giroux, "Modeling of a generic high-penetration no-storage wind-diesel system using Matlab/Power System Blockset," *Global Windpower Conference*, France, April 2002.
112. B. Saulnier, A.O. Barry, B. Dube, and R. Reid, "Design and development of a regulation and control system for the high-penetration no-storage wind/diesel scheme," *European Community Wind Energy Conference 88*, Denmark, June 1988.
113. R. Sebastian and J. Quesada, "Distributed control system for frequency control in an isolated wind system," *Renewable Energy*, 31, 285–305, 2006.
114. R. Gagnon, G. Sybille, S. Bernard, D. Pare, S. Casoria, and C. Larose, "Modeling and real-time simulation of a doubly-fed induction generator driven by a wind turbine," *Proceedings of the International Conference on Power Systems Transients* IPST'05, Montreal, June 2005.
115. A. Mirecki, X. Roboam, and F. Richardeau, "Comparative study of maximum power strategy in wind turbines," *IEEE International Symposium on Industrial Electronics*, Vol. 2, pp. 993–998, May 2004.
116. M.R. Dubois, H. Polinder, and J.A. Ferreira, "Generator topologies for direct-drive wind turbines, and adapted technology for turbines running in cold climate," *Proceedings of the Conference on Wind Energy in Cold Climates*, Matane, Canada, 2001.
117. M.R. Dubois, H. Polinder, and J.A. Ferreira, "Comparison of generator topologies for direct-drive wind turbines," *IEEE Transactions on Energy Conversion*, 21, 2006, 725–733, 2006.
118. M.R. Dubois, H. Polinder, and J.A. Ferreira, "Comparison between axial and radial-flux permanent magnet generators for direct-drive wind turbines," *Proceedings of the European Wind Energy Conference and Exhibition*, Copenhagen, pp. 1112–1115, July 2001.
119. J. Morren, S.W.H. de Haan, and J.A. Ferreira, "Ride through of DG units during a voltage dip," *IEEE General Meeting of Power Engineering Society*, Denver, Colarado, June 2004.
120. R. Jones, "Power electronic converters for variable speed wind turbines," *IEE Colloquium on Power Electronics for Renewable Energy*, [digest no. 1997/170], June, pp. 1–8, 1997.
121. N.R. Ullah and T. Thiringer, "Variable speed wind turbines for power system stability enhancement," *IEEE Transactions on Energy Conversion*, 22, 52–60, 2007.
122. X. Zhang, W. Wang, U. Liu, and J. Cheng, "Fuzyy control of variable speed wind turbine," *World Congress on Intelligent Control and Automation*, 1, 3872–3876, 2006.
123. F.D. Kanellos and N.D. Hatziargyriou, "A new control scheme for variable speed wind turbines using neural networks," *IEEE Power Engineering Society Winter Meeting*, Vol. 1, pp. 360–365, January 2002.
124. J.L.R. Amendo, S. Arnalte, and J.C. Burgos, "Automatic generation control of a wind farm with variable speed wind turbines," *IEEE Power Engineering Review*, 22, 65, 2002.
125. J.G. Slootweg, S.W.H. de Haan, H. Polinder, and W.L. Kling, "Voltage control methods with grid connected wind turbines: A tutorial review," *Wind Engineering*, 25, 353–365, 2001.
126. S. Li and A.T. Haskew, "Analysis of decoupled d–q vector control in DFIG back-to-back PWM converter," *IEEE Power Engineering Society General Meeting*, pp. 1–7, June 2007.
127. F.D. Kanellos, S.A. Papathanassiou, and N.D. Hatziargyriou, "Dynamic analysis of a variable speed wind turbine equipped with a voltage source AC/DC/AC converter interface and a reactive current control loop," *Proceedings of the Tenth Mediterranean Electrotechnical Conference*, Vol. 3, pp. 986–989, May 2000.
128. C. Batlle and A.D.-Cerezo, "Energy-based modeling and simulation of the interconnection of a back-to-back converter and a doubly-fed induction machine," *American Control Conference*, p. 6, June 2006.
129. S. Wijnbergen, S.W.H. de Haan, and J.G. Slootweg, "Plug 'n' play interface for renewable generators participating in voltage and frequency control," *IEEE 6th Africon Conference*, Vol. 2, pp. 579–584, October 2002.
130. M. Dubois, "Optimized permanent magnet generator topologies for direct drive wind turbines," PhD thesis, Delft University of Technology, Delft, The Netherlands, 2004.
131. International Energy Agency, available at: http://www.iea.org/

3

Tidal Energy Harvesting

3.1 Introduction

The generation of electrical power from ocean tides is very similar to traditional hydro-electric power generation. The simplest generation system for tidal plants involves a dam, known as a barrage, across an inlet. Usually, a tidal power plant consists of a tidal pond created by a dam, a powerhouse containing a turbo-generator, and a sluice gate to allow the bidirectional tidal flow. The rising tidal waters fill the tidal basin after opening the gate of the dam, during the flood tide. The gates are closed, when the dam is filled to capacity. After the ocean water has receded, the tidal basin is released through a turbo-generator. Power can be generated during ebb tide, flood tide, or both. Ebb tide occurs when the water is pulled back, and flood tide occurs when the water level increases near the shore [1]. Studies demonstrate that the tidal power will be economical at sites where mean tidal range exceeds 16 ft [1,2].

One of the advantages of tidal energy harvesting is that the tidal current is regular and predictable. Furthermore, tidal current is not affected by climate change, lack of rain, or snowmelt. Environmental and physical impacts and pollution issues are negligible. In addition, tidal power can be used for water electrolysis in hydrogen production and desalination applications [3]. However, tidal power generation is a very new technology, which needs further investigations and developments.

Similar to the wind turbines, tidal turbines can be used for tidal energy harvesting. Tidal turbines and wind turbines are similar in both appearance and structure, to some extent. Tidal turbines can be located in river estuaries and wherever there is a strong tidal flow. Since water is about 800 times as dense as air, tidal turbines have to be much stronger than wind turbines. They will be heavier and more expensive; however, they will be able to capture more energy at much higher densities [4]. A typical tidal turbine is shown in Figure 3.1.

Various tidal turbines' structures and the critical components implemented will be discussed in detail in Sections 3.2.4 and 3.3.3.2.

Usually, tidal fences are mounted in the entrance of ocean channels. In this way, tidal water is forced to pass through a fence structure, which is called caisson. Unlike barrage stations, tidal fences do not require basins and they can be used in a channel between the mainland and a nearby offshore island, or between two islands. Tidal fences have much less impact on the environment, since they do not require flooding the basin. They are also significantly cheaper to install; however, the caisson may disrupt the movement of large marine animals and shipping [4].

Tidal fences can be mounted at the entrance of channels where ocean water enters the land via a bay (Figure 3.2a), or between the main land and an island (Figure 3.2b), or simply between two islands (as shown in Figure 3.2c).

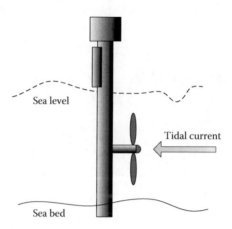

FIGURE 3.1 Tidal turbine.

Sometimes, even the top of the fence could be simultaneously used as roads connecting two islands. The structure of a sample tidal fence with a road on it is shown in Figure 3.3. This fence structure can be used in channel blocking locations illustrated in Figure 3.2.

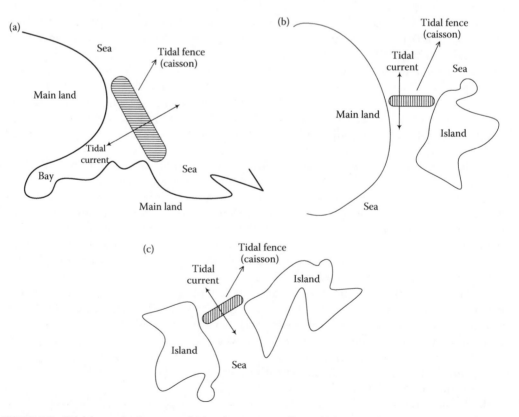

FIGURE 3.2 Tidal fences can be mounted (a) at the entrance of bays, (b) between the main land and an island, and (c) between two islands.

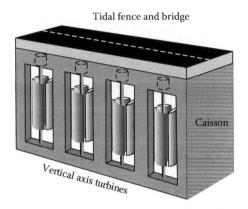

FIGURE 3.3 Tidal fence and bridge structure. (Redrawn from "Ocean Energy," Technical Report of the US Department of Interior, Minerals Management Service.)

3.1.1 History of Tidal Energy Harvesting

The earliest evidence of the usage of the oceans' tides for power conversion dates back to about 900 A.D., in Britain and France [5]. Much later, American colonists built tidal-powered mills in New England. Naturally occurring tidal basins are used by Americans to construct tidal power plants by building a barrage across the opening of the basin and allowing the basin to fill with the rising tide, impounding the water as the tide falls. The impounded water could be released through an energy conversion device like a waterwheel or a paddlewheel [5], as shown in Figures 3.4 and 3.5. The power was used quite widely for grinding grains and corn.

The idea of generating electric power from exploiting the power of the tides in estuaries was proposed in 1920 [6]. In 1967, the world's first tidal electric plant was successfully completed on the Rance Estuary in Brittany, France, which is shown in Figure 3.6. The enclosed estuary of the Rance River has very large tides with 13.5 m difference between high and low tides during the equinox [7]. The 740-m-long barrage is simultaneously used

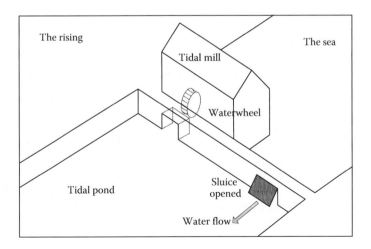

FIGURE 3.4 Water flows inside to the tidal pond.

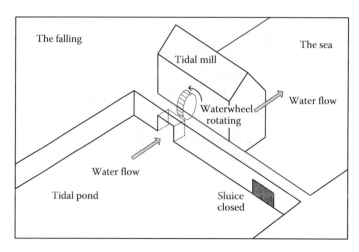

FIGURE 3.5　Water flows from tidal pond to the sea.

as a road and it also contains a ship lock. This 240 mW power plant has 24 two-way 10 mW turbines, and it is sufficient to power 4% of the homes in Brittany [7]. So far it is still the largest operating commercial tidal facility in the world. It attracts hundreds of thousands of visitors and students every year.

3.1.2　Physical Principles of Tidal Energy

Gravitational interaction between the moon, the sun, and the earth attributes the rhythmic rising and lowering of tidal height. The attraction force exerted by the moon or the sun on

FIGURE 3.6　**(See color insert following page 80.)** La Rance tidal power station. (From Dam of the tidal power plant on the estuary of the Rance River, Bretagne, France, *image licensed under the Creative Commons Attribution 2.5 License.* With permission.)

a molecule of water can be calculated as

$$f = \frac{KMm}{d^2},$$ (3.1)

where f is the attraction force, K is the universal constant of gravitation, M is the mass of the moon or sun, m is the mass of a water molecule, and d is the distance from a water molecule to the moon/sun.

Due to the less distance between the moon and the earth, the moon exerts 2.17 times greater force on the tides as compared to the sun [8]. As a result, the tide closely follows the moon during its rotation around the earth, bulging along the axis pointing directly at the moon. When the sun and the moon are in line, whether pulling on the same side or on the opposite side, their gravitational forces are combined together and this results in a high "spring tide" [6]. When the moon and the sun are located at 90° angle to each other, their gravitational force pulls the water in different directions, causing the bulges to eliminate each other's effect. This results in a smaller difference between high and low tides, which is known as a "neap tide" [6]. The concepts of spring tide and neap tide are shown in Figure 3.7.

The period between the two spring tides or neap tides is around 14 days, half of the lunar cycle. The range of a spring tide is commonly about twice as that of a neap tide, whereas the longer period cycles impose smaller perturbations [6].

The amplitude of the tide wave is very small in the open ocean. The typical tidal range is about 50 cm in the open ocean. However, the tide height can increase dramatically, when it

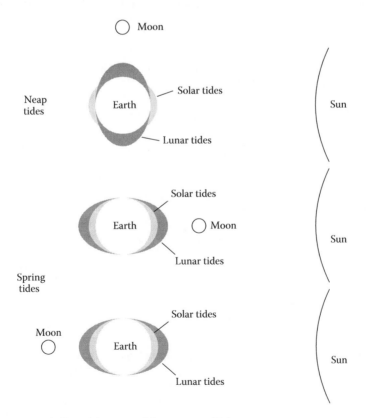

FIGURE 3.7 Gravitational effect of the sun and the moon on tidal range.

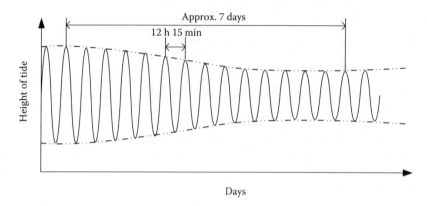

FIGURE 3.8   Tidal variation during a week.

reaches coasts or continental shelves, bringing huge masses of water into narrow bays and river estuaries along a coastline. At some sites, the tidal flow can be heightened to more than 10 m by the complex resonance effects. Bay of Fundy, in Canada, where the greatest tides in the world can be found, and the Severn Estuary, in England, are famous examples of this effect [6]. In these areas, two high tides occur in one day, called semidiurnal tide, with a tidal cycle of about 12 h and 25 min, as shown in Figure 3.8 [9]. This daily variation is quasisinusoidal.

Tidal wave can also be reinforced by reflections between the coast and the shelf edge, and funneling effect (due to the shape of the coastline as the tidal bulge progresses into a narrowing estuary), as demonstrated in Figure 3.9 [8].

In addition, tide can also be diurnal (one tidal cycle per day with a period of 24 h and 50 min) or mixed (with a tidal cycle intermediate between a diurnal tide and a semidiurnal tide).

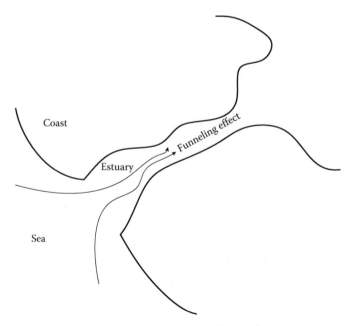

FIGURE 3.9   Funneling effect in an estuary entrance.

3.2 Categories of Tidal Power and Corresponding Generation Technology

Utilization of tidal power can be classified into two main types: potential energy harvesting and kinetic energy harvesting.

3.2.1 Potential Energy

A traditional approach is exploiting the potential energy from the difference in height between high and low tides, by constructing a barrage to make a basin, as shown in Figure 3.10.

Given a basin, the theoretical potential energy can be calculated as

$$E = g\rho A \int z \, dz = \frac{1}{2} g\rho A h^2, \tag{3.2}$$

where E is the energy (J), g is the acceleration of gravity (m/s²), ρ is the seawater density (approximately 1022 kg/m³ for seawater), A is the surface area of the basin (m²), z is the vertical coordinate of the ocean surface (m), and h is the water head (m).

The average value for the acceleration of gravity is $g = 9.8 \, \text{m/s}^2$. Therefore, for seawater

$$(g\rho) = 10.0156 \, \text{kN/m}^3.$$

Thus, potential energy for a tide cycle per square meter of ocean surface would be

$$\frac{E}{A} = 5.0078 h^2,$$

which is [10]

$$\frac{E}{A} = 1.39 h^2 \, \text{W-h}.$$

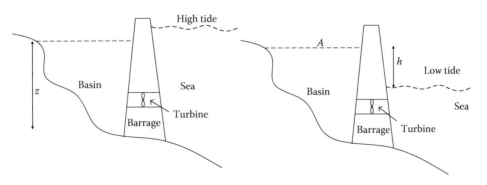

FIGURE 3.10 Difference in height between high and low tides.

3.2.2 Tidal Barrages Approach

Tidal barrages approach is a conventional method, where a barrage is placed across an estuary with a large tidal range to create a pressure difference, and operate a hydroelectric power plant with intermittent flow, as shown in Figure 3.11. The Rance River Estuary plant is a well-known example of this approach. The basic elements of a barrage are caissons, embankments, sluices, turbines, and ship locks. Sluices, turbines, and ship locks are housed in caisson (very large concrete blocks). Embankments seal a basin where it is not sealed by caissons.

There are three main operation patterns in which power can be generated from a barrage: ebb generation, flood generation, and two-way generation.

3.2.2.1 Ebb Generation

In ebb generation, first the basin is filled by the incoming tide through the sluice gate, and then the sluice gates are closed. The turbine gates are kept closed until the sea level falls to create sufficient head across the barrage, and then they are opened so that the water passes through the turbines to generate power until the water level inside the basin is again low. This process is expressed in Figure 3.12. Ebb generation takes its name because generation occurs as the tide ebbs.

3.2.2.2 Flood Generation

The flood generation method uses incoming tide to generate power, when water passes through turbines. When the tide is coming, the sluices and the turbine passages are closed so that the water level on the ocean side of the barrage rises. When sufficient head is created, the turbine gates open and the generators start to work. The process of this approach is presented in Figure 3.13. Generally, its efficiency is a little lower than that of ebb generation method, because the surface area of the estuary decreases with depth, which means less water can be contained [11]. Flood generation is less flavored due to problems such as more severe potential ecological problems for the basin and reduced access of shipping.

3.2.2.3 Two-Way Generation

Two-way generation approach combines the ebb generation and flood generation methods. Generation occurs as the tide both ebbs and floods in every cycle. The process of this

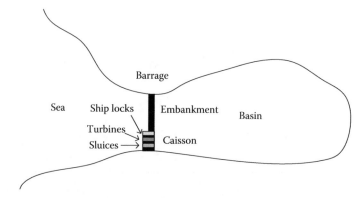

FIGURE 3.11 Barrage components.

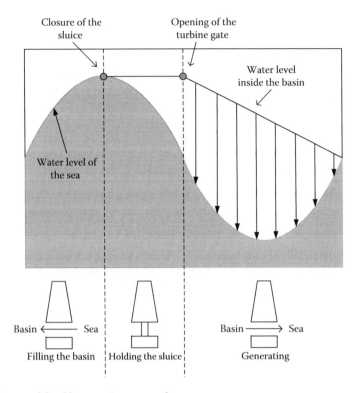

FIGURE 3.12 Process of the ebb generation approach.

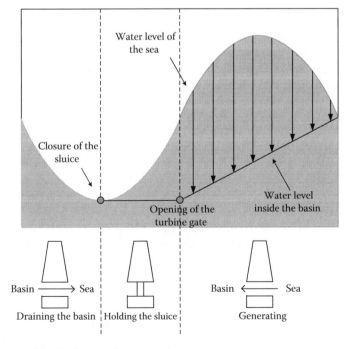

FIGURE 3.13 Process of the flood generation approach.

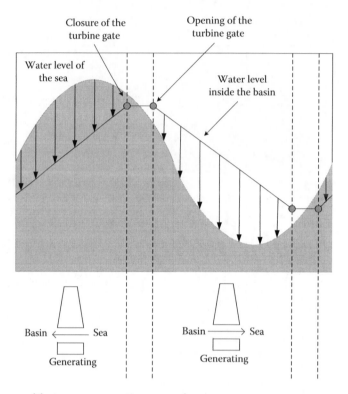

FIGURE 3.14   Process of the two-way generation approach.

approach is expressed in Figure 3.14. This method can extract more energy than ebb or flood generation. However, practically, it cannot significantly improve the efficiency due to several facts. First, the turbines have to be designed to operate in both directions, which results in less efficiency and more cost. Second, the water passage has to be longer, in order to capture energy from both water flow directions, which is another issue in two-way generation [11]. In addition, it also has the potential ecological and shipping problems associated with the flood generation method.

3.2.3   Tidal Lagoons Concept

The tidal lagoon, as shown in Figure 3.15, is an offshore tidal power conversion approach. It may relieve the environmental and economic problems of the tidal barrage method. This concept was proposed by the Tidal Electric Ltd. Tidal lagoons can be built of loose rock, sand, and gravel. A 60 MW project has been proposed for Swansea Bay, UK, which covers 5 km^2 in one mile offshore area [12].

In comparison to barrages constructed at estuary, offshore tidal lagoon can utilize both the ebb and the flood tides for generation, since the environmental and shipping problems are much smaller. Its generation cycle is illustrated in Figure 3.16.

Conventional low-head hydroelectric generation equipment and control systems are proposed to be utilized in this concept. This concept uses mixed-flow reversible bulb turbines.

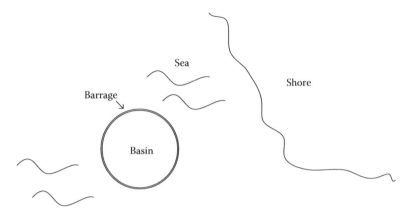

FIGURE 3.15 Schematic diagram of a tidal lagoon.

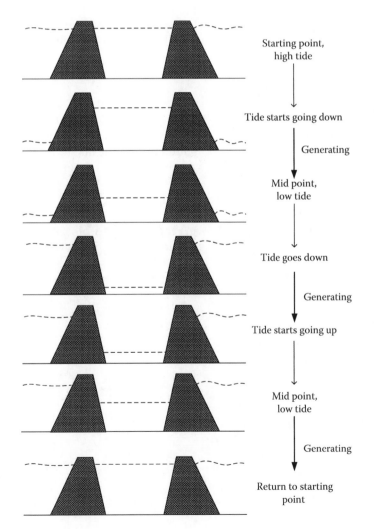

FIGURE 3.16 Operation process of a tidal lagoon.

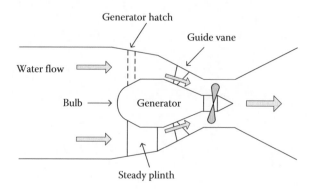

FIGURE 3.17   Bulb turbine.

3.2.4 Tidal Turbines Used in Barrages

The power generating technology used in barrage approach is very similar to the conventional low-head hydro systems. The following three types of turbines are commonly used in tidal power generating [13].

3.2.4.1 Bulb Turbine

Bulb turbine is one of the most popular turbines utilized in tidal barrages. The La Rance tidal plant in France uses bulb turbines. It is similar to a bulb, as shown in Figures 3.17 and 3.18. All the essential turbine components, such as turbine bearing and shaft seal box, as well as the generator are all placed inside a bulb. During its operation, water flows around the turbine, passing the generator, guide vanes, blades, and draft tube into tail channel. It has normally three to five blades made of stainless steel.

Since water must be prevented from flowing through the turbine during maintenance, its maintenance is relatively difficult. There is a hatch located above the generator. It provides access to the generator for components maintenance and some assembly or dismantling tasks inside the bulb. So far, bulb turbines are still the most popular turbines used by low-head barrage designers.

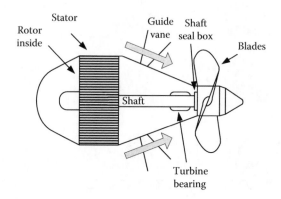

FIGURE 3.18   Bulb turbine inside.

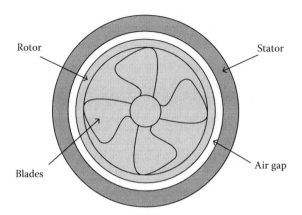

FIGURE 3.19   Generator structure of a rim turbine.

3.2.4.2   Rim Turbine

The generator of a rim turbine is designed so that the rim of the generator rotor is attached directly to the periphery of the blades of the turbine, as shown in Figure 3.19.

The general structure of a rim turbine, also known as the Straflo (straight flow) turbine, is shown in Figure 3.20.

The generator stator is mounted on the barrage, and therefore only the turbines are in the water flow. As a result, the maintenance problem for generator part in the bulb turbines is alleviated in rim turbines. The sealing between the blades and generator (at the edge of the rotor rim) is an issue in the rim turbines. The water passage configuration is essentially the same as for a bulb turbine. These turbines are used in Annapolis Royal in Canada [7].

3.2.4.3   Tubular Turbine

In tubular turbine, the generator is on top of the barrage and the blades are connected to a long shaft, as shown in Figure 3.21. This design gives some room to a gearbox, which can allow for more efficient operation of generators. There are some issues about the vibration

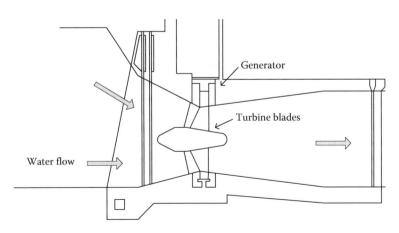

FIGURE 3.20   Rim turbine.

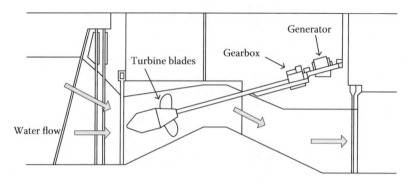

FIGURE 3.21 Tubular turbine.

of the long shaft. The tubular turbines have been used in some hydro plants in the United States and they are proposed for the Severn tidal project in the United Kingdom.

3.3 Turbine and Generator's Control

The hydraulic turbine drives an electric generator that takes the energy of the flowing water. The water flow is directed to the turbine's blades, creating a torque on the blades. Hence, the energy is transferred to the turbine from water flow. Since the electric generator should be driven by a specific speed, the turbine speed should be controlled. In modern hydraulic turbine models, electric–hydraulic speed control has been commonly included. In hydraulic turbine model, two main sections can be provided: (a) prime movers including water supply conduit and (b) speed controls of the prime mover.

The block diagram given in Figure 3.22 depicts the basic elements of a hydro turbine speed control model [14].

Figure 3.23 gives a more detailed model of a hydro turbine including components such as turbine control dynamics, load, rotor, turbine, and conduit dynamics. The rotor dynamics calculate the rotor speed with respect to the difference between the electromagnetic torque and the mechanical torque. Turbine dynamics are associated with the conduit dynamics through the relationship between the pressure and the flow. At the same time, turbine control dynamics decides the gate opening with respect to the assigned generation value and the rotor speed.

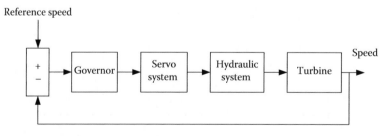

FIGURE 3.22 Hydro turbine speed control model.

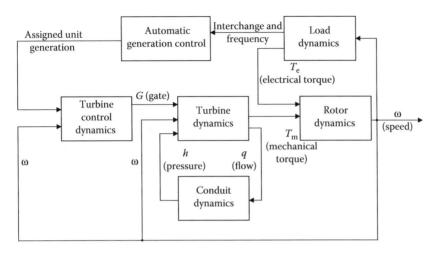

FIGURE 3.23 Block diagram of hydraulic turbine and its control system.

3.3.1 Modeling of Hydraulic Turbine Conduit Dynamics

In Figure 3.24, the dynamic characteristics of a hydraulic turbine are shown. The turbine has a penstock, unrestricted head, and tail race. Depending on the design, the turbine may or may not have a surge tank [15,16].

The fluid, for example, water, which is assumed to be incompressible, and a nonelastic water column (conduit) has a length of L and a cross-sectional area of A. Losses in the penstock are represented with h_1, which is proportional to flow squared q^2 and head loss coefficient is represented as f_p, which can be generally ignored.

Using the momentum law, the flow rate change in the conduit can be expressed as

$$\frac{d\bar{q}}{dt} = \left(\bar{h}_0 - \bar{h} - \bar{h}_1\right)\frac{gA}{L}, \tag{3.3}$$

where $\bar{q}$ is the turbine flow rate (m³/s), A is the penstock cross-sectional area (m²), L is the length of the penstock (m), g is the gravitational acceleration (m/s²), $\bar{h}_0$ is the static head of

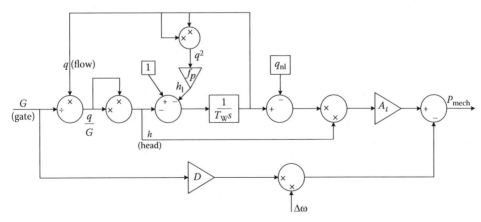

FIGURE 3.24 Nonlinear turbine model.

the water column (m), $\bar{h}$ is the head of the turbine admission (m), and $\bar{h}_1$ is the head loss to conduit friction (m).

The water time constant is defined as

$$T_W = \frac{L}{A}\frac{q_{base}}{h_{base}g} \text{ (s),} \tag{3.4}$$

where q_{base} is the static head of the water column above the turbine and the maximum flow rate chosen, which occurs when the gates are fully open ($G = 1$). At the same time, h_{base} is the head at the turbine. In other words, the base head is the total available static head, and the base gate is the maximum gate opening. The base quantities can be anything else.

The per unit representation of Equation 3.3 becomes

$$\frac{dq}{dt} = \frac{(1 - h - h_l)}{T_W}, \tag{3.5}$$

where h is the head at the turbine and h_l is the head loss in per unit.

The water flow rate to the turbine is a function of the gate opening and the head:

$$q = f(G, h). \tag{3.6}$$

In per unit system, flow rate through the turbine is

$$q = G\sqrt{h}. \tag{3.7}$$

The mechanical power equals the flow multiplied by head with necessary conversion factors in an ideal turbine. The efficiency of the turbine is naturally less than 100%, thus the effective flow is the difference between the no-load flow and the actual flow. Then, the effective flow should be multiplied by the head. Gate opening also has an effect on speed deviation damping and results in mechanical power reduction. The mechanical power can be expressed as

$$P_m = A_t h(q - q_{nl}) - DG\Delta\omega. \tag{3.8}$$

Here, q_{nl} is the per unit no-load flow, reports of the fixed power losses. A_t is the proportionality constant given in Equation 3.9, which can be calculated using turbine power rating (MW) and generator base power (MVA).

$$A_t = \frac{\text{turbine power rating (MW)}}{h_r(q_r - q_{nl}) \times \text{generator power rating (MVA)}}, \tag{3.9}$$

where h_r is the per unit head at the turbine at rated flow, while q_r is the per unit flow at rated load. At the rated load, per unit gate is generally less than unity (1).

Equation 3.9 defines the parameter A_t, which represents the conversion relationship between gate opening and per unit turbine power on the MVA base of the generator.

If the friction losses are neglected, the nonlinear turbine model (Figure 3.24) can be simplified and linearized as shown in Figure 3.25.

According to this linearized and simplified model, the time constants can be defined as

$$T_1 = \frac{q_o - q_{nl}}{T_W} \tag{3.10}$$

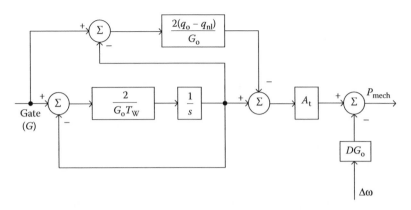

FIGURE 3.25 Linearized turbine model.

and

$$T_2 = \frac{G_o T_W}{2},\tag{3.11}$$

where G_o is the per unit gate opening at the operating point and q_o is the per unit steady-state flow rate at the opening point. Based on these time constants, according to Figure 3.25, the mechanical power output becomes

$$\Delta P_m = \frac{A_t(1 - T_1 s)\Delta G}{(1 + T_{2s})} - DG_o \Delta \omega.\tag{3.12}$$

It should be noted that more gate opening results in more flow rate. In per unit representation,

$$G_o = q_o.\tag{3.13}$$

The damping term in Equation 3.12 can be neglected to obtain the commonly used linear transfer function of the penstock/turbine as

$$\frac{\Delta P_m}{\Delta G} = A_t \times \frac{1 - G_o T_W s}{1 + G_o T_W s/2},\tag{3.14}$$

where $G_o T_W$ is the approximate effective water starting time for small perturbations around the operating point.

Detailed hydro turbine models can be implemented by considering other parameters such as the traveling wave models, surge tank effects, elastic water column conditions, multiple penstocks, and other mechanical parameters [14–17]. However, the model presented here is adequate for most of the analysis and control design procedures.

3.3.2 Hydro Turbine Controls

Figure 3.26 presents a basic hydraulic turbine speed control system, which consists of a governor, a servo system, a hydraulic system, and the hydraulic turbine.

For stable control performance, hydro turbines require transient droop features in the speed control. This is because of the initial inverse response characteristics of power to gate

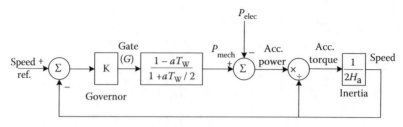

FIGURE 3.26 Hydro turbine and speed control linear model.

changes. The transient droop stands for the governor regulation. For slow changes and in the steady state, the governor exhibits the normal low regulation (high gain), while for fast deviations in frequency, the governor exhibits high regulation (low gain).

The main duty of the governor is to control the gate opening for the hydraulic system. A governor is generally composed of a pilot valve, servomotor, speed limit, droop compensation systems, and a gate servomotor [14,15,17]. This complete system can also be expressed by a linear transfer function. Generally, a governor can be represented by a proportional (P), proportional–integral (PI), or a proportional–integral–derivative controller (PID). On the other hand, the servomotor can be represented by a first-order transfer function. Figure 3.27 shows a block diagram of a typical governor including the servo system and servo motor along with permanent and transient droop compensations.

The turbine gate is controlled by a two-stage hydraulic position servo. The parameters in the model are as follows:

T_p is the pilot valve and servo motor time constant.

Q is the servo gain.

T_g is the main servo time constant.

R_p is the permanent droop.

R_t is the transient droop.

T_R is the reset time or dashpot time constant.

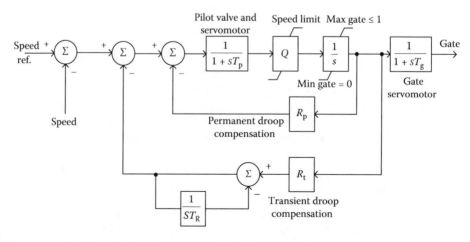

FIGURE 3.27 A typical hydro turbine governor model.

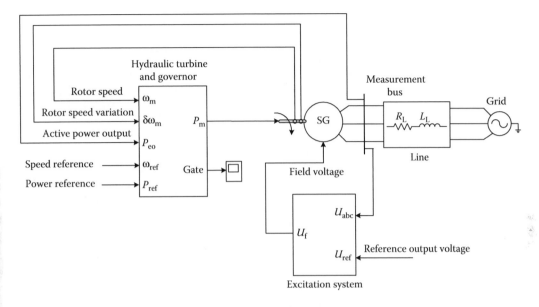

FIGURE 3.28 Hydraulic turbine and governor coupled to a synchronous generator.

The turbine, governor, and speed control models can be implemented in MATLAB/Simulink for dynamic analyses. The general block diagram of the hydraulic turbine, governor system, synchronous machine, excitation voltage control, transmission line, and three-phase grid is shown in Figure 3.28.

In the model, synchronous generator is a salient-pole machine with 202 mW nominal power and 13.8 kV phase-to-phase nominal voltage. The nominal frequency is 60 hz and number of pole pairs is 32. Thus, the nominal revolutions per unit is

$$n_s = \frac{60 \times 60}{32} = 112.5 \, \text{rpm}.$$

The hydraulic turbine and governor system consists of a hydraulic turbine, a servomotor, and a PID-based governor. The hydraulic turbine is modeled as Figure 3.29. This model is the Simulink implementation of Figures 3.25 and 3.26, and Equation 3.8. The parameters *beta* and t_w in this model are speed deviation damping coefficient and water starting time,

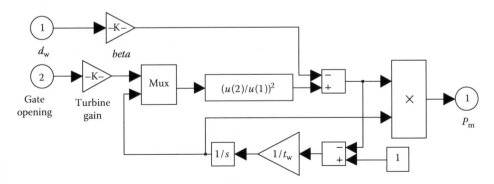

FIGURE 3.29 Hydraulic turbine model.

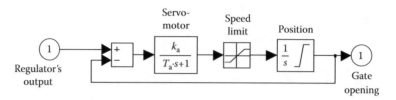

FIGURE 3.30 Servomotor model.

respectively. The servomotor system is controlled by a regulator (PID governor), which controls the gate opening for the turbine. The servomotor system is shown in Figure 3.30.

The parameters for the servomotor are $K_a = 3.33$ and $T_a = 0.07$s. Gate opening maximum and minimum values are set at 0.01 and 0.985, respectively. The gate opening and closing speed values are set between $[-0.5, 0.5]$ p.u./s, respectively. The permanent droop for the turbine is 0.05 p.u. and governor regulator gains are $K_p = 9.304, K_i = 0.84, K_d = 0$ (PI type governor), and $T_d = 0.01$s.

Figure 3.31 demonstrates a system composed of a servomotor, a regulator, and the permanent droop that are connected with the hydraulic turbine.

The excitation voltage control system shown in Figure 3.28 is a PI controller. The measured ABC phase voltages and the reference ABC phase voltages are first converted to dq-axis voltage quantities. Then, the reference and the measured dq-axis voltages are compared, and based on the error signal value, the appropriate excitation voltage level is applied to the generator in order to maintain a fixed generator output voltage.

The reference power generation of the turbine is selected as 0.8 and the reference speed is taken as unity (1). From Figure 3.28, it is seen that the hydraulic turbine and governor system produces a mechanical power and this mechanical power is applied to the generator's rotor shaft. The output of the synchronous generator is connected to the grid through an RL circuit, representing the transmission line. For 22 s of simulation the rotor speed settles at 1, whereas the rotor speed deviation is damped to 0 after some small oscillations, as seen from Figures 3.32 and 3.33. The generator voltage is shown in Figure 3.34, which is controlled by the excitation voltage controller. The mechanical power supplied to the generator's shaft,

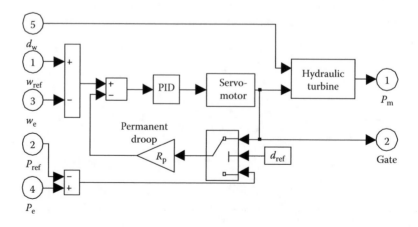

FIGURE 3.31 Complete turbine, governor, and servomotor model.

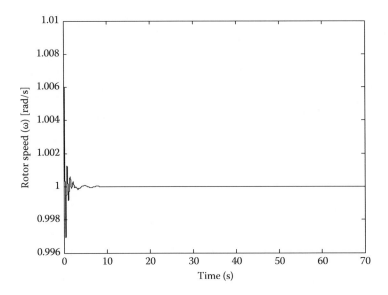

FIGURE 3.32 Rotor speed.

shown in Figure 3.35, is increasing to the reference value of 0.8, which is commanded by the gate opening, shown in Figure 3.36.

Due to the efficiency of the electromechanical energy conversion of synchronous generator, the generator output active power has a waveform similar to Figure 3.37. Finally, the real value of the generator output phase-to-phase voltage is shown in Figure 3.38. It should be noted that the rated voltage of the generator is 13.8 kV, phase to phase.

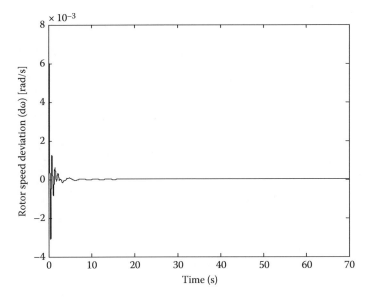

FIGURE 3.33 Rotor speed deviation.

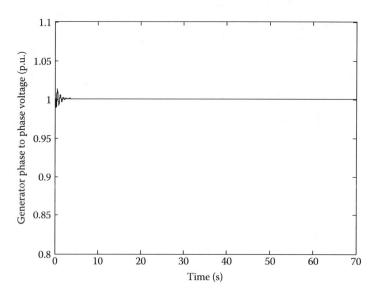

FIGURE 3.34   Generator voltage.

3.3.3   Kinetic Energy

Instead of using costly dams, which potentially occupy too much space, the other approach is to take advantage of the horizontal flow of tide current. In this way, the kinetic energy of the moving unconstrained tidal streams is utilized to drive the turbines, without dams, in a similar way to harvest wind energy from wind. In this concept, instead of collecting the ebb tidal water in a dam and using its potential energy, the energy of moving tidal current is harvested.

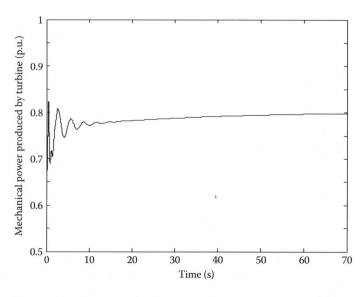

FIGURE 3.35   Turbine mechanical power.

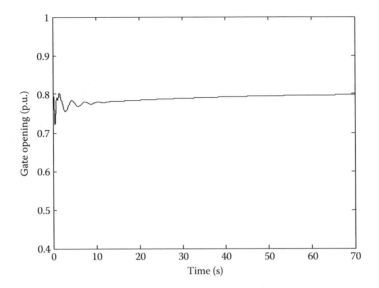

FIGURE 3.36 Gate opening.

3.3.3.1 Energy Calculation for Tidal Current Energy Harvesting Technique

A schematic model of a tidal turbine is presented in Figure 3.39.

The ideal kinetic energy for a tidal turbine model can be calculated as

$$E = \frac{1}{2}mV^2, \tag{3.15}$$

where V is the speed of the water, m is the mass of water, which is given by

$$m = \rho v, \tag{3.16}$$

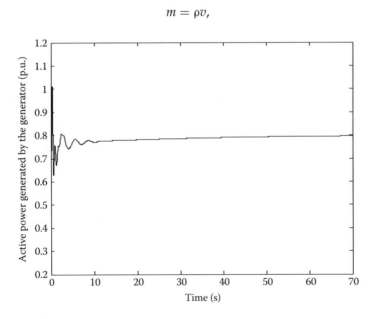

FIGURE 3.37 Generator active power.

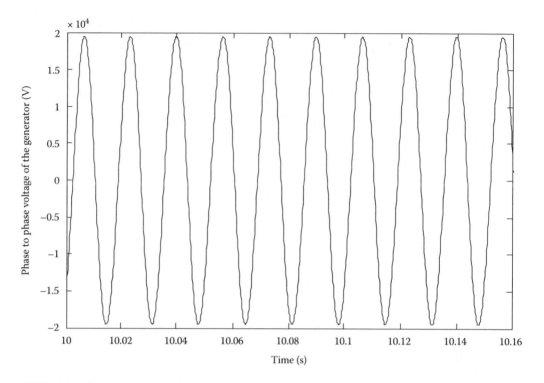

FIGURE 3.38   Generator output voltage.

where ρ is the seawater density (approximately $1022\,\text{kg}/\text{m}^3$) and v is the water volume. Thus, the kinetic energy can be expressed as

$$E = \frac{1}{2}\rho v V^2. \tag{3.17}$$

For a mass of water passing through a rotor with a cross-sectional area of A, the ideal power can be expressed as

$$P = \frac{1}{2}\rho A V^3. \tag{3.18}$$

Considering the efficiency of the turbine, the power can be expressed as

$$P_T = \frac{1}{2}\rho A V^3 C_p, \tag{3.19}$$

where P_T is the power developed by the rotor (W), A is the area swept out by the turbine rotor (m^2), V is the stream velocity (m/s), and C_p is the power coefficient of the turbine.

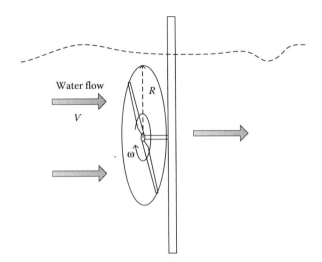

FIGURE 3.39 Conceptual tidal current turbine used for tidal energy harvesting.

At sea level and at 20°C, dry air has a density of 1.2 kg/m³, and seawater has a density of approximately 1022 kg/m³. Assuming the tidal flow rate is one-fifth of that of the wind flow, the tidal turbine would have a rotor with a diameter about half of a wind turbine of the same rated power.

The power coefficient, C_p, is the percentage of power that the turbine can extract from the water flowing through the turbine. According to the studies carried out by Betz [18], the theoretical maximum amount of power that can be extracted from a fluid flow is about 59%, which is referred to the Betz limit.

The thrust force applied to the rotor can be expressed by

$$F = \frac{1}{2}\rho A V^2. \tag{3.20}$$

Hence, the ideal maximum torque on the rotor is

$$T = \frac{1}{2}\rho A V^2 R, \tag{3.21}$$

where R is the radius of the rotor.

The ratio between the actual torque developed by the rotor and the theoretical torque is called torque coefficient (C_T), which is given as

$$C_T = \frac{T_T}{T} = \frac{2T_T}{\rho A V^2 R}. \tag{3.22}$$

The ratio of the rotor velocity to the water current velocity is called tip speed ratio (TSR) (λ), which is given as

$$\lambda = \frac{R\omega}{V}, \tag{3.23}$$

where ω is the angular speed of the rotor.

Both the power coefficient and torque coefficient of a rotor depend on the tip ratio. From Equation 3.19,

$$C_P = \frac{2P_T}{\rho A V^3} = \frac{2T_T \omega}{\rho A V^3}.$$

(3.24)

Dividing Equation 3.24 by Equation 3.22 gives

$$\frac{C_P}{C_T} = \frac{R\omega}{V} = \lambda.$$

(3.25)

Thus, the TSR (λ) is equal to the ratio between the power coefficients (C_P) and the torque coefficient (C_T) [19]. C_T is a function of β, which is the pitch angle, so C_P is a function of λ and β. Thus, the power coefficient is a function of λ and β.

3.3.3.2 Tidal Turbines Used in Tidal Current Approach

To harvest tidal current energy, energy conversion systems are required to convert water kinetic energy into the motion of mechanical systems, which can drive generators. Generally, these devices can be characterized into three fundamental types (shown in Figure 3.40) (a) horizontal axis systems; (b) vertical axis systems; and (c) linear lift-based systems.

3.3.3.2.1 Horizontal Axis Turbine

One simple way to capture tidal stream energy is using horizontal axis tidal turbines, which are similar to the horizontal axis wind turbines and have been developed over the last 30 years.

Horizontal axis turbines have their blades rotating around a main shaft pointed into the direction of tidal current flow. Between blades and the shaft, a hub works to transfer and convert captured stream energy to the rotational energy. The hub is mostly designed to accommodate the pitch control that allows turning of the blades to their optimum orientation in the tidal stream. Gearbox is another critical component commonly used in horizontal axis turbine system. The gear increases the low speed of the turbine shaft to the desired operating speed of the generator shaft, while the generator converts the mechanical rotation energy into electric power. The projects based on horizontal axis turbines are described in the following paragraphs.

1. *Hammerfest Strom's E-Tide Project*   As shown in Figure 3.41, the so-called *Blue or E-Tide Concept* was developed by Hammerfest Strom in 2003. A 300 kW prototype has been

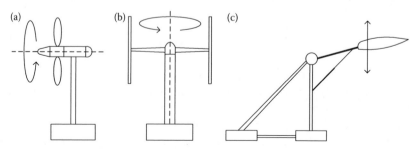

FIGURE 3.40   Three tidal turbine fundamental types: (a) horizontal axis system; (b) vertical axis system; (c) linear lift based system.

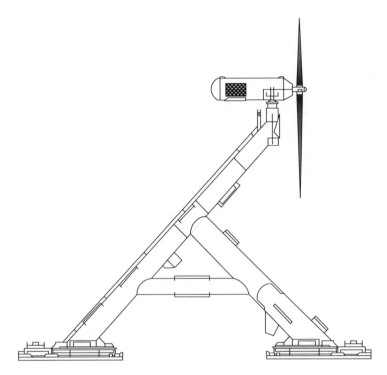

FIGURE 3.41 E-tide turbine.

tested, and its estimated energy output is about 0.7 GWh/yr. Now the first generation of commercial mills with much larger output is under design. It consists of steel pipes and gravity footing, with a height of 20 m, and the gravity weight is 200 tons.

2. The Marine Current Turbine's Seaflow and SeaGen turbines Figure 3.42 demonstrates a tidal turbine, called SeaGen, developed by the Marine Current Turbine Ltd. (MCT). Its swiveling rotor is mounted on a fixed steel pile, 2.1 m in diameter, driven into seabed, and tall enough to keep its top above the surface of the sea. They can be physically raised up above the sea level for maintenance or repairs.

The length of the blades is 7.5 m (24.6 ft), from the hub center to the tip. The optimum operating point of the turbines is for coastal current speeds between 4 and 5.5 mph. In this current speed, each turbine in the farm can generate the rated power of 300 kW. This is the amount of energy that a 30-m (98.5-ft)-long wind turbine can generate. Ideal locations for tidal turbine farms are close to shore in water depths of 20–30 m (65.5–98.5 ft).

As shown in Figure 3.43, SeaGen is another turbine system developed by MCT. Based on the MCT reports, SeaGen is the world's first commercial scale tidal energy turbine, which is going to be installed in North Ireland. Its biggest difference from Seaflow turbine is that each SeaGen has two rotors mounted at the both ends of a pair of streamlined wing-link arms, which also can be physically raised up to the sea surface for maintenance or repairs. Each power train is driven by a rotor, and consists of a gearbox and a generator rated about 500 kW, so each SeaGen's rated power is around 1 mW, about three times the power of Seaflow.

3. The Lunar Energy's turbines Figure 3.44 demonstrates another concept of horizontal turbine developed by Lunar Energy Inc. In this design, a bidirectional horizontal axis turbine is housed in a symmetrical duct, which is fixed to seabed by its gravity foundation.

FIGURE 3.42   (**See color insert following page 80.**) A SeaGen turbine installed at Strangford Lough. (Courtesy of Seaflow and SeaGen Turbines, Marine Current Turbines Ltd., available at: http://www.marineturbines.com, March 2008.)

FIGURE 3.43   (**See color insert following page 80.**) SeaGen turbine farm. (Courtesy of Seaflow and SeaGen Turbines, Marine Current Turbines Ltd., available at: http://www.marineturbines.com, March 2008.)

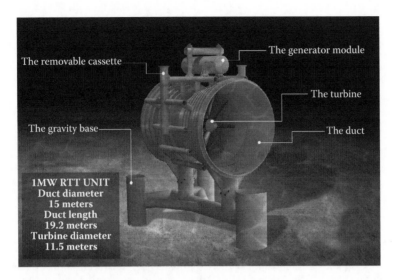

FIGURE 3.44 **(See color insert following page 80.)** A generating system with a horizontal axis turbine, Rotech Tidal Turbine. (Courtesy of Rotech Tidal Turbine (RTT), Lunar Energy™, Harnessing Tidal Power, available at: http://www.lunarenergy.co.uk/)

The cassette that contains the generator module and the turbine is removable. It can be lifted up over the sea when maintenance is required.

The duct is designed as shown in Figure 3.45, which can capture a large area of tidal stream and accelerate the current speed through a narrow channel and can work at a tidal flow offset of up to 40°, eliminating the need for the yaw control mechanism. This would keep the design simple and cost effective.

4. Tidal Stream Energy's Semi-Submersible Turbine   This system is designed for Pentland Firth, which separates the Orkney Islands from Caithness in the north of Scotland, with a water depth of 60 m. Their latest design is shown in Figure 3.46, and it is anchored by a gravity base. The total power output of four 20 m rotors is rated at 4 mW [22].

Easy maintenance is considered as one of the key features of this Semi-Submersible Turbine (SST) design concept. The swing arm is hinged at its upper end to the main spar buoy. When maintenance is needed, water is pumped out of the main body spar buoy tanks, so

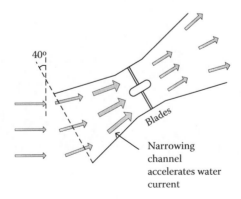

FIGURE 3.45 Ducted turbine cross section.

FIGURE 3.46 (**See color insert following page 80.**) Semisubmersible turbine. (Courtesy of Semi-Submersible Turbine, TidalStream Limited *"The Platform for Tidal Energy,"* available at: http://www.tidalstream.co.uk/)

that the swing arm will start to roll and rotors will rise until they reach a stable position on the surface of the sea. This process is expressed in Figure 3.47.

Figure 3.48 shows the SST's comparison against an offshore wind turbine with a same power rating. According to Tidal Stream Energy's statements [22], the SST in this figure

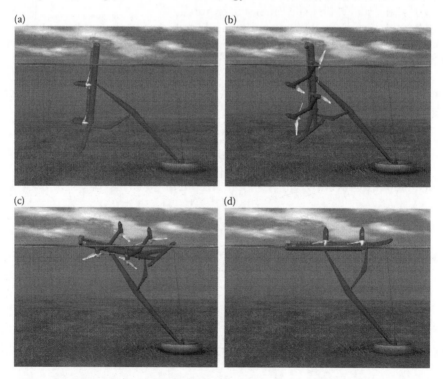

FIGURE 3.47 (**See color insert following page 80.**) Position transformation of SST to its maintenance position. (a) Water is started to be pumped out of the main body, (b) swing arm started to roll, (c) rotors are rising to the water surface, (d) rotors reach a stable position on the sea surface. (Courtesy of Semi-Submersible Turbine, TidalStream Limited *"The Platform for Tidal Energy,"* available at: http://www.tidalstream.co.uk/)

FIGURE 3.48 **(See color insert following page 80.)** Comparison of SST with a wind turbine for the same power rating. (Courtesy of Semi-Submersible Turbine, TidalStream Limited *"The Platform for Tidal Energy,"* available at: http://www.tidalstream.co.uk/)

is at 60 m water depth, and the wind turbine is located at 25 m water depth for offshore applications. Each rotor of SST has a diameter of 20 m, whereas the wind turbine with a 100 m diameter rotor captures around the same energy in a typical 10 m/s wind speed. The gravity base needed for the wind turbine is 25% larger than that needed by the tidal turbine, which has to deal with higher direct loads but not overturning moment.

3.3.3.2.2 Vertical Axis Turbines

As shown in Figure 3.49, horizontal axis turbines can harness bidirectional tidal flow. This requires a more complex two-direction generator.

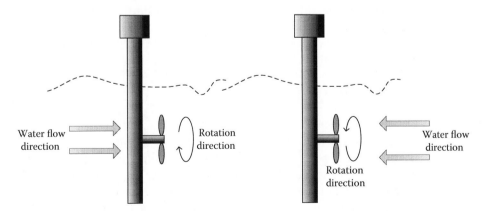

FIGURE 3.49 Operation of a bidirectional horizontal axis turbine.

Vertical axis turbines are cross-flow turbines, with the axis positioned perpendicular to current flow. The idea of the vertical axis turbines is derived from the early Darrieus turbine—a windmill turbine developed decades ago. Compared to horizontal turbines, vertical axis turbines permit the harnessing of tidal flow from any direction, facilitating the extraction of energy not only in two directions, but making use of the full tidal ellipse of the flow [23].

1. Blue Energy's vertical axis turbine Ref. [24] provides a vertical axis turbine system developed by Blue Energy Inc. In this design, four fixed hydrofoil blades are connected to a rotor, which drives an electrical generator through a gearbox. The gearbox and generator sit at the top, above the surface of the water, for easy access. Each turbine is expected to have approximately 200 kW output power. Power is transmitted through submersible DC cabling buried in the ocean sediments with power drop points for coastal cities and connections to the continental power grid.

For large-scale power production, more numbers of these kinds of turbines could be linked in series to create a tidal fence across an ocean passage or inlet to harness the energy of the tides [4], as also provided in Ref. [24].

A 2.2 GW tidal fence project was proposed by the Blue Energy Inc. for the San Bernadino Strait in the Philippines [24]. However, due to political and capital issues, the proposed $2.8 billion project is still on hold [24].

Commercial technologies of tidal fence are expected to be suitable for off-grid power supply in islands and other remote regions, as well as grid-connected applications [25]. They have the potential to provide a predictable, near-continuous source of power in many areas of the world, where moderate tidal currents exist.

2. Gorlov's helical turbine Gorlov's helical turbine (GHT) is another kind of vertical axis turbine. In GHT design, two, three, or more air foil blades are twisted around the axis along a cylindrical surface like a screw thread as shown in Ref. [26].

The GHT can produce power from water current flow as low as 1.5 m/s. It solves the vibration problem of the straight airfoil and is very efficient. It is self-starting and can be installed either vertically or horizontally to the water current flow direction [27]. In a standard GHT (1 m in diameter, 2.5 m in length), rated power is 1.5 kW for 1.5 m/s water current speed and 180 kW for 7.72 m/s.

3. KOBOLD turbine Since 2001, an ENERMAR plant, developed by Ponte di Archimede Co., has been anchored in Italy's Strait of Messina about 150 m (490 ft) offshore in waters 18–25 m (59–82 ft) deep [28]. It is fixed with four concrete anchoring blocks and chains. Ponte di Archimede Co.'s data indicate that the plant can produce 25 kW energy from a marine current of only 2 m/s (4 knots), and about 22,000 kWh of useful energy can be extracted each year.

KOBOLD turbine is implemented in this system. It can not only rotate independent of the direction of the current, but also has a high torque that permits spontaneous starting even under intense conditions without the need of an ignition device. In this turbine, a three-blade rotor is mounted on a vertical shaft.

3.3.3.2.3 Linear Lift-Based Device

The Stingray is a device that converts tidal energy into electrical energy. It is a seabed-mounted device consisting of a hydroplane with variable pitch on a pivoted arm. The tidal stream exerts a resultant force as the tidal stream flows over the hydro turbine. The Stingray is developed by Engineering Business Ltd. and is shown in Figure 3.50. Its hydroplane has an attack angle relative to the approaching water stream, which is varied using a simple mechanism. This causes the supporting arm to oscillate, which in turn forces hydraulic

FIGURE 3.50 (See color insert following page 80.) Stingrays mounted on the seabed. (Courtesy of Engineering Business Ltd., UK, "Stingray tidal stream generator," available at: http://www.engb.com, March 2008.)

cylinders to extend and retract [29]. This produces high-pressure oil that is used to drive a hydraulic motor, which is cross-coupled to an electric generator [29]. Engineering Business Ltd. has a project to design, build, test, and install a full-scale offshore demonstration of its Stingray tidal stream generator.

3.4 Tidal Energy Conversion Systems

The major components of a tidal energy conversion system are blades, shaft, gearbox, generator, sensors, and control system.

3.4.1 Generators

Tidal turbine generators operate based on the same basic principles as generators utilized in wind energy harvesting. Synchronous generator (PMSG and BLDC generator) and DFIG are common in tidal energy systems.

3.4.1.1 Synchronous Generator

Most generators used in barrages are synchronous generators. They are widely used in generating hydraulic electric energy. Nowadays over 97% of all electric power around the world is generated by synchronous generators, because synchronous generators are very efficient and convenient to operate. Their efficiencies vary from 75% to 90%.

The simplest form of the generator, as shown in Figure 3.51, is composed of a stator and a rotor, with the same number of poles. In Figure 3.51, the stator circuit has three pairs of coils,

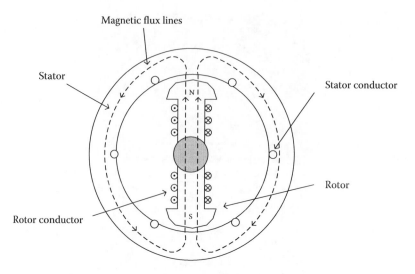

FIGURE 3.51   Conceptual representation of a two-pole synchronous machine.

with 120° phase angle. The rotor can be either PM (PMSM) or electromagnet excited by a DC source. A synchronous generator's rotor and magnetic field run at a same constant speed, which is the so-called synchronous speed. The synchronous speed can be expressed by

$$N_s = \frac{60f}{p} \text{(rpm)},$$ (3.26)

where f is the frequency (Hz) and p is the numbers of pole pairs

$$\omega_m = \frac{\omega}{p} = \frac{2\pi f}{p} \text{(rad/s)},$$ (3.27)

where ω_m is the machine angular velocity and ω is the electric angular velocity. Based on Equations 3.26 and 3.27, speed regulation is achievable through controlling the frequency. The equivalent circuit and the phasor diagram of the synchronous generator are shown in Figure 3.52. In this equivalent circuit, E_0 is the equivalent induced voltage (back emf), Z is the load impedance, E is the output voltage, X is the synchronous reactance, and I is the stator current for one phase. Equation 3.28 is the general voltage equation for the synchronous generator:

$$\vec{E} = \vec{E}_0 - \vec{I}R - \vec{I}X.$$ (3.28)

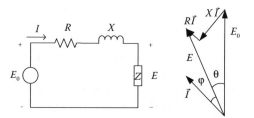

FIGURE 3.52   Equivalent circuit of synchronous generator.

Output power and electromagnetic torque can be calculated based on the following equations:

$$P = 3UI \cos \varphi = \frac{3E_0 E}{X} \sin \theta \qquad (3.29)$$

and

$$T_e = \frac{3E_0 E}{\omega_m X} \sin \theta, \qquad (3.30)$$

where φ is the power coefficient angle, θ is the torque angle, and ω_m is the machine angular velocity.

PMSM can be actively controlled by vector control of rotor magnetic field orientation to get better torque characteristics. Figure 3.53 shows the vector control of a PMSM. PMSM is a synchronous machine with a rotor made of PM, so its vector control is much simpler than that of asynchronous machines [30].

In ideal conditions, mathematical equations of vector control of PMSM can be expressed as

$$u_d = Ri_d + L_d \frac{di_d}{dt} - \omega L_q i_q, \qquad (3.31)$$

$$u_q = Ri_q + L_q \frac{di_q}{dt} + \omega L_d i_d + \omega \psi_r, \qquad (3.32)$$

$$T_e = \frac{3p}{2} \left[\psi_r i_q + \left(L_d - L_q \right) i_q i_d \right], \qquad (3.33)$$

where u_d and u_q are the stator winding voltage in d–q coordinate system, i_d and i_q are the stator winding current in d–q coordinate system, R is the stator winding resistance, L_d and L_q are the stator winding inductance, ψ_r is the rotator magnetic flux linkage, ω is the angular speed of the stator, and T_e is the torque of the rotator.

A commonly used method is fixing the vector of stator current on q-axis, so that the part of d-axis does not exist. In other words, the rotating magnetic flux linkage is not variable

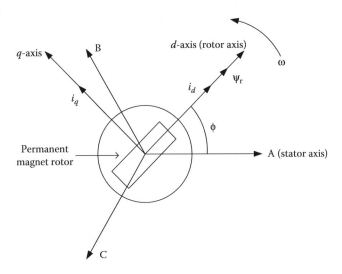

FIGURE 3.53 PMSM vector control scheme.

and it does not affect the electromagnetic torque. Thus, the electromagnetic torque is

$$T_e = \frac{3p}{2}\psi_r i_q.$$
(3.34)

Since the parameters of excitation and torque are decoupled, the electromagnetic torque can be controlled by regulating the corresponding q-axis current of the stator.

In a system of kinetic tidal energy harvesting with synchronous generator, the operating speed changes randomly with fluctuations in the tidal current velocity and, therefore, the output voltage and frequency would be variable. Therefore, a power electronic interface is necessary between the generator and grid to control the output voltage and frequency of the system. It will rectify the AC output of the generator to DC. Then the DC current will be inverted back to AC current with the desired frequency and amplitude of the grid, as shown in Figure 3.54.

3.4.1.2 Asynchronous Generator

Compared to synchronous generators, asynchronous generators are more rugged and usually have a lower capital cost. Like synchronous generators, they are composed of a stator and a rotor, and the stator circuit has three pairs of coils. The difference is that the rotor in an induction generator is composed of short-circuit coils. Its synchronous speed is the same as that of a synchronous machine, as given in Equation 3.26.

The difference between the rotational speed and the synchronous speed in percent of synchronous speed is called the generator's slip. The slip is defined as

$$s = \frac{N_s - N}{N_s},$$
(3.35)

where N_s is the synchronous speed (in revolutions per minute) and N is the rotor's speed.

3.4.1.2.1 Doubly Fed Induction Generator

DFIG has wider working speed range (about ±30% of the synchronous speed). This feature of DFIG creates the opportunity to be widely used in wind and tidal current energy harvesting, where variable speed systems are required.

Compared to the induction generator, the stator circuit of the DFIG is connected to the grid directly, but the rotor circuit is connected to a converter via slip rings, as shown in Figure 3.55.

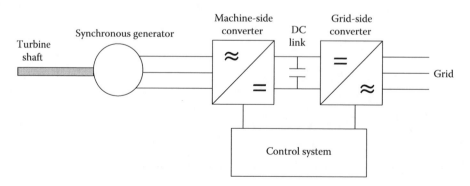

FIGURE 3.54 Variable speed operation of a synchronous generator.

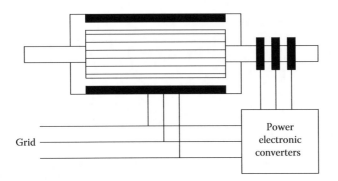

FIGURE 3.55 Circuit connection of a DFIG.

The rotor winding is connected to the grid through "back-to-back" converters. A DC-link capacitor is placed, as energy storage, between these two converters to decrease the voltage ripple in the DC link. The machine side converter is used to control the torque or the speed of the DFIG and also the power factor at the stator terminals. The main objective for the grid side converter is to keep the DC-link voltage constant [26], as shown in Figure 3.56.

Figure 3.57 presents one-phase equivalent circuit of the DFIG.

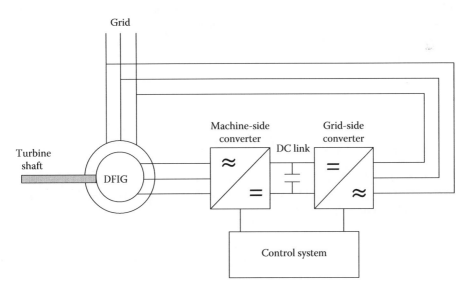

FIGURE 3.56 Variable speed generating with DFIG.

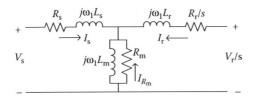

FIGURE 3.57 Equivalent circuit of DFIG. (Modified from S. Tnani et al., "A generalized model for double fed induction machines," *IASTED Conference*, 1995.)

Based on Kirchhoff's voltage law,

$$V_s = R_s I_s + j\omega_1 L_s I_s + j\omega_1 L_m (I_s + I_r + I_{Rm}), \tag{3.36}$$

$$\frac{V_r}{s} = \frac{R_r}{s} I_r + j\omega_1 L_r I_r + j\omega_1 L_m (I_s + I_r + I_{Rm}), \tag{3.37}$$

$$0 = R_m I_{Rm} + j\omega_1 L_m (I_s + I_r + I_{Rm}), \tag{3.38}$$

where V_s is the stator voltage, R_s is the stator resistance, V_r is the rotor voltage, R_r is the rotor resistance, I_s is the stator current, L_s is the stator leakage inductance, I_r is the rotor current, L_r is the rotor leakage inductance, I_{Rm} is the magnetizing resistance current, R_m is the magnetizing resistance, ω_1 is the stator frequency, L_m is the magnetizing inductance, and s is the slip, which is

$$s = \frac{\omega_1 - \omega_r}{\omega_1} = \frac{\omega_2}{\omega_1}, \tag{3.39}$$

where ω_r is the rotor speed and ω_2 is the slip frequency.

3.4.1.2.2 Dynamic Model of the DFIG

The Park model is a commonly used model for the doubly fed generators [16]. As the vector control theory proposed by Blaschke in 1971, an asynchronous machine can be described by a differential equations system in d, q components. Ramuz et al. [16] chose axes shifted 90° behind the vector of stator voltage ($V_{sd} = 0$ and $V_{sq} = V_s$) as the reference frame, as shown in Figure 3.58.

Equations of the machine are

$$V_s = R_s I_s + \frac{d\Phi_s}{dt} + j\omega_s \Phi_s, \tag{3.40}$$

$$V_r = R_r I_r + \frac{d\Phi_r}{dt} + j\omega_r \Phi_r, \tag{3.41}$$

where ω_s is the angular velocity of the axes system (d, q), ω is the angular velocity of the rotor compared to the stator, and R_s and R_r are the stator and rotor resistances, respectively.

Equations of the magnetic flux are

$$\Phi_s = L_s I_s + M I_r, \tag{3.42}$$

$$\Phi_r = L_r I_r + M I_s, \tag{3.43}$$

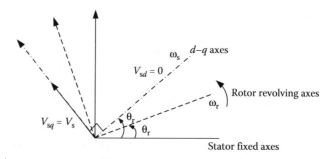

FIGURE 3.58 Rotating dq-axis reference frame system of a DFIG.

where L_s is the stator cyclic inductance, L_r is the rotor cyclic inductance, and M is the cyclic mutual inductance.

Projecting Equations 3.40 through 3.43, for the d–q axes system, they can be represented as

$$V_{sd} = R_s I_{sd} + \frac{d\Phi_{sd}}{dt} - \omega_s \Phi_{sq}, \tag{3.44}$$

$$V_{sq} = R_s I_{sq} + \frac{d\Phi_{sq}}{dt} + \omega_s \Phi_{sd}, \tag{3.45}$$

$$V_{rd} = R_r I_{rd} + \frac{d\Phi_{rd}}{dt} - \omega_r \Phi_{rq}, \tag{3.46}$$

$$V_{rq} = R_r I_{rq} + \frac{d\Phi_{rq}}{dt} + \omega_r \Phi_{rd}. \tag{3.47}$$

Equations of flux become

$$\Phi_{sd} = L_s I_{sd} + M I_{rd}, \tag{3.48}$$

$$\Phi_{sq} = L_s I_{sq} + M I_{rq}, \tag{3.49}$$

$$\Phi_{rd} = L_r I_{rd} + M I_{sd}, \tag{3.50}$$

$$\Phi_{rq} = L_r I_{rq} + M I_{sq}. \tag{3.51}$$

3.4.2 Gearbox

Gearbox is one of the most vulnerable parts of the generation system in the tidal turbines. Since the tidal current velocity is typically in the range of 5–30 rpm [32], if no gearbox is utilized, the rotor of the tidal turbine will experience low rotational speeds. Generators are generally designed for higher rotational speeds. A gearbox is placed between the turbine shaft and generator shaft to increase the speed at the shaft of the generator, so that value of tip speed can be expressed as

$$b = \omega \frac{R_T}{R}, \tag{3.52}$$

where ω is the rotational speed of a rotor, R_T is the turbine radius, and R is the transmission ratio.

In order to eliminate energy loss associated with gearbox, gearless generator systems (direct-drive generators) were proposed. These systems not only improve the efficiency, but also prevent failures in gearboxes and simplify the maintenance. Figure 3.59 shows a system-level configuration of a gearless variable-speed generation system configuration.

These low-speed high-torque synchronous generators are already manufactured and utilized in wind energy harvesting industry. However, a directly driven generator will be relatively heavy and expensive, due to the fact that the mass of the rotor of the generator has to be roughly proportional to the torque.

3.4.3 Optimal Running Principle of Water Turbine

Figure 3.60 illustrates that at a water current speed of V_1, when the generator operates at a speed of ω_1 the maximum output power of the generator is P_1. However, when the

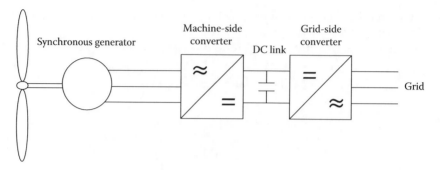

FIGURE 3.59　Variable-speed direct-driven (gearless) turbine.

water current changes to V_2 at the same operating speed of ω_1, the generator can only produce an output power of P_2. On the other hand, based on the output power versus speed characteristics of the generator, the maximum output power for speed V_2 is obtained at ω_2 and is equal to P_3. Therefore, in the tidal energy generation system, the variable-speed generation system is more efficient than the fixed-speed generation system, due to the variable tidal current speed. The angular speed of the turbine ω and the power coefficient of the C_p relation is expressed earlier in Equations 3.19 and 3.24 [33].

3.4.4　Maximum Power Point Tracking

In a variable speed generating system, the MPPT is one of the most important parts of the system. MPPT methods are implemented by controlling the speed of the tidal generator to ensure that the maximum available energy from the tidal current is captured. It can be seen in Figure 3.61 that with variable current speed (from V_1 to V_n) we can get the maximum value (from P_1 to P_n) with variable turbine speed (from ω_1 to ω_n). By connecting all these

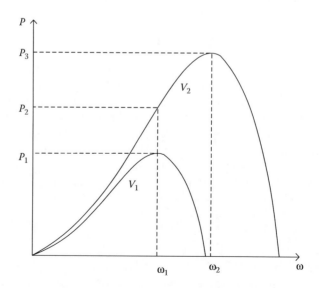

FIGURE 3.60　Output power versus rotating speed for a tidal energy generator.

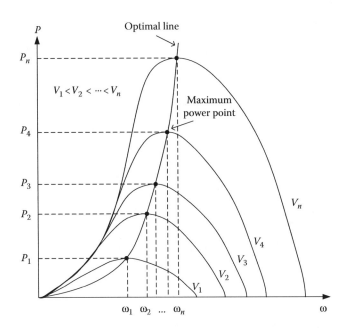

FIGURE 3.61 Generator power curves at various tidal speeds.

top points, we can get the optimal line. The goal is to keep the operating point on that line dynamically.

The variable speed generation system is more effective and hence suitable than the fixed-speed generation system in tidal energy harvesting applications [34]. The main reason of this is the fact that the variable-speed generation system can be operated at the most efficient rotating speed when the tidal current speed changes. The variable speed operation in tidal applications can be controlled by using the MPPT control method.

So far, many control methods of MPPT have been proposed for wind turbines and tidal turbines. Basically, they can be categorized into two control techniques: table look-up method and calculation method.

3.4.4.1 MPPT Method Based on Look-Up Table

A traditional method used to track the maximum power point is to predetermine a look-up table or an equation describing the loading required to achieve the MPP, and then measure the current speed. The speed is used to calculate the optimal load required for that particular current speed in this proposed open-loop-type system, as shown in Figure 3.62. In this method, the optimal load for various tidal current speeds needs to be preprogrammed into the control unit [35].

In the control unit, a PID controller is commonly used, as shown in Figure 3.63. Assuming that the water current speed c is known, it is multiplied by the desired TSR yielding the result of c_1. Thus desired generator rotor tip speed is obtained. Real-time tip speed b is proportional to turbine's rotating speed ω_t. Then difference between the desired tip speed c_1 with the real tip speed b results in the tip speed error signal e. This error signal is then used as an input for the PID controller to set the duty cycle of the converter.

However, there are some drawbacks in this method. First, the speed of the tidal current cannot be measured very accurately due to the variable current faced by the turbine blades.

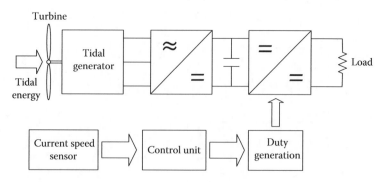

FIGURE 3.62   Control method based on current speed measurements.

Second, due to the measuring errors, system reliability decreases, and due to the necessity of additional algorithms and equipments for sensing the speed, the system becomes more complex and expensive [36,37].

3.4.4.2   MPPT Method Based on Current Speed Calculation

This method is mostly applicable to turbines with synchronous machines. In a synchronous machine, the electrical frequency produced by the generator is proportional to the angular speed of the rotor. Therefore, by measuring the electrical frequency, the rotor speed can be calculated [35]. The power versus rotor speed profiles is unique for each current speed, so for one particular rotor speed and one particular tidal current speed, there is a unique deliverable power.

Because the power delivered by the current turbine is measurable and the electrical frequency can also be determined, it is possible to calculate the instantaneous current speed responsible for the system's power generation at any time. The system level structure is expressed in Figure 3.64. Since the power profiles might be only available for a few of the current speeds, interpolation is used to determine the speeds, where the measured power does not exactly correspond to a given power profile [35].

3.4.5   P&O-Based MPPT Method

In this MPPT method, water current speed is neither measured by a sensor nor calculated by a processor. In this method, the MPPT is achieved by controlling the duty cycle of the buck converter, as shown in Figure 3.65 [35,38].

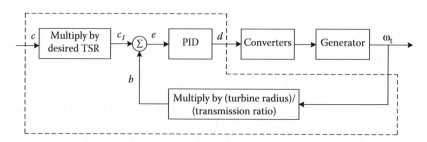

FIGURE 3.63   Block diagram of a MPPT controller.

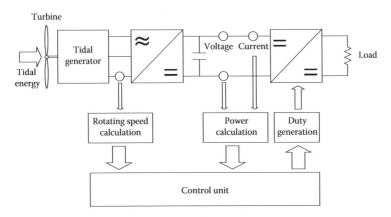

FIGURE 3.64 Control method based on current speed calculation.

The tidal turbine characteristics corresponding to tidal speed values features a second-order equation curve as shown in Figure 3.66. The output power of tidal generator is maximized at a particular rotation speed of tidal generator in various tidal speeds.

The maximum power can be generated from the tidal generator when the variation of generator output power versus the variation of rotation speed of generator is equal to zero:

$$\frac{dP}{d\omega} = 0. \tag{3.53}$$

Applying the chain rule [39], Equation 3.53 can be expressed by

$$\frac{dP}{d\omega} = \frac{dP}{dD} \times \frac{dD}{dV_{in}} \times \frac{dV_{in}}{d\omega} = 0. \tag{3.54}$$

In this way, the information of the tidal turbine characteristics, tidal speed measurement, and the tidal turbine speed are not necessary anymore. Therefore, this method can overcome

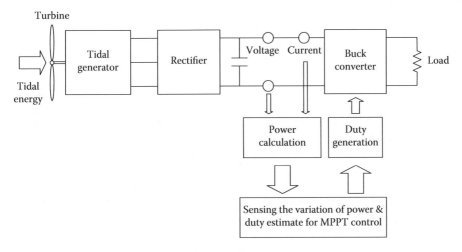

FIGURE 3.65 Block diagram of a MPPT control system.

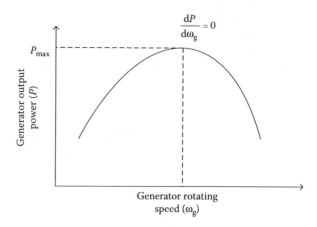

FIGURE 3.66 Tidal turbine speed–power characteristics.

the drawbacks of MPPT methods that are based on look-up table or water current speed calculation.

The duty ratio (D) of a buck converter, in steady-state operation, can be expressed as

$$D = \frac{V_{\text{out}}}{V_{\text{in}}}, \tag{3.55}$$

where V_{out} is the output voltage and V_{in} is the input voltage of the converter.

The variation of duty ratio of this converter in terms on its input voltage is as

$$\frac{dD}{dV_{\text{in}}} = -\frac{V_{\text{out}}}{V_{\text{in}}^2} < 0. \tag{3.56}$$

The three-phase rectifier output voltage is equal to converter input voltage. The voltage variation over the speed variation is positive. In other words, they vary through the same direction (increasing or decreasing):

$$\frac{dV_{\text{in}}}{d\omega} > 0. \tag{3.57}$$

Using Equations 3.75, 3.76, and 3.73, it can be concluded that

$$\frac{dP}{dD} = 0. \tag{3.58}$$

Therefore, the MPPT control of the tidal energy generation system can be achieved through the regulation of duty ratio. Thus, the zero point of the output power variation over duty cycle is equal to the zero point of the output power variation over the generator rotation speed.

From Figure 3.72, it is seen that the generator speed should be increased for output power of the generator to reach the maximum power point, if the current generator speed is on the left side of the optimal rotation speed [40]. This means that the $dP/d\omega$ should be greater than zero:

$$\frac{dP}{d\omega} > 0. \tag{3.59}$$

Using Equations 3.58 and 3.59, the relationship between the generator rotation speed variation and the duty ratio of converter versus output power of generator is

$$\frac{\mathrm{d}P}{\mathrm{d}D} < 0. \tag{3.60}$$

From Figure 3.72, it can be observed that if the current operation point is on the right side of the optimal rotation speed, the duty cycle of the converter should be decreased in order to reach the maximum power point. In this case, $\mathrm{d}P/\mathrm{d}\omega$ has to be less than zero. In this condition, Equation 3.60 becomes

$$\frac{\mathrm{d}P}{\mathrm{d}D} > 0. \tag{3.61}$$

Based on Equations 3.60 and 3.61, the location (on the left or right) of rotation speed of the generator in Figure 3.72 can be estimated. Then the duty ratio will be adjusted and the rotation speed of the turbine will reach its optimal point. The process of output power measurement and duty cycle adjustment are repeated continuously. Thus MPPT is achieved through controlling the duty cycle of the interface DC/DC converter, as shown in Figure 3.66.

As a result, this method does not rely on the turbine characteristics and speed–power curve directly since it requires only the output power variation, but not the tidal current and turbine speed measurements.

3.5 Grid Connection Interfaces for Tidal Energy Harvesting Applications

3.5.1 Grid Connection Interfaces for Tidal Turbine Applications

Due to the unpredictable speed behavior of tidal currents, the tidal current generators may produce varying and unstable alternating currents and voltages. Since the density of tidal current energy is varied depending on the tidal current velocity, the output power needs further conditioning prior to the grid connection. In addition, the frequency of the output voltage should be regulated to be the same as grid frequency. Output power conditioning, amplitude, phase, and frequency regulation of the conversion system are done by utilizing power electronic converters. AC/DC and DC/AC power electronic interfaces can be used for frequency and voltage regulation.

Block diagram of a power conditioning system for a grid-connected ocean tidal energy conversion system is shown in Figure 3.67. The rise and fall of tide water causes the differences between water levels. This difference rotates the water turbine. The flow rate variation of the tide water results in variable speed of the turbine. The AC power produced by the generator is converted into DC power via three-phase bridge rectifiers followed by a DC/AC inverter. The output of the DC/AC inverter generally contains harmonics, which should be filtered. Finally, output power can be connected to grid and transmitted to consumers after its voltage is increased and isolated by a power transformer.

The tide water has a natural periodicity and power generated by a tidal current generator has intermittencies. During these intermittencies, power cannot be generated. Thus, an energy storage system must be connected in parallel at the generator output or output of

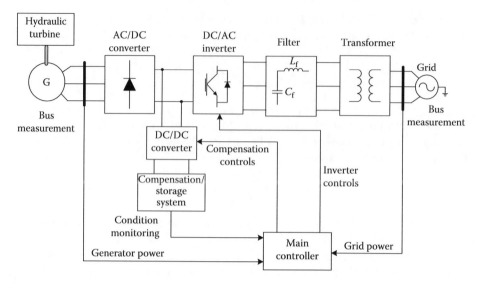

FIGURE 3.67   Grid connection and controls of tidal current power conditioning system.

a conversion stage in the power conditioning system. During the intermittency periods, the stored energy can be supplied to the grid. Hence, it is ensured that continual power is supplied to the grid side. Contrarily, this power source will be charged, if the available energy from tides is more than the demand energy.

Tidal current energy conversion systems can operate either as grid-connected or stand-alone (off-network) systems.

3.5.1.1   Grid-Connected Systems

In grid-connected systems, the grid and tidal current power generation systems should be synchronized. In addition, if a fault, grid power outage, or any unexpected event occurs, the problems of islanding (disconnection and reconnection) should be considered. Moreover, to ensure efficient and reliable operation, grid-connected power converters are required.

For the voltage and frequency synchronization, the phase-locked-loop (PLL) method can be utilized. The operation schematic of the PLL method is depicted in Figure 3.68. PLL is a control system that has an output signal with a fixed relation to the phase of the input reference signal. The PLL mechanism can be realized using analog or digital circuits. In both mechanisms, a phase detector, a variable electronic oscillator, and a feedback sensor are required.

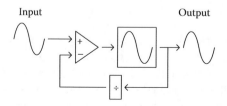

FIGURE 3.68   A simplified PLL principle.

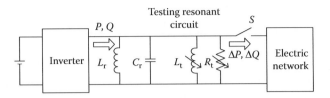

FIGURE 3.69 Islanding detection circuit.

In the case of any fault such as under- or overvoltage of the grid, overcurrent or overheat of the power electronic devices, and undervoltage of the generator output, the grid-connected generation system may not detect the power outage immediately and self-cut from the grid [41]. As a result, the tidal current power generation system and near-by loads form a stand-alone island that is not fed by the grid [42]. In general, the islanding may harm the entire power system and the consumer side. Islanding affects voltage and frequency stability. It can also contribute to phase asynchrony. Therefore, detection of islanding by appropriate detection circuit is very essential for the tidal energy conversion systems. Figure 3.69 demonstrates an islanding detection system.

The detection circuit is located between the inverter and the grid. It consists of an RLC resonant circuit. In grid-connected systems, over-/underfrequency and over-/undervoltage detection circuits are essential parts of the system. The internal protection of the power electronic interface stops the supply in case of any fault. In case of any fault, even though the voltage and frequency requirements are satisfied, in order to prevent islanding, additional detection circuits are required [41,42].

In a grid-connected system, inverter is controlled to produce a stable sinusoidal current. This current should have the same frequency as the grid and should be in phase with grid voltage. A synchronous phase-lock-loop (SPLL) system can be used to ensure these requirements [41]. A grid-connected current tracking and synchronous phase-lock control system can be employed for a grid connection system as shown in Figure 3.70.

In the control strategy shown in Figure 3.70, the inner loop is employed for real-time current track. The purpose of the additional outside loop is to reduce the error. Using zero cross detectors, the amplitude and phase of the grid-connected system is determined and detected values are processed in a PI controller to produce the compensative current to

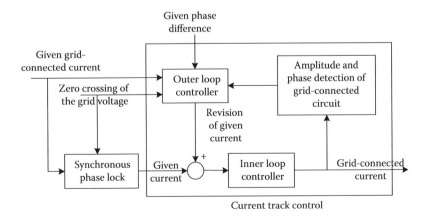

FIGURE 3.70 Grid-connected current control.

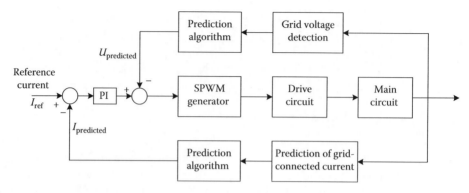

FIGURE 3.71   Control system of the inner control loop.

decrease the amplitude and phase of the error. The summation of the corrected values of the given current and the current signal produced by the SPLL forms the reference current of the inner loop. Thus, the inner and outer loops cooperate to track the given current [41]. The reference point to produce the sinusoidal signal is provided by the zero-crossing of the grid voltage. Consequently, the SPLL is used to generate the given current in phase with the grid voltage. The control system block diagram is shown in Figure 3.71 [42], where I_{ref} is the reference current injected to the grid, $I_{predicted}$ is the predicted grid current, and $U_{predicted}$ is the prediction of the grid voltage.

In this strategy, the reference current is compared with the feedback of the measured real-time grid-connected current. It should be noted that the reference current has the same frequency as the grid and is synchronized with the grid. The SPWM signals are obtained by modulating the output by a triangle wave. These signals are amplified to produce a sinusoidal current that has the same phase and frequency with the grid.

Outer loop control system is shown in Figure 3.72. Employing only inner loop cannot alleviate the phase and amplitude errors for the AC input signals. In this case, the error signal is a cosine function in steady state. An additional outer loop should be used to eliminate

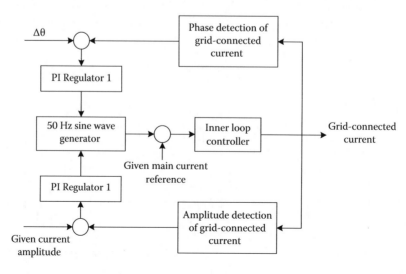

FIGURE 3.72   Block diagram of the outer loop control system.

this steady-state error to control the grid-connected current's phase and amplitude. The feedback input of the phase control loop is obtained using the phase difference between the grid-connected current produced by the inner loop and the grid voltage. On the other hand, the feedback input of the amplitude control loop is obtained using the amplitude of the grid-connected current. The PI regulators are utilized as controllers to track the reference current, as shown in Figure 3.72.

The grid connection technique described above can be used for tidal current power applications with variable speed hydraulic turbines coupled to the variable speed generators. DFIG is one of the variable speed generation techniques. Control of DFIGs as variable speed generators is generally divided into two methods: speed regulation method and power regulation method [43–46]. A grid-connected DFIG using bidirectional AC/DC/AC power electronic converters is shown earlier in Figure 3.56 [43,47].

The stator windings are directly connected to the grid. Hence, the magnetic flux induced by the stator side current component is constant. The d- and q-axis flux can be expressed as

$$\lambda_{ds} = L_0 i_{ms} = L_s i_{ds} + L_0 i_{dr}, \tag{3.62}$$

$$\lambda_{qs} = 0, \tag{3.63}$$

where λ_{ds} and λ_{qs} are stator fluxes in the stator reference frame (Wb), i_{ms} is the magnetizing current in the stator reference frame (A), i_{ds} is the stator current in the stator reference frame (A), and i_{dr} is the rotor current in the stator reference frame (A).

The q-axis flux equation would be

$$\lambda_{qs} = L_{ls} i_{qs} + L_0 (i_{qs} + i_{qr}) = L_s i_{qs} + L_0 i_{qr}. \tag{3.64}$$

Since the stator side q-axis flux is zero, the q-axis rotor current can be expressed as

$$i_{qs} = -\frac{L_0}{L_s} i_{qr}, \tag{3.65}$$

where L_{ls} is the leakage inductance (H), i_{qs} is the stator current in the stator reference frame (A), L_0 is the mutual inductance (H), i_{qr} is the rotor current in stator reference frame (A), and L_s is the stator winding inductance (H).

The active and reactive output power of the generator are

$$P_s = \frac{3}{2}(v_{qs} i_{qs} + v_{ds} i_{ds}), \tag{3.66}$$

$$Q_s = \frac{3}{2}(v_{qs} i_{ds} - v_{ds} i_{qs}), \tag{3.67}$$

where v_{qs} and v_{ss} are the stator voltage in the stator reference frame (V), P_s and Q_s are the stator active power (W) and stator reactive power (VAr). Equations 3.66 and 3.67 are used to determine the stator d- and q-axis currents.

Stator d-axis voltage v_{ds} is 0 due to the three-phase symmetry of the stator windings, thus stator current is a function of only the rotor q-axis current as given in Equation 3.68 [44,47,48]:

$$P_s = -\frac{3}{2}\left(\frac{L_0}{L_s} v_{qs}\right) i_{qr}. \tag{3.68}$$

The q-axis stator voltage can be expressed as

$$v_{qs} = r_s i_{qs} + \omega_e \lambda_{ds} + \frac{d\lambda_{qs}}{dt}. \tag{3.69}$$

Assuming a small stator resistance and applying Equations 3.63 through 3.69, the stator-side current component to retain magnetic flux (i_{ms}) can be expressed as

$$i_{ms} = \frac{v_{qs}}{\omega_e L_0}, \tag{3.70}$$

where ω_e is the synchronous speed in rad/s. Equation 3.69 can be applied to Equation 3.62 and the result can be solved for i_{ds} to obtain the relation given in Equation 3. 71:

$$i_{ds} = \frac{L_0}{L_s} \left(\frac{v_{qs}}{\omega_e L_0} - i_{dr} \right). \tag{3.71}$$

The rotor d-axis current can be used to control the stator reactive power output as given in Equation 3.72.

$$Q_s = \frac{3}{2} v_{qs} \frac{L_0}{L_s} \left(\frac{v_{qs}}{\omega_e L_0} - i_{dr} \right). \tag{3.72}$$

Thus, the active and reactive power (P_s^* and Q_s^*) can be controlled using i_{qr}^* and i_{dr}^*.

3.5.2 Grid Connection and Synchronization for Tidal Energy Harvesting with Basin Constructions

Generally, synchronous generators are more appropriated for basin- or dam-based tidal energy harvesting applications, since the water can be collected in reservoirs and water flow through the gate and governor can be controlled. This is very similar to the systems of hydroelectric power plants.

In basin-based tidal applications, due to the natural periodicity of ocean tides, the water flow may not be continuous. Thus, the generator may be disconnected from the grid bus or shutdown for maintenance purposes.

The synchronization conditions must be satisfied in order to re-connect the generator to the grid bus. The synchronization conditions are as follows:

1. The rms output voltage of the generator must be equal to the grid bus rms voltage.
2. The frequency of the generator must be equal to the grid frequency.
3. The phase sequences of the generator and the grid must be same.
4. The grid and the generator voltages must be in phase.

Figure 3.73 shows all the satisfied synchronization conditions for generator and grid. V_m and V_g are the grid and generator voltages, ω_m and ω_g are the frequencies of the grid and the generator, and α_m and α_g are the phase angles of the grid and the generator, respectively.

There are several methods for grid synchronization. One of the synchronization techniques is "zero crossing" method, which obtains the phase information by detecting the

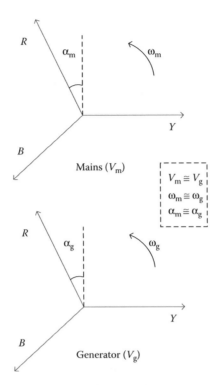

FIGURE 3.73 Phasor diagram for synchronized grid and generator.

zero crossing of the utility voltage. However, in every period, there are only two zero crossings, thus in some cases the performance might be low.

For the frequency and voltage regulation, back-to-back converters can be used for grid connection. In this method, the output of the generator is rectified and then inverted again with desired amplitude, phase, and frequency specifications. In this method, Clarke or Park transformations can be used. In Clarke transformations, the ABC phase quantities are converted to $\alpha\beta$ stationary frame. The reference and measured $\alpha\beta$ quantities are compared, and using controllers, the appropriate PWM signals can be generated for the converters. The Park transformations involve the similar procedure. The difference is the conversion of ABC phase quantities into dq axis quantities instead of $\alpha\beta$ quantities. By filtering the input signals such as three-phase voltages, the phase angle of the utility grid voltage can be estimated.

The other method of synchronization is the PLL technique, where one signal tracks the other. The details of PLL are aforementioned in Section 3.5.1.1.

3.6 Potential Resources

Worldwide, approximately 3000 GW of energy is continuously available from tides. Table 3.1 provides 26 potential tidal energy-generating sites. It has been estimated that only 2% (60 GW) of this power can be recovered as electrical energy [49,50].

TABLE 3.1

Prospective Sites for Tidal Energy Projects

Country	City	Mean Tidal Range (m)	Basin Area (km^2)	Capacity (MW)	Approximate Annual Output (TWh/yr)	Annual Plant Load Factor (%)
Argentina	San José	5.8	778	5040	9.4	21
	Golfo Nuevo	3.7	2376	6570	16.8	29
	Rio Deseado	3.6	73	180	0.45	28
	Santa Cruz	7.5	222	2420	6.1	29
	Rio Gallegos	7.5	177	1900	4.8	29
Australia	Secure Bay (Derby)	7.0	140	1480	2.9	22
	Walcott Inlet	7.0	260	2800	5.4	22
Canada	Cobequid	12.4	240	5338	14.0	30
	Cumberland	10.9	90	1400	3.4	28
	Shepody	10.0	115	1800	4.8	30
India	Gulf of Kutch	5.0	170	900	1.6	22
	Gulf of Khambat	7.0	1970	7000	15.0	24
Korea (Rep.)	Garolim	4.7	100	400	0.836	24
	Cheonsu	4.5	—	—	1.2	—
Mexico	Rio Colorado	6–7	—	—	5.4	—
UK	Severn	7.0	520	8640	17.0	23
	Mersey	6.5	61	700	1.4	23
	Duddon	5.6	20	100	0.212	22
	Wyre	6.0	5.8	64	0.131	24
	Conwy	5.2	5.5	33	0.060	21
USA	Pasamaquoddy	5.5	—	—	—	—
	Knik Arm	7.5	—	2900	7.4	29
	Turnagain Arm	7.5	—	6500	16.6	29
Russian Fed.	Mezen	6.7	2640	15,000	45	34
	Tugur	6.8	1080	7800	16.2	24
	Penzhinsk	11.4	20,530	87,400	190	25

3.7 Environmental Impacts

Tidal energy has significant environmental benefits. It is nonpolluting and replaces coal and hydrocarbon fuels. A barrage, sometimes, can also provide protection over coastal areas during very high tides by acting as a storm surge barrier or breakwater. However, any proposed large-scale energy project has environmental impacts that must be weighed and considered.

3.7.1 Sediment

Estuaries often have high volume of sediments moving through them, from the rivers to the sea. The introduction of a barrage into an estuary may result in sediment accumulation within the barrage, which requires more frequent dredging.

3.7.2 Fish

Fish migration is also affected heavily by tidal barrages. Some fishes such as the salmon must migrate from saltwater to freshwater to spawn and migrate back, multiple times in their lifespan. When the sluices are open, they can get through. Otherwise they have to seek way out through the turbines, which may end their life. Marine mammal's migration may also be affected in some regions and, therefore, fishing industry may be affected, too.

3.7.3 Salinity

Barrages may also cause less water exchange between the sea and the basin. A result of this will be the decrease in the average salinity inside the basin, which may affect the food chain of the creatures inside the basin.

3.8 Summary

Tidal energy is a result of the orbital kinetic energy of the moon, earth, and sun. It provides zero gas, solid, or radiation pollution and is an inexhaustible supply of energy. As a kind of green energy, tidal energy has some significant merits. Tidal power energy does not depend on the season or the weather type, so it is more predictable, compared to other sources of energy such as wind or solar energy. In this chapter, various features of the ocean tidal energy harvesting are explained, which have a great potential; however, they are not widely utilized yet. Different energy generation technologies, their optimal operation principles, and possible utilization techniques are described throughout this chapter.

References

1. Ocean Energy: Technology Overview, Renewable Development Initiative, available at: http://ebrdrenewables.com
2. D.A. Dixon, "Fish and the energy industry," EPRI (Electric Power Research Institute) Research Report, ASMFC Energy Development Workshop, October 2006.
3. P.O'Donnell, "Update '05: Ocean wave and tidal power generation projects in San Francisco," *Proceedings of the IEEE Power Engineering Society General Meeting*, Vol. 2, pp. 1990–2003, June 2005.
4. "Ocean Energy," Technical Report of the US Department of Interior, Minerals Management Service.
5. Tidal Electric Ltd., London, UK, "Technology: History of tidal power," available at: http://www.tidalelectric.com/History.htm, March 2008.
6. G. Boyle, *Renewable Energy; Power for a Sustainable Future*, Oxford: Oxford University Press, 1996.
7. S. Sheth, "Tidal energy in electric power systems," *IEEE Power Engineering Society General Meeting*, Vol. 1, pp. 630–635, 2005.
8. T.J. Hammons, "Tidal power," *Proceedings of the IEEE*, 8 (7), 1997.

9. S.E. Ben Elghali, M.E.H. Benbouzid, and J.F. Charpentier, "Marine tidal current electric power generation technology: State of the art and current status," *Electric Machines & Drives Conference, IEMDC '07, IEEE International*, Vol. 2, pp. 1407–1412, 2007.

10. A.M. Gorlov, "Tidal energy," *Encyclopedia of Science and Technology*, London: Academic Press, 2001, pp. 2955–2960.

11. G.N. Tiwari and M.K. Ghosal, *Renewable Energy Resources: Basic Principles and Applications*, Oxford, UK: Alpha Science International Ltd., April 2005.

12. Sustainable Development Commission, "Tidal power in the UK, Research Report 2—Tidal Technologies Overview, Contract 2, Final Report," October 2007.

13. E.M. Wilson, "Tidal energy and its development," *Ocean*, 4, 48–56, 1972.

14. IEEE Committee Report, "Hydraulic turbine and turbine control models for system dynamic studies," *IEEE Transactions on Power Systems*, 7, 167–179, 1992.

15. D.G. Ramey and J.W. Skooglund, "Detailed hydrogovernor representation for system stability studies," *IEEE Transactions on Power Apparatus and Systems*, 89, 106–112, 1970.

16. D. Ramuz, M. Camara, H. Clergeot, and J.M. Kauffmann, "Simulation of a doubly fed induction generator used in an autonomous variable speed hydro turbine with maximum power point tracking control," *Power Electronics and Motion Control Conference, EPE-PEMC 2006, 12th International*, pp. 1620–1624, 2006

17. L.N. Hannet and B. Fardanesh, "Field tests to validate hydro turbine-governor model structure and parameters," *IEEE Transactions on Power Systems*, 9, 1744–1751, 1994.

18. A. Betz, "Windmills in the light of modern research," U.S. National Advisory Committee for Aeronautics, Technical Memorandum No. 474, Vol. 18, no. 46, p. 27, November 1927.

19. S. Mathew, *Wind Energy Fundamentals, Resource Analysis and Economics*, New York: Springer, February 2006.

20. Seaflow and SeaGen Turbines, Marine Current Turbines Ltd., available at: http://www.marineturbines.com, March 2008.

21. Rotech Tidal Turbine (RTT), Lunar Energy™, Harnessing Tidal Power, available at: http://www.lunarenergy.co.uk/

22. Semi-Submersible Turbine, TidalStream Limited *"The Platform for Tidal Energy,"* available at: http://www.tidalstream.co.uk/

23. S. Kiho, M. Shiono, and K. Suzuki, "Power generation from tidal currents by darrieus turbine," *Renewable Energy*, 9 (1–4), 1242–1245, 1996.

24. Blue Energy Vertical Axis Turbine, Blue Energy Inc., Canada, available at: http://www.bluenergy.com, March 2008.

25. Rise Research Institute for Sustainable Energy, "Tidal resources," available at: http://www.rise.org.au, March 2008.

26. A.N. Gorban', A.M. Gorlov, and V.M. Silantyev, "Limits of the turbine efficiency for free fluid flow," *Journal of Energy Resources Technology*, 123, 311–317, December 2001.

27. A.M. Gorlov, "The helical turbine and its applications for tidal and wave power," *OCEANS 2003 Proceedings*, Vol. 4, p. 1996, 2003.

28. G. Colcagno, F. Salvatore, L. Greco, A. Moroso, and H. Eriksson, "Experimental and numerical investigation of a very promising technology for marine current exploitation: the Kobold turbine," *Proceedings of 16th International Offshore and Polar Engineering Conference (ISOPE)*, Vol. 1, part 1, pp. 323–330, San Francisco, USA.

29. Engineering Business Ltd., UK, "Stingray tidal stream generator," available at: http://www.engb.com, March 2008.

30. A. Petersson, "Analysis, modeling and control of doubly-fed induction generators for wind turbines," PhD dissertation, Chalmers University of Technology, Gothenburg, Sweden, 2005.

31. S. Tnani, S. Diop, A. Berthon, and J.M. Kauffmann, "A generalized model for double fed induction machines," *IASTED Conference*, 1995.

32. Z. Du, F. Zhao, and X. Du, "Investigation of the control strategy for tracking maximum power point in VSCF wind generation," *Chinese Control Conference*, pp. 1585–1589, August 2006.

33. L.L. Freris, *Wind Energy Conversion Systems*, Englewood Cliffs, NJ: Prentice-Hall, pp. 182–184, 1990.
34. G.D. Moor, H.J. Beukes, "Maximum power point trackers for wind turbines," *2004 35th Annual IEEE Power Electronics Specialists Conference*, Aachen, Germany, 2004.
35. L.L. Freris, *Wind Energy Conversion Systems*, Englewood Cliffs, NJ: Prentice-Hall, pp. 182–184, 1990.
36. S. Morimoto, H. Nakayama, M. Sanade, and Y. Takeda, "Sensorless output maximization control for variable-speed wind generation system using IPMSM," *IEEE Transactions on Industry Applications*, 41 (1), 60–67, 2005.
37. J.S. Choi, R.G. Jeong, J.H. Shin, C.K. Kim, and Y.S. Kim, "New control method of maximum power point tracking for tidal energy generation system," *Proceeding of International Conference on Electrical Machines and Systems*, Seoul, Korea, pp. 165–168, October 2004.
38. E. Koutroulis and K. Kalaitzakis, "Design of a maximum power tracking system for wind energy conversion applications," *IEEE Transactions on Industrial Electronics*, 53, (2) 486–494, 2006.
39. J.S. Choi, R.G. Jeong, J.H. Shin, C.K. Kim, and Y.S. Kim, "New control method of maximum power point tracking for tidal energy generation system," *Proceedings of the International Conference on Electrical Machines and Systems*, Korea, pp. 165–168, October 2007.
40. H.-da Liu, D.-pu Li, and Z.-li Ma, "The grid-connection control system of the tidal current power station," *Proceedings of the 33rd Annual Conference of the IEEE IECON*, Taiwan, pp. 1293–1298, November 2007.
41. J. Stevens, R. Bonn, J. Ginn, and S. Gonzales, "Development and testing of an approach to anti-islanding in utility interconnected photovoltaic systems," 2000.8, SAND 2000-1939, Sandia National Laboratories Report.
42. G. Iwanski and W. Koczara, "Control system of the variable speed autonomous doubly fed induction generator," *Proceedings of the International Power Electronics and Motion Control Conference*, PEMC 2004, Latvia.
43. P. Pena, J.C. Clare, and G.M. Asher, "Doubly fed induction generator using back-to-back PWM converters and its application to variable speed wind-energy generation," *IEE Proceedings of the Electric Power Applications*, Vol. 143, pp. 231–241, May 1996.
44. S. Muller, M. Deicke, and R.W. De Doncker, "Adjustable speed generators for wind turbines based-on doubly-fed induction machines and 4-quadrant IGBT converters linked to the rotor," *IEEE Industry Applications Conference*, Vol. 4, pp. 2249–2254, 2000.
45. M. Mohan, T.M. Undeland, and W.P. Robbins, *Power Electronics*, 3rd ed., New York: Wiley, pp. 164–171, 2003.
46. J.W. Park, K.W. Lee, and H.J. Lee, "Wide speed operation of a doubly-fed induction generator for tidal current energy," *Proceedings of the 30th Annual Conference of the IEEE Industrial Electronics Society*, Vol. 2, pp. 1333–1338, November 2004.
47. F.M. Gardner, *Phase Lock Techniques*, New York: Wiley, 1979.
48. A.C. Baker, "The development of functions relating cost and performance of tidal power schemes and their application to small-scale sites," *Tidal Power*, London: Thomas Telford, 1986.
49. R.H. Charlier and J.R. Justus, *Ocean Energies: Environmental, Economic, and Technological Aspects of Alternative Power Sources (Elsevier Oceanographic Series)*, Amsterdam: Elsevier, p. 534, 1993.
50. L.B. Bernshtein, "Strategy of tidal power stations utilization to provide a number of planet regions with ecologically pure energy," *Proceedings of International Symposium on Ocean Energy Development*, Muroran, Hokkaido, Japan, pp. 65–77, 1993.

4

Ocean Wave Energy Harvesting

4.1 Introduction to Ocean Wave Energy Harvesting

Uneven heating around the globe results in winds that generate ocean waves. As a result, water particles adopt circular motions as part of the waves [1,2]. This motion carries kinetic energy and the amount of this energy is related to the speed and duration of the wind, the depth, and area of the water body affected by the wind, and the seabed conditions. Waves occur only in the volume of the water closest to its surface, whereas in tides, the entire water body moves, from the surface to the seabed [2]. The total power of waves breaking around the world's coastlines is estimated at 2–3 million megawatts. The west coasts of the United States and Europe and the coasts of Japan and New Zealand are good sites for harnessing wave energy [1].

Ocean wave energy systems convert the kinetic and potential energy contained in the natural oscillations of ocean waves into electricity. There are a variety of mechanisms for the utilization of this energy source. One method to harness wave energy is to bend or focus the waves into a narrow channel of concentration, to increase their power and size. Then waves can be channeled into a catch basin or used directly to spin turbines [1]. The energy of the waves is converted into electricity by means of surge devices and oscillating column devices. Wave energy conversion (WEC) is one of the most feasible future technologies; however, since it is not mature enough, construction cost of wave power plants are considerably high. These energy systems are not developed and matured commercially due to the complexities, sea conditions, and difficulty of interconnection and transmission of electricity through turbulent water bodies. In addition, they may require higher operation and maintenance costs in comparison to the land power systems due to their location at sea.

A typical ocean wave energy harvesting system consists of a wave power absorber, a turbine, a generator, and power electronic interfaces. The absorber captures the kinetic energy of the ocean waves. The absorbed mechanical kinetic energy of the waves is either conveyed to turbines or the absorber directly drives the generator. The turbine is used to drive the shaft of the electric generator. Turbines are generally used with systems that feature rotational generators. In other methods, a linear motion generator is used, and this can be directly driven by the power absorber or movement of the device. Both linear and rotational generators produce variable frequency and variable amplitude AC voltage. This AC voltage is rectified to DC voltage in order to take advantage of the DC energy transmission capabilities of salty ocean water. DC transmission in salty water does not require an additional cable for the negative polarity. Thus, it will be more cost-effective than transmitting the power in AC form, which requires three-phase cables. Transmission cable length varies depending on the location of the application, which is either nearshore or offshore. However, the principles are the same for both types of applications. After the DC

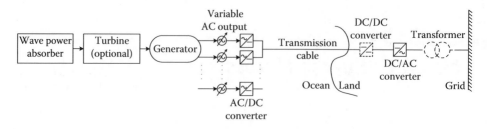

FIGURE 4.1 System-level diagram of ocean wave energy harvesting.

power is transmitted from ocean to land, a DC/DC converter or a tap-changing transformer is used for voltage regulation. Depending on the utilized voltage regulation system, a DC/AC inverter is used before or after the voltage regulator. The voltage synchronization is provided by the inverter and the output terminals of the inverter can be connected to the grid.

Figure 4.1 shows a system-level diagram of the ocean wave energy harvesting. WEC devices (including the absorber, turbine, and generator) are interconnected within an in-water substation. The substation consists of the connection equipments and controllers for individual devices. The outputs of the generators are connected to a common DC bus using DC/AC converters for transforming power before transmission to the shore. A transmission line connects the cluster to the shore. An onshore inverter converts the DC voltage to a 50 or 60 Hz AC voltage for grid connection. An optional shore transformer with tap changer or a DC/DC converter compensates the voltage variations. A group of an absorber, a turbine, and generators could be used in a farm structure to increase the amount of captured energy.

Alternatively, the land converters might be moved offshore to overcome in-land space limitations. The converters can be placed on a platform or enclosed in a buoy anchored to the seabed. However, this adds a level of complexity to the system and may require more maintenance than in-land converters. Figure 4.2 shows a land DC/DC converter moved offshore.

Another option is installing an offshore transformer as seen in Figure 4.3. This would increase the power transmission capability, since the higher voltage transmission will result in less transmission losses, since, for the same power rating, the current will be lower with a higher transmission voltage level. However, this case does not allow for DC transmission.

As an alternative, boost DC/DC converters can be installed after the AC/DC converter of the generator. This allows a high-voltage DC transmission link as illustrated in Figure 4.4. In this case, both transmission losses will be kept at a minimum and only the single-line DC transmission through the ocean water will be needed. The disadvantage of this configuration is that more power components will be used with some additional losses and additional cost will be required to install the boost converter.

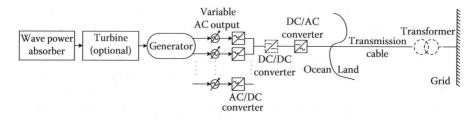

FIGURE 4.2 System-level diagram; the land DC/DC converter is moved offshore.

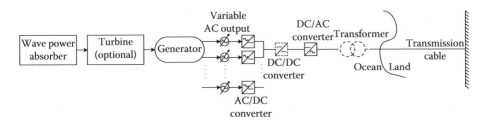

FIGURE 4.3 System-level diagram with the offshore transformer.

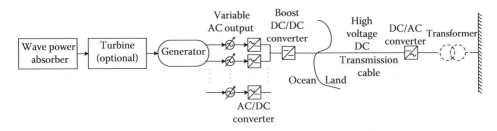

FIGURE 4.4 System-level diagram with the high voltage DC transmission link.

4.2 The Power of Ocean Waves

The physics of ocean waves and wave spectra are explained by Van Dorn in [3]. In Figure 4.5, the characteristics of two types of waves with different periods are presented. These two types of waves represent the main categories of ocean waves.

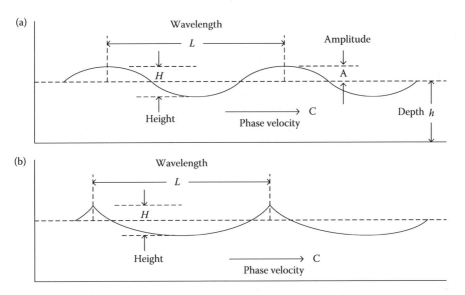

FIGURE 4.5 Characteristics of a wave. (Modified from W.J. Jones and M. Ruane, "Alternative electrical energy sources for Maine, Appendix I, Wave Energy Conversion by J. Mays," Report No. MIT-E1 77-010, MIT Energy Laboratory, July 1977.)

Figure 4.5a shows periodic, progressive surface waves resembling a sinusoidal curve, which is symmetrical around the equilibrium water level. In Figure 4.5b, the waves have higher periods and no longer are sinusoidal or symmetric. Their crests are peaked, and extend higher above the equilibrium level than the long and smooth troughs, which fall below it.

The total energy in an ocean wave can be calculated in joules per unit of width of wave front by summing up the potential and kinetic energy together. The potential energy in a wave of length L is generated by the displacement of the water away from the mean sea level. The kinetic energy of a wave is a result of both horizontal and vertical water particle motions [4].

The total potential and kinetic energy of an ocean wave can be expressed as

$$E = \frac{1}{2}\rho g A^2,$$ (4.1)

where g is the acceleration of gravity (9.8 m/s^2), ρ is the density of water (1000 kg/m^3), and A is the wave amplitude (m). To obtain the average energy flux or power of a wave period, energy E is multiplied by the speed of wave propagation, v_g.

$$v_g = \frac{L}{2T},$$ (4.2)

where T is the wave period (s) and L is the wavelength (m) [4].

$$P_w = \frac{1}{2}\rho g A^2 \frac{L}{2T}.$$ (4.3)

The dispersion relationship describes the connection between the wave period T and the wavelength L as

$$L = \frac{gT^2}{2\pi}.$$ (4.4)

If Equation 4.4 is substituted into Equation 4.3, the power or energy flux of an ocean wave can be calculated as

$$P_w = \frac{\rho g^2 T A^2}{8\pi}.$$ (4.5)

Instead of using the wave amplitude, wave power can also be rewritten as a function of wave height, H. Considering that the wave amplitude is half of the wave height, the wave power becomes

$$P_w = \frac{\rho g^2 T H^2}{32\pi}.$$ (4.6)

See Refs. [3] and [5] for additional calculations.

4.3 Wave Energy Harvesting Technologies

Various methods have been proposed for converting wave energy into practically usable electrical energy. Some of the early examples of the wave power systems are the Salter Cam developed in England [6,7], a hinged floating system developed in England [8], a pressure-activated submerged generator developed by Kayser [9,10] in Germany, the wave-powered pump developed at Scripp's Institution of Oceanography [10,11], and a pneumatic wave converter originally developed by Masuda [12,13] in Japan. The first four systems are large-scale energy generation systems and the last one is used as a low-level power supply for navigation buoys.

Current experimental and theoretical researches indicate that up to 90% of the wave's power can be extracted given certain conditions. Thus, the ocean wave power can be efficiently converted into electrical energy [4]. The total conversion efficiency of wave energy is around 35% when considering all the ancillary conversion processes throughout the year [4]. The major issue with wave energy production is integrating this unsteady and unpredictable energy source into the grid. Although the wave energy has zero fuel cost, it has a relatively high capital cost and an output that is ever-changing and mostly unpredictable.

Generally, there are two types of wave energy generation sites with respect to their distance from the shore, which are discussed in detail in the following subsections.

4.3.1 Offshore Energy Harvesting Topologies

Offshore applications are located away from the shore and they generally use a floating body as wave power absorber and another body that is fixed to the ocean bottom. Salter Cam and buoys with air-driven turbines are the only applications involving rotational generators in offshore applications. Generally, linear generators with buoys are used in offshore applications. Linear generators are directly driven by the movement of a floating body on the ocean. Before classifying the offshore applications, dynamics of fixed and floating bodies are discussed in this section.

4.3.1.1 Dynamics of Fixed Bodies in Water

When waves hit a body fixed to the ocean bottom, such as a pier or an offshore platform, the waves will dispatch around the body relative to the geometrical shape of the wave and body, which dictate the reflection of the incident waves from the body [4].

If the body does not extend continuously to the bottom below the free surface, some of the wave energy may leak underneath the structure. The energy of a wave is distributed from the surface down to the bottom exponentially and is not localized only at the surface. At a depth of approximately one-half wavelength, the wave effect can be considered to be negligible [4].

A two-dimensional (2D) case of a fixed cylinder is shown in Figure 4.6. The incident wave of the height H_I is partially reflected by the cylinder, while the rest of the energy in the wave is transmitted through the cylinder to the other side, where it can behave as a wave again and continue on its way. In an ideal case with no power loss, the amount of energy in the incident wave must be equal to the sum of the energies of the transmitted wave and the reflected wave:

$$\frac{1}{2}\rho g \left(\frac{H_I}{2}\right)^2 = \frac{1}{2}\rho g \left(\frac{H_T}{2}\right)^2 + \frac{1}{2}\rho g \left(\frac{H_R}{2}\right)^2. \tag{4.7}$$

Here, H_T is the height of the transmitted wave, H_R is the height of the reflected wave, and H_I is the height of the incident wave.

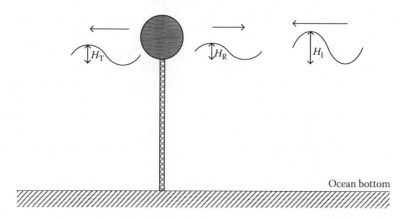

FIGURE 4.6 Side view of a fixed body in waves showing incident, reflected and transmitted waves. (Modified from W.J. Jones and M. Ruane, "Alternative electrical energy sources for Maine, Appendix I, Wave Energy Conversion by J. Mays," Report No. MIT-E1 77-010, MIT Energy Laboratory, July 1977.)

This wave interaction can cause considerable forces on the body, although it is assumed that the waves are not of the braking kind. During the breaking of the waves on shore, a significant amount of energy is lost in the resulting turbulence [4]. If we replace the cylinder with a sphere, the behavior is similar, with some differences in the reflection of the waves around the sides. The diffracted waves appear as a complicated pattern of the interfering waves. Also, in this case, more wave energy leaks under the sphere compared to the cylinder case [4].

4.3.1.2 Dynamics of Floating Bodies in Water

In the floating body case, the same cylinder is assumed to be freely floating on the ocean. The motions of a floating body can be defined as: heave, sway, and roll corresponding to vertical, horizontal, and rotational motions as shown in Figure 4.7 according to the terminology of naval architecture.

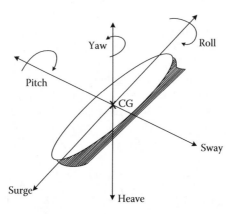

FIGURE 4.7 Directions of motion of a floating body. (Modified from W.J. Jones and M. Ruane, "Alternative electrical energy sources for Maine, Appendix I, Wave Energy Conversion by J. Mays," Report No. MIT-E1 77-010, MIT Energy Laboratory, July 1977.)

If the cylinder is pushed up and down, waves are created, which will spread out from either side. Similar behavior is expected if we cause it to sway by pushing it from side to side. No disturbance to the water would be caused by rolling the object, if it is a perfect circular cylinder without roughness. If the shape is generalized to a box, it is expected to create waves when disturbed [4].

Thus, a combination of waves is expected as a result of causing the body to move in different manners. Three additional types of motion are the surge, yaw, and pitch [4] as shown in Figure 4.7.

Floating devices can generate electricity from the bobbing action of a floating object. The object can be mounted to a floating raft or to a device fixed on the ocean floor. These types of devices can be used to power lights and whistles on buoys.

4.3.1.3 Air-Driven Turbines

WEC is generally composed of several stages. In air-driven turbine systems for nearshore and offshore applications, the primary conversion is from wave to the pressurized air. The secondary stage is extracting the mechanical energy of the pressurized air to the rotation of the shaft of the turbine. The last stage is converting the mechanical rotation into electric power through electric generators.

Figure 4.8 shows the operating principle of an offshore application, which consists of a floating buoy with an air chamber and an air-driven generator. In this system, when the waves hit the body, the water level inside the channel of the buoy increases. This increase in water level applies a pressure to the air in the air chamber. When the air is pressurized, it applies a force to the ventilator turbine and rotates it. This turbine drives the electric generator, creating electricity at its output terminals. When the waves are pulled back to the ocean, the air in the air chamber is also pulled back as the water level in the buoy channel decreases. Due to the syringe effect, the turbine shaft rotates into the contrary direction, still producing electricity.

In order to achieve maximum efficiency during the syringe effect, the air chamber and the ventilating generator should be carefully mechanically insulated. However, this adds design complexity and further cost to the system.

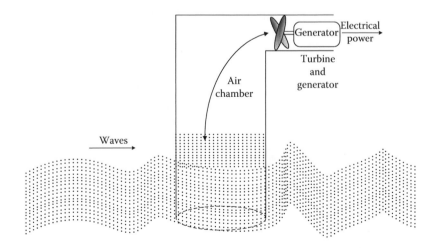

FIGURE 4.8 Spinning the air-driven turbines using wave power. (Modified from "Ocean Energy," Report of the U.S. Department of Interior Minerals Management Service.)

4.3.1.4   Fixed Stator and Directly Driven PM Linear Generator–Based Buoy Applications

The idea is based on the height difference of the wave from top and bottom. The buoy follows the motion of the wave when it is floating on the ocean surface. The buoy can move vertically on a pillar, which is connected to a hull. PMs are mounted on the surface of the hull, and coil windings on the outside of the hull. The pillar and stator are connected on a foundation standing on the seabed of the ocean. The hull with the mounted magnets is called the rotor or the piston of the generator, and they are the moving parts of the generator. Since the motion is linear, this generator is called a linear generator.

The illustration of linear generator in use with the floating buoy and fixed pillar is shown in Figure 4.9.

Due to the varying amplitude and period of the ocean waves, the amplitude and frequency of the output voltage of the generator varies. Thus, an AC/DC rectifier followed by a DC/AC converter is required to make the grid connection possible.

One drawback of linear generators might be the fact that the associated large and expensive electromagnetic converters may generate flux changes that are too slow due to the low velocities. However, the latest developments in linear generators have increased the interest in using linear generators for wave energy applications.

The preferable geometry of a buoy is a cylindrical shape since it can act as a point absorber and intercept waves coming from different directions. When the wave rises, the buoy pulls the generator piston by the rope. When the wave subsides, the generator will be drawn back by the spring that stores the mechanical energy in the first case. Thus, the electric generation is provided during both up and down motion.

The generator AC voltage starts at zero when the buoy is in its lowest position, increases until the buoy reaches its highest position at the top of the wave and descends back to zero as the buoy stops.

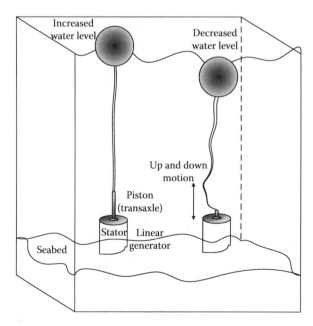

FIGURE 4.9 Linear generator–based buoy-type wave energy harvesting method.

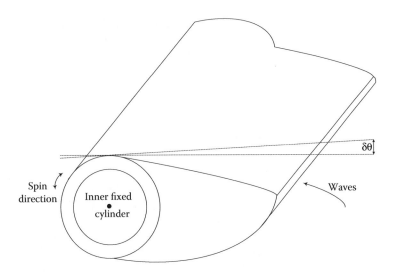

FIGURE 4.10 Side view of Salter Cam.

4.3.1.5 *Salter Cam Method*

Salter Cam is shown in Figures 4.10 and 4.11. Salter Cam, also known as "the nodding duck," features an outer shell that rolls around a fixed inner cylinder that is activated by the incoming waves. The power can be captured through the differential rotation between the cylinder and the cam. In this application, the motion of the cam is converted from wave into a hydraulic fluid, and then the hydraulic motor is used to convert the pressurized hydraulic fluid into rotational mechanical energy. Consequently, the rotational mechanical energy is converted to electricity by utilizing electric generators. As an intermediate step,

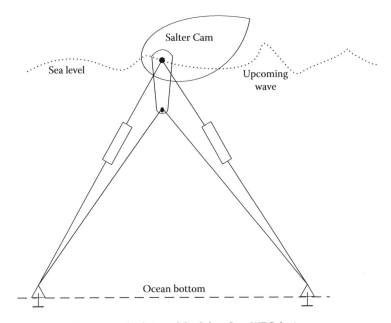

FIGURE 4.11 A schematic illustrating the fixing of the Salter Cam WEC device.

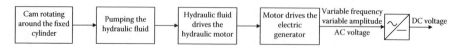

FIGURE 4.12   Operating principle of the Salter Cam.

flywheels or pressurized liquids can be used in order to reduce the intermittencies of the wave power in conversion to electric power.

The inner cylinder is fixed to the ocean bottom through hawser lines as shown in Figure 4.11.

The shape of the Salter Cam allows a large interaction with the incident wave. The shape should not generate a wave on its backside during the roll over due to its cylindrical backside [4].

The Salter Cam WEC generates electric power through the harmonic motion of the free part of the device around the fixed inner cylinder. The outer part rises and falls with respect to the motion caused by the ocean waves. This motion pumps a hydraulic fluid that drives a hydraulic motor. This motor is coupled to the shaft of an electric generator that generates electricity. The block diagram explaining the operation principle of the Salter Cam is shown in Figure 4.12. The Salter Cam is extremely efficient in energy production and its applications have gained a lot of interest recently.

The cam acts as a damped harmonic oscillator in that it exhibits a resonant behavior at certain wavelengths. This damped pendulum consists of a mass located either at the end or distributed along the length of a rod (body of the cam), a restoring force due to gravity (hydrostatic force on the cam), and a damping force in the opposite direction of the motion caused by the viscosity of the medium (internal energy absorber of the cam) as shown in Figure 4.13.

When displaced away from the current stall position, it is expected that the pendulum (cam) oscillates for a few cycles until its motion has been thoroughly damped. If, however, a force is applied in a periodic manner, then the pendulum (cam) is expected to oscillate at its driven period. The amplitude of the motion not only depends on how hard it is pushed, but also on the natural period of the device, which is determined by gravity, cam geometry, and wave force. This period may be in the order of seconds [4].

In Salter Cam [6,15,16], there is a "string" or common spine, which provides a stable frame of reference along the roll. The cams run on rollers, which are the bodies of rotary pumps.

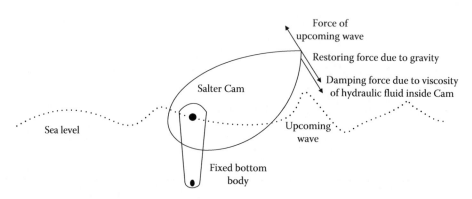

FIGURE 4.13   Forces affecting a Salter Cam.

High-pressure oil drives hydraulic swash-plate motors coupled to electric generators. Electrical transmission is provided by undersea cables.

In deepwater, the wave power spatial flux (in kW/m of wave front crest) is given by the wave height (H_S in m) and the peak wave period (T_p in s). Based on these two parameters, the incident wave power (J in kilowatts per meter of wave crest length, or kW/m) associated with each sea state record, assuming 84% efficient wave power turbine, is estimated by the following equation:

$$J = 0.42 \times (H_S)^2 \times T_p. \qquad (4.8)$$

It should be noted that wave power varies with the square of wave height—in other words, if the height of the wave doubles, the generated power will be increased four times [2].

4.3.2 Nearshore Energy Harvesting Topologies

Nearshore topologies are applied to the shore or within the surfing zone of the ocean. Nearshore applications have some advantages and disadvantages in comparison to the offshore applications with respect to the selected method. The different nearshore applications are described in the following section.

4.3.2.1 *Nearshore Wave Energy Harvesting by the Channel/Reservoir/Turbine Method*

In order to harness the wave energy, wave currents can be funneled into a narrow channel to increase their power and size. The waves can be channeled into a catch basin and used directly to rotate the turbine as shown in Figure 4.14. This method is more expensive in comparison to the other offshore applications, since it requires building a reservoir to collect the water carried by the waves to drive the turbine. However, all the components of the WEC system are located inland, and this results in easier and less maintenance in comparison to the offshore applications. Additionally, since a reservoir collects the ocean water, the intermittencies can be eliminated with respect to the size of the reservoir. This will create a convenient platform for voltage and frequency regulation. However, it will be more advantageous to build this type of plants in the locations where they have regular and sustaining wave regimes.

4.3.2.2 *Air-Driven Turbines Based on the Nearshore Wave Energy Harvesting Method*

Another way to harness wave energy is with an oscillating water column (OWC) that generates electricity from the wave-driven rise and fall of water in a cylindrical shaft or pipe. The rising and falling water drives air into and out of the top of the shaft, powering an air-driven turbine as shown in Figure 4.15.

Figure 4.15a shows the general structure of the nearshore air-driven turbine. When the wave fills in the wave chamber, it pushes the air through the ventilator, which drives the electrical machine as shown in Figure 4.15b. In Figure 4.15c, the wave retreats from the wave chamber causing the pressure of the air inside the channel to decrease.

This method is advantageous because it is able to use not only wave power but also the power from the tidal motions. However, mechanical isolation should be provided within wave and air chambers in order to obtain better efficiencies. This will also add some more cost and design complexity to the system.

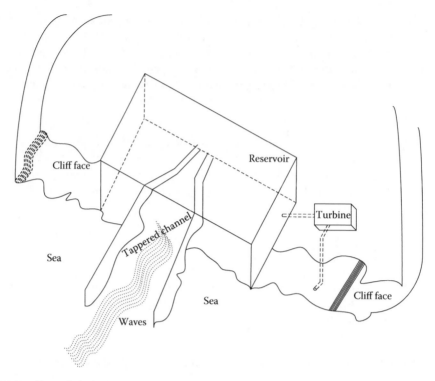

FIGURE 4.14 Channeled ocean wave to a reservoir to spin the turbines.

4.3.3 Wave Power Absorbers

A wave power absorber absorbs the energy of a wave and converts it into mechanical energy that drives the turbines coupled to generators or directly moves the pistons of linear generators. Consequently, the mechanical power is converted to electrical power

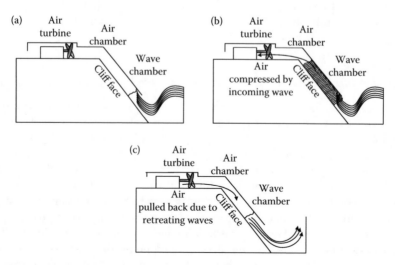

FIGURE 4.15 Air-driven turbines using the wave power [14]. (a) Upcoming wave starts filling the chamber, (b) air is compressed by rising water, and (c) air is pulled back by retreating waves.

by generators. Wave power absorbers used in different applications are discussed in this subsection.

When a floating body is displaced above or below the water position, a restoring force tends to bring the floating body to its original equilibrium position. The force that brings the body back to its equilibrium position results in potential energy. The kinetic energy is associated with the motion of the body [4].

In order to efficiently extract the wave system energy, the mechanical energy given in Equation 4.1 should be considered. The basic principle of wave power absorbers is to take the mechanical energy of the waves, which can be then converted into electrical energy. Pumps, flywheels, compressors, turbines, and linear or rotational generators are the devices that could be used for wave energy extraction as mechanical energy converters [4].

There are several studies reported in the literature to harness the power of ocean waves [17–19]. Isaacs [13] outlines some of the proposed modes of operation for those devices including the Scripps wave pump. In [20], Richards reviews the buoy-shaped wave-activated turbine generators. The designs of Masuda, McCormic, Isaacs, Kayser (Figures 4.16 and 4.17), and Falnes are one of those designs.

In these devices, the heave motion of a cylindrical buoy is used to provide a pressure head, which then drives a turbine generator by hydraulic or pneumatic methods. In Figure 4.16, Masuda units of 70 and 120 W are shown, which is used by the Japanese Maritime Safety Agency. Masuda has also proposed a floating, octahedral shaped buoy (Figure 4.17) with an outside diameter of 120 m, which is expected to produce 3–6 MW in the high seas in Japan.

A wave contouring raft whose joints are formed by hydraulic pumps operating on the differential motions of the linked rafts as shown in Figure 4.18 was proposed by Wooley and Platts. The details of this method were discussed in Ref. [4]. This method has been experimented by Wavepower Ltd of Southampton, England. The individual rafts are built at a length of one-quarter of the wavelength of the waves that, on the average, contribute the greatest amount of energy. The rafts would be somewhat wider than their length in the direction of the advancing wave.

Cylindrical buoys are preferable due to their ability to extract energy from the wave system that is in excess of their diameters and they are not sensitive to the wave direction.

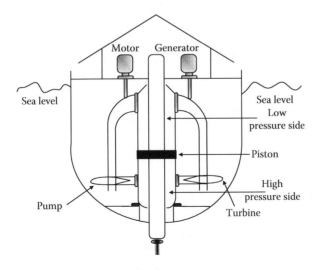

FIGURE 4.16 Falnes buoy-shaped wave-activated turbine.

(a)

(b)

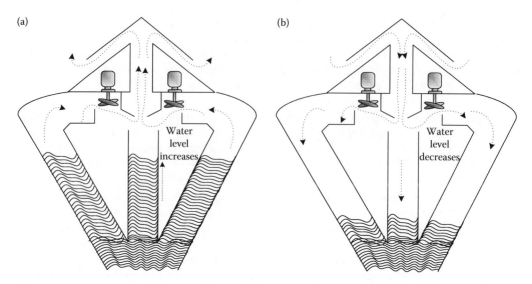

FIGURE 4.17 Air pressure ring buoy. (a) Water level increases and air is taken out from the upper outlets. (b) Water level decreases and air is pulled back from the upper inlets.

The power or energy flux in a wave is about H^2T of the wave front. The floating body can extract an amount of power greater than that of intercepted waves by the diameter of the body because of the diffracted and radiated waves [4].

An array of these devices can capture more of the wave energy with the proper devices to control their motions with respect to one another and the incident waves. In [21], it is reported that an array of floating bodies spaced a wavelength apart are capable of coordinating operation in two modes such as heave and roll and they may extract 100% of the incident wave energy.

4.3.4 Wave Power Turbine Types

In this subsection, some of the most common wave power turbines that are coupled to the rotational generators for wave energy applications are discussed. These devices are generally installed within fixed structures to the shoreline. Since they are fixed, they provide an appropriate frame for wave forces that come against them so that they can have high conversion efficiencies.

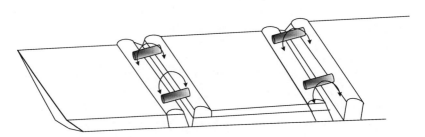

FIGURE 4.18 Wave contouring raft.

4.3.4.1 The "Wells" Air Turbines

The Wells turbines, which are low-pressure air turbines, are commonly used in OWC applications. Using Wells turbines eliminates the need for expensive and delicate valve systems to rectify the direction of the airflow.

OWCs are used as shore-mounted structures and they have fixed frames of references to meet the wave forces. The OWCs are essentially resonant devices and they operate in response to the incoming wave activity [22]. The pressure of the water inside the column increases due to the incident wave crest. The internal water level rises and in turn it pushes air out from the top of the column [22]. This airflow rotates the air turbine. Then the airflow reverses and flows into the column again when the waves are dragged back to the ocean.

OWCs use Wells turbines, which are self-rectifying air turbine as shown in Figure 4.19 [22], and do not need additional rectifying valves. Wells can extract energy from incoming or outgoing airflows to/from the air column. The Wells turbine has a low drag and can be driven at high rotational speeds of several hundred rpm without the need for a gearbox. The efficiency of Wells rotor is the highest when the air pressure is corresponding to 2–3 m of water rise in the column, which is the typical ocean wave height.

The airflow in both directions produces a forward thrust on the blades. Even though the airflow is bidirectional, the turbine spins unidirectionally.

The block diagram of the Wells turbine, which drives the generator of the wave power system, is given in Figure 4.20. The input of the turbine model is air velocity, which is caused by the pressure increase in the column and the output of the model is the mechanical torque that drives the generator [23,24].

The output mechanical torque is

$$T_{\mathrm{m}} = \frac{P_{\mathrm{shaft}}}{\omega_{\mathrm{m}}}, \tag{4.9}$$

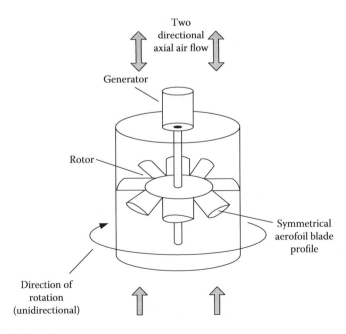

FIGURE 4.19 The Wells turbine.

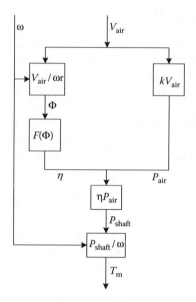

FIGURE 4.20   Block diagram of the Wells turbine.

where P_{shaft} is the power of the turbine shaft and ω_m is the angular speed of the turbine shaft. The shaft power P_{shaft} can be calculated by

$$P_{shaft} = \eta P_{air}, \qquad (4.10)$$

where η and P_{air} are the efficiency of the turbine and air power, respectively. The turbine efficiency is a function of flow coefficient, Φ. Turbine efficiency versus flow coefficient is shown in Figure 4.21.

The air power is proportional to the air velocity as

$$P_{air} = kV_{air}, \qquad (4.11)$$

where k is the proportion constant.

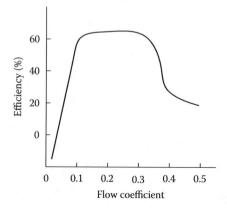

FIGURE 4.21   Turbine efficiency as a function of flow coefficient.

The flow coefficient Φ can be determined as

$$\Phi = \frac{V_{air}}{\omega_m r}, \tag{4.12}$$

where r is the turbine radius.

The turbine output torque has a pulsating form. In other words, the output waveform of the torque has oscillations with respect to the airflow. Since the air velocity at the turbine input is related to the wavelength and wave height, coupling a flywheel with high inertia can make the turbine output torque smoother. The flywheel location is along the shaft between the turbine and the induction generator [23,24].

The block diagram of a Wells turbine given in Figure 4.20 can be realized using MATLAB and Simulink. An example of a pulsating waveform of a Wells turbine is given in Figure 4.22 for the inertia $J = 0.02\,\mathrm{kg\,m^2}$, power conversion proportion constant $k = 0.5$, and radius of the turbine $r = 1\,\mathrm{m}$, for a small-scale Wells turbine.

4.3.4.2 Self-Pitch-Controlled Blades Turbine for WEC

The working principle, characteristics, and behavior of the self-pitch-controlled blades turbine is explained in this subsection [25]. The operating principle of the turbine using self-pitch-controlled blades is explained through Figure 4.23. This turbine is suitable for WEC and it can be used as an alternative for Wells turbines.

The turbine blade can oscillate between two predefined angles of $\pm\gamma$, which are set on the hub by a pivot near the leading edge [25]. The turbine blades can flip by themselves when an airfoil experiences a pitching moment, M, at a certain angle of incidence about the pivot. They can flip $+\gamma$ or $-\gamma$ depending on the flow direction. Therefore, the turbine's higher torque and higher operating efficiency is obtained with a lower rotating speed.

The torque coefficient C_T and the input power coefficient C_A versus the flow coefficient Φ are used for evaluating the turbine's characteristics under steady-state flow conditions.

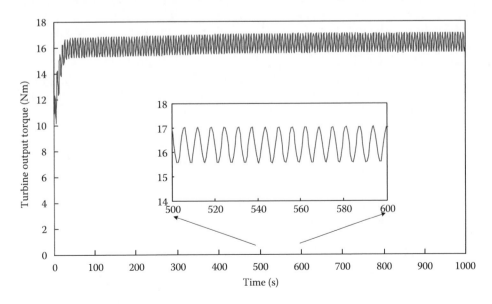

FIGURE 4.22 Output torque waveform of the Wells turbine with sinusoidal input airflow.

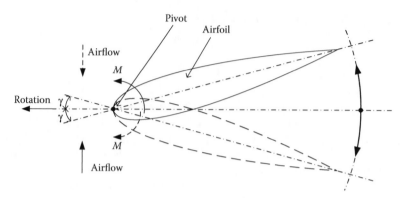

FIGURE 4.23 Turbine using self-pitch-controlled blades for WEC.

These coefficients are defined as

$$C_T = \frac{T_0}{\left[\rho(v^2 + U_R^2)blzr_R/2\right]},\tag{4.13}$$

$$C_A = \frac{\Delta p Q}{\left[\rho(v^2 + U_R^2)blzv/2\right]},\tag{4.14}$$

$$\phi = v_a/U_R,\tag{4.15}$$

where ρ is the air density, v is the mean axial flow velocity, U_R is the circumferential velocity at mean radius (r_R), b is the rotor blade height, l is the chord length, Δp is the total pressure drop between before and after the turbine, T_0 is the output torque, and z is the number of rotor blades [25].

The C_T–Φ characteristics vary with respect to the different blade angles. The C_T value decreases if γ increases in the stall-free region. This shows that the larger γ results in better starting conditions. Besides, the stall point increases with γ as well as the flow coefficient at no-load condition. The input coefficient C_A is higher at $\gamma = 0$ than nonzero γ values. This is caused by the rotor geometry. If γ is different from zero, the C_A value is negative at small flow coefficients. This means that the rotor acts as a fan at smaller inlet angles.

Thus, the averaged performance of the rotor goes down under an oscillating flow condition if the setting angle is fixed. It is possible to overcome this problem using the self-pitch-controlled blades because the torque at a small relative inlet angle becomes zero [25]. The overall view of a wave energy turbine with self–pitch-controlled blades is shown in Figure 4.24.

According to the previously constructed wave power plants, the axial flow velocity during exhalation such as from the air chamber to the outside is higher than that during inhalation [26–29].

Thus, the optimum setting pitch angle will be different, according to the direction of the airflow. For instance, authors investigated the effect of setting angles on the turbine performances under real sea conditions [25]. The analytical procedure for this performance is explained in [30].

The behavior of rotating systems can be described as

$$I\frac{d\omega}{dt} + T_L = T_0.\tag{4.16}$$

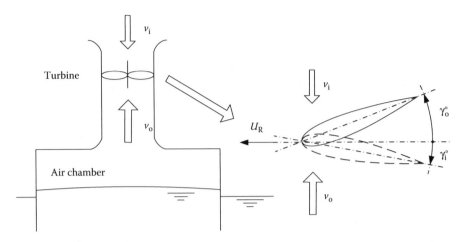

FIGURE 4.24 Overall view of wave energy turbine with self-pitch-controlled blades.

Here, I is the inertia moment of the rotor, t is the time, and T_L is the loading torque. For given values of I, T_L, and T_0, Equation 4.16 can be numerically solved. At the beginning, this gives the starting characteristics of the turbine and provides the running characteristics at the asymptotic condition. The turbine performance can be calculated as a mean efficiency when the solution is in the asymptotic condition:

$$\bar{\eta} = \frac{\left((1/T) \int_0^T T_0\omega \, dt\right)}{\left((1/T) \int_0^T \Delta pQ \, dt\right)}. \tag{4.17}$$

$\bar{\eta}$ is evaluated over one wave period, when the rotor is rotating at constant speed. In Equation 4.17 T is the wave period, T_0 is the output torque, ω is the angular rotor speed, and Q is the flow rate (m^3/s). In [25], the effect of different pitch angle variations on the turbine performance (mean efficiency) is presented.

4.3.4.3 Kaplan Turbines for WEC

In this subsection, Kaplan-type hydraulic turbines are described with their structure and operating principles. Kaplan turbines are suitable for both tidal and wave energy applications as well as tapered channel and other hydroelectric applications. These turbines are propeller-type water turbines with adjustable blade angles for speed regulation and torque control.

The fluid changes pressure as it moves through the turbine and transfers its energy to mechanical energy, which makes the turbine an inward flow reaction turbine. The efficiency of the Kaplan turbine is around 90%; however, it might decrease for very low head conditions. The inlet is a tube located around the wicket gate of the turbine. Water is directed tangentially through the wicket gate, which spins the turbine due to the spirals on to a propeller-shaped runner. The outlet is a draft tube used to decelerate the water and recover its kinetic energy.

The turbine does not need any minimum water flow rate since the draft tube remains full of water. A higher turbine location may increase the suction; however, this may result in faster cavitations of the turbine.

Double-regulated turbines have adjustable runner blades, because double regulation involves both wicket gate opening and angle controlling of the runner blades. The most efficient operation can be obtained under varying head, flow, or load conditions by controlling the blade angles and the wicket gate opening. Actual head values and wicket gate opening are used for the determination of the runner blade angle. Combinational dependence of $y_R(y_W, H_b)$ is specified in order to run the turbine at maximum efficiency, regardless of the operating conditions [31]. y_R is the runner blades' servomotor piston position corresponding to runner blades angle, y_W is the guide vanes' servomotor piston position corresponding to wicket gate opening, and H_b is the gross head diameter in meters. This relation is provided by the combinator (cam), called special device, in the turbine control system.

A hydroelectric power conversion system consists of a hydraulic turbine, a governor and a generator, which are shown in Figure 4.25 with double-regulated turbine [31].

Double-regulated hydraulic turbines are used in low-head hydraulic power plants such as WEC plants involving rotational turbines, which are driven by the water flow from the reservoir with short pipelines. Thus, water and cable channel can be considered as incompressible and the inelastic water column equation can be used as follows:

$$\frac{q - q_0}{h - h_0} = -\frac{1}{T_W s}, \tag{4.18}$$

where q and q_0 are the turbine discharge and its initial value (p.u.), respectively, h and h_0 are the turbine head and its initial value, respectively (p.u.), T_W is the water starting time (in seconds) and s is the Laplace operator.

The turbine that drives the electric generator has the mechanical torque at the coupling point, which can be calculated as

$$m = \frac{qh\eta_t}{\omega}, \tag{4.19}$$

where m is the turbine torque (p.u.), η_t is the turbine efficiency, and ω is the speed (p.u.). Head, speed, guide vanes, and runner blades are the parameters affecting the discharge and efficiency of the turbine as

$$q = q(h, \omega, y_W, y_R), \tag{4.20}$$

$$\eta_t = \eta_t(h, \omega, y_W, y_R). \tag{4.21}$$

Hence, the turbine mechanical torque is also affected by these four variables as

$$m = m(h, \omega, y_W, y_R). \tag{4.22}$$

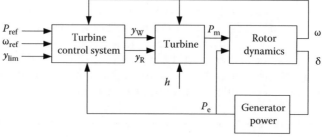

FIGURE 4.25 Double-regulated turbine model with control system and rotor dynamics.

The turbine characteristics define these complex nonlinear functions. The turbine speed is around a fixed value particularly if the power system is connected to the grid. If the speed variations are ignored, the turbine discharge and efficiency characteristics could be written independently from the speed [31] as

$$q = q(h, y_W, y_R), \tag{4.23}$$

$$\eta_t = \eta_t(h, y_W, y_R). \tag{4.24}$$

These characteristics are defined for h_{char}, which is the specific turbine head. For any h value, the turbine discharge determined from the $q(y_W, y_R)$ can be recalculated as

$$q = q(y_W, y_R)\sqrt{(h/h_{char})}. \tag{4.25}$$

Additional to the head variations, the turbine efficiency may vary and the efficiency characteristics for specific head h_{char} can be calculated as

$$\eta_t = \eta_t(y_W, y_R). \tag{4.26}$$

Discharge and efficiency functions with Equations 4.23 and 4.24 are used to build the nonlinear model of double-regulated Kaplan turbine. The block diagram of a Kaplan turbine is presented in Figure 4.26 [31].

4.3.4.4 Other Types of Turbines Used for WEC

In this subsection, other turbine types used for WEC applications are reviewed. These turbines are

- Biplane wells turbine with guide vanes (BWGVs), Figure 4.27 [32].
- Impulse turbine with self-pitch-controlled guide vanes (ISGVs), Figure 4.28 [33].
- Impulse turbine with fixed guide vane (IFGV), Figure 4.29 [34].

The different wave turbines are experimentally tested in "Performance comparison of turbines for wave power conversion" [35]. The test rig in this study consists of a settling chamber, a large piston-cylinder, and a test section with 3 m diameter with a bell-mouthed entry and a diffuser exit.

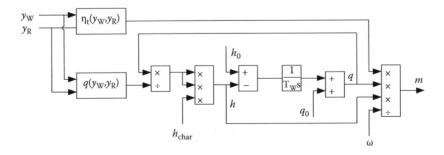

FIGURE 4.26 Nonlinear model of hydraulic turbine.

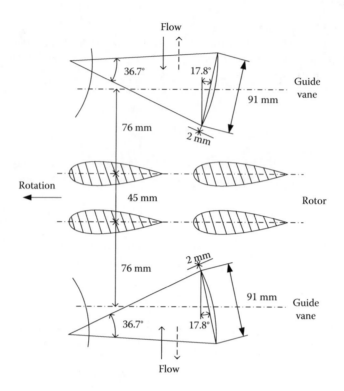

FIGURE 4.27 Biplane wells turbine with guide vanes.

The turbine rotor with $v = 0.7$ hub-to-tip ratio is placed at the center of the test section. Then the turbine rotor is tested at a constant rotational speed under steady-state conditions. The performance of the turbine is evaluated in terms of the turbine angular speed, ω, turbine output torque, T_0, flow rate, Q, and total pressure drop between the inside and outside of the air chamber, Δp. It should be noted that all of the turbines are self-starting [35].

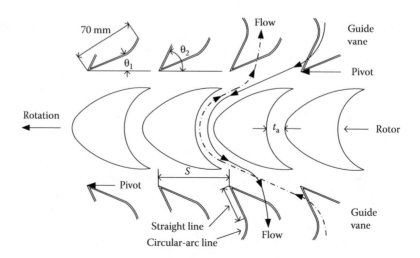

FIGURE 4.28 Impulse turbine with self-pitch-controlled guide vanes by link motion (ISGV).

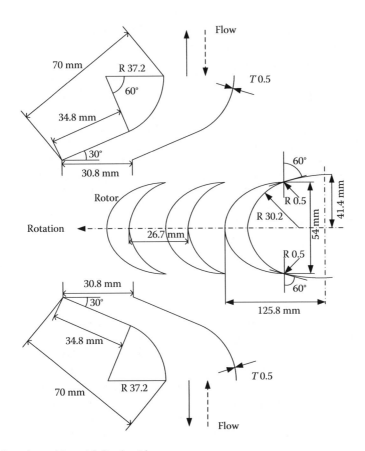

FIGURE 4.29 Impulse turbine with fixed guide vanes.

The turbines used for the experimental tests have the following specifications:

- BWGV: AR $= 0.5, l_r = 90$ mm, $\sigma_{rR} = 0.45, \sigma_{gR} = 1.25$.
- ISGV: $t_a/S_r = 0.4, l_r = 54$ mm, $\gamma = 60°, \sigma_{rR} = 2.02, \sigma_{gR} = 2.27, \theta_1 = 17°, \theta_2 = 72.5°$, and $\lambda = -7.5°$.
- IFGV: $t_a/S_r = 0.4, l_r = 54$ mm, $\gamma = 60°, \sigma_{rR} = 2.02, \sigma_{gR} = 2.27, \theta = 30°$, and $\lambda = -7.5°$,

where AR is the aspect ratio, l_r is the rotor chord length, σ is the solidity, t_a is the width of the flow path at r_R, S_R is the rotor blade space at r_R, γ is the blade inlet angle for impulse turbine rotor, θ is the setting angle of guide vane, and λ is the sweep angle.

Turbine characteristics were evaluated using the torque coefficient C_T, the input power coefficient C_A, and the flow coefficient ϕ [35]:

$$C_T = \frac{T_0}{\{\rho_a w^2 b l_r z r_R/2\}},$$
(4.27)

$$C_A = \frac{\Delta p Q}{\{\rho_a w^2 b l_r z v_a/2\}},$$
(4.28)

$$\phi = \frac{v_a}{U_R},$$
(4.29)

where T_0 is the output torque, ρ_a is the air density, b is the rotor blade height, l_r is the rotor's chord length, w is the relative inflow velocity, z is the number of rotor blades, r_R is the mean radius, Δp is the total pressure drop between the settling chamber and atmosphere, v_a is the mean axial flow velocity, and U_R is the circumferential velocity at r_R.

According to the torque coefficient versus flow coefficient characteristics of these turbines, the rotor stall causes rapid decreases in C_T in Wells-type rotors such as BWGV. For impulse-type rotor turbines such as ISGV and IFGV, the increase in ϕ also increases the C_T value. Furthermore, the C_T value is larger at the larger ϕ region than other Wells turbines.

According to the C_A–ϕ characteristics of the turbines, the C_A value is the largest in WTGV; this means that the air chamber pressure is higher and may require more frequent bearing maintenance, due to the greater thrust force. ISGV and IFGV have smaller pressure increases in the air chamber when they are used as wave power generator devices [35].

The turbine efficiency can be obtained using C_T–ϕ and C_A–ϕ characteristics. The efficiency is defined as

$$\eta = \frac{C_T}{C_A\phi}. \tag{4.30}$$

However, it should be noted that efficiency is not the only important factor for optimal turbine selection in WEC. In addition, turbine characteristics vary with the efficiency of the air chamber, which is the ratio of the power of the OWC and the incident wave power [35]. Airflow generated by the OWC is irregular due to the irregular natural behavior of ocean waves. Thus, it is important to clarify the turbine characteristics under irregular flow conditions. The irregular test wave is based on The International Ship Structure Congress spectrum, which is specific for irregular wave behavior and is used typically in ocean applications [36].

A WEC system with an air turbine is presented in Figure 4.30, which shows the upcoming waves applying pressure to the air chamber, resulting in the rotation of the air turbine.

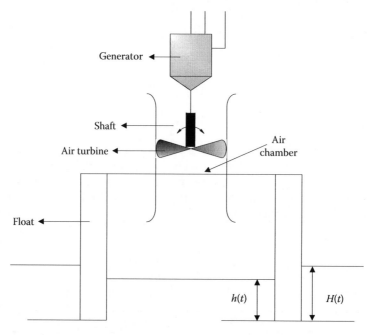

FIGURE 4.30 Schematic of the air turbine-type wave power generator system.

In this type of WEC system, the incident wave height and the wave height in the air chamber are interrelated as seen in the following equation [34]:

$$\frac{d}{dt}\left(\rho_s h A_C \frac{dh}{dt}\right) = \left\{\rho_s g(H - h) - \Delta p\right\} A_c,$$ (4.31)

where ρ_s is the seawater density, h is the wave height in the air chamber, H is the incident wave height, A_c is the cross-sectional area of the air chamber, and g is the gravity. The mean axial flow velocity is $v_a = (1/m)(dh/dt)$ and Δp is a function of (dh/dt) for a given rotational speed U_R. Extracting the relationship between Δp and (dh/dt) from the C_A–ϕ curve, Equation 4.31) can be rewritten as [35]

$$\rho_s A_c \left\{\left(\frac{dh}{dt}\right)^2 + h\frac{d^2h}{dt^2}\right\} = A_c \left\{\rho_s g(H - h) - \Delta p\right\},$$ (4.32)

if it is assumed that $\Delta p/\rho_s \equiv F(dh/dt)$.
Then,

$$h\frac{d^2h}{dt^2} + \left(\frac{dh}{dt}\right)^2 + F\left(\frac{dh}{dt}\right) - g(H - h) = 0.$$ (4.33)

The Runge–Kutta–Gill method can be used to solve this equation to obtain the wave height in the air chamber. The power of the incident wave is W_i and the power of the OWC W_o can be defined as

$$\bar{W}_i = \frac{\sum_{i=1}^{N}(1/32\pi)\rho_s g^2 H_i^2 T_i^2}{\sum_{i=1}^{N} T_i},$$ (4.34)

$$\bar{W}_{oi} = \frac{\sum_{i=1}^{N}(1/32\pi)\rho_s g^2 h_i^2 T_i^2}{\sum_{i=1}^{N} T_i}.$$ (4.35)

The efficiency of the air chamber is the ratio of the OWC power to the incident wave power as

$$\tilde{\eta}_c = \frac{\bar{W}_o}{\bar{W}_i}.$$ (4.36)

The axial velocity is directly proportional to the wave height if the flow is assumed to be incompressible. The axial flow velocity through the turbine can be expressed as

$$v_a^* = \frac{d\left(H/H_{1/3}\right)}{d\left(t/\overline{T}\right)} = \frac{dh^*}{d\bar{t}^*}.$$ (4.37)

The motion of the turbine's rotating system under irregular flow can be described as

$$K^2 X_I \frac{d\bar{\omega}^*}{d\bar{t}^*} + X_L = C_T(\phi)\frac{(K\bar{\omega}^*)^2 + v_a^{*2}}{2}\sigma_{rR}\frac{4(1-v)}{1+v},$$ (4.38)

where

$$\phi = v_a^*/(K\bar{\omega}^*), \qquad\qquad (4.39)$$

$$K\bar{\omega}^* = \omega m r_R \overline{T}/H_{1/3}, \qquad\qquad (4.40)$$

$$v_a^* = \frac{m\overline{T}v_a}{H_{1/3}}. \qquad\qquad (4.41)$$

Here, K is the period, X_I is the inertia moment, X_L is the loading torque, and σ_{rR} is the solidity at $r_R = l.z/(2\pi r_R)$.

In Equation 4.38, the first and second terms on the left-hand side are the inertia and loading terms and the right-hand side is the torque produced by the turbine. Using Equation 4.38, the turbine behavior at the starting conditions can be calculated as a function of $K\bar{\omega}^*$ and v_a^* if the loading characteristics, torque coefficient, and geometrical specifications of the rotor are given. The operation characteristics after the start-up can be obtained assuming a constant rotor speed. The mean output and input coefficients C_o and C_i can be calculated, respectively, as

$$\bar{C}_o = \frac{1}{t^*}\int_0^{t^*} C_T(\phi)\frac{(K\bar{\omega}^*)^2 + v_a^{*2}}{2} \times \sigma_{rR}\frac{4(1-v)}{1+v}\bar{\omega}^* \, d\bar{t}^*, \qquad\qquad (4.42)$$

$$\bar{C}_i = \frac{1}{t^*}\int_0^{t^*} C_A(\phi)\frac{(K\bar{\omega}^*)^2 + v_a^{*2}}{2K} \times \sigma_{rR}\frac{4(1-v)}{1+v}v_a^* \, d\bar{t}^*. \qquad\qquad (4.43)$$

The mean turbine efficiency can be calculated as

$$\tilde{n}_t = \frac{\bar{C}_o}{\bar{C}_i}\bar{C}_o/\bar{C}_i. \qquad\qquad (4.44)$$

Using the turbine and air chamber efficiencies, the overall conversion efficiency of the wave energy device is

$$\tilde{n} = \tilde{n}_c \cdot \tilde{n}_t. \qquad\qquad (4.45)$$

The conversion efficiencies of different turbine types are shown in Figure 4.31. For impulse-type turbines, the efficiency is higher at the large $1/(K\bar{\omega}^*)$ region in comparison to the Wells-type turbines. The rotor stall does not occur for the impulse-type turbines so the torque can be obtained in a comprehensive region of flow coefficient. The disadvantage of ISGV is the maintenance of pivots; however, it has the largest maximum efficiency, even larger than the IFGV in which the guide vanes are fixed [35]. Therefore, the running characteristics of the impulse turbine are better than the Wells turbine.

According to the starting characteristics of the turbines, very short time is required for the start of the impulse-type turbine. This means that impulse turbines can start more quickly than Wells turbines and can generate more power at the same start-up duration. In addition, the operational speed is lower than the Wells-type turbines. These are due to the higher torque coefficient of the impulse turbines than the torque coefficient of the Wells turbines. Moreover, flow coefficient is lower for the impulse turbines in no-load conditions. Thus, in terms of mechanical advantages and noise reduction, impulse turbines seem to be more suitable to design better applications for WEC systems.

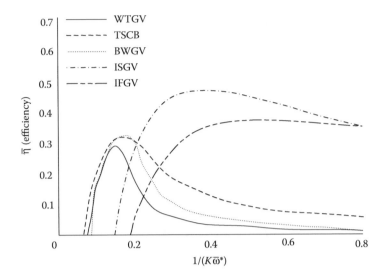

FIGURE 4.31 Comparison efficiency of wave energy converters with different turbines.

4.3.5 Wave Power Generators

There are several types of generators used to convert the wave energy into electrical energy. Generally, rotating machines such as synchronous generators or induction generators are used within the system employing a turbine to provide the mechanical driving for the generator shaft. These systems are very common for nearshore applications. Rarely rotating generators are found in buoy-shaped air turbine–driven generator systems. In other offshore applications including buoys, linear PM generators are common. Cam-shaped cylindrical ocean wave energy converters are exceptional, although they are offshore applications, rotating machines are employed with these devices.

4.3.5.1 A Wave-Activated Linear Generator Model

A wave-activated electric generator, which operates on the same principles as alternating current generators, is presented in [37]. The device forms a two degree of freedom mechanical system producing power by electromagnetic induction.

The proposed generator consists of an electric power generator enclosed in a buoy. The generator has a rectangular wire loop of N turns, directly supported by the buoy frame and a system with two PMs that are connected by helical springs to the buoy as shown in Figure 4.32. The wire loop moves between the magnets and magnets exhibit a two degree of freedom system excited by the wave-induced motion of the buoy. Since the system has no rotating parts, there will not be any fouling.

The electric power is generated as the reason of induced current in the conductor due to the moving conductor relative to a magnetic force.

The electric power generated in this type of linear generator can be expressed as

$$P = \frac{N^2 B^2 L^2 V_r^2}{R}. \tag{4.46}$$

Here, V_r is the relative motion (distance) of the wire loop with respect to the magnet, B is the magnetic induction, R is the load resistance in the wire loop, and L is the wire length

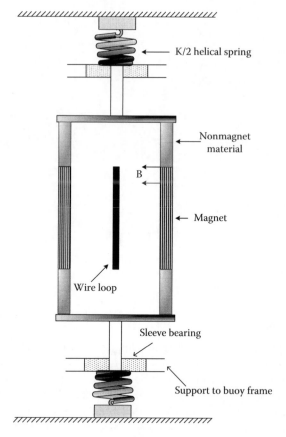

FIGURE 4.32 A linear wave activated generator system. (Modified from T. Omholt, *Oceans*, 10, 585–589, 1978.)

within the magnetic field in the plane perpendicular to the plane surface [37]. As a result of the current flow, a force will be applied on the loop opposing its motion, which can be written as in Equation 4.47. An equal but opposite force also applies on the magnets:

$$F = \frac{N^2 B^2 L^2 V_r}{R}. \tag{4.47}$$

A simple schematic diagram of a buoy-generator system is shown in Figure 4.33 [37].

The sum of inertia, damping, and restoring forces for a buoy in this type of generator equals the excitation forces.

$$\text{Inertia forces} + \text{Damping forces} + \text{Restoring forces} = \text{Excitation forces.} \tag{4.48}$$

The following second-order equations describe the motion of the system by considering only vertical motion of a buoy-generator with infinitesimal waves:

$$(M + M_a)\ddot{Z} = \rho g K_p a_a \cos(\omega t) - \frac{B^2 N^2 L^2}{R}(\dot{Z} - \dot{X}) - K(Z - X) - b\dot{Z} - \rho g A Z, \tag{4.49}$$

$$M_m \ddot{X} = -\frac{B^2 N^2 L^2}{R}(\dot{X} - \dot{Z}) - K(X - Z), \tag{4.50}$$

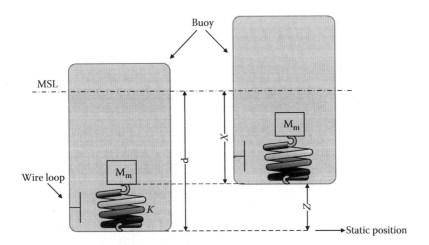

FIGURE 4.33 Schematic diagram of buoy generator system.

where $(m + m_a)\ddot{Z}$ represents the inertia forces, $b\dot{Z}$ stands for damping forces, and $\rho g AZ$ is the restoring force. The right-hand side of Equation 4.50 shows the sum of the excitation forces. Here, m is the mass of the buoy (mass of the spring-supported magnetic system is not included), m_a is the added mass, b is the damping coefficient, ρ is the water density, K_p is the pressure response factor ($K_p = e^{-kd}$), a is the wave amplitude, K is the total spring constant, ω is the wave frequency, and Z and X are the displacements shown in Figure 4.33, respectively. The derivations of these displacements give the corresponding velocity and acceleration. It is assumed that the buoy diameter is small in comparison to the wavelength [37].

Equation 4.50 is based on the assumption that the mass of the PMs is m_m and the system with PMs is excited by the induced force (F) and this force is transmitted through two helical springs [37].

From Equations 4.49 and 4.50, it can be seen that the $X = Z$ or $\dot{X} = \dot{Z}$ cases are not energy-generating motions. So, displacement of X and Z should not be synchronized or, in other words, the phase angle between two motions must be greater than $0°$ but less than $180°$.

The system with PMs is excited by the force that is transmitted through two helical springs. The electric power generation can be expressed as

$$P = \frac{N^2 B^2 L^2}{R}(\dot{Z} - \dot{X})^2. \tag{4.51}$$

The average power over one wave period T can be obtained by integrating the power Equation given in Equation 4.51.

$$\overline{P} = \frac{1}{T}\int_0^T P\,\mathrm{d}t. \tag{4.52}$$

Recalling that the power that is carried by an ocean wave is expressed by $P_W = \rho g^2 a^2 TD/8\pi$, the conversion rate of the system can be expressed as the ratio of the electric power generation over the total wave power, which is also the system efficiency:

$$\eta = \frac{\overline{P}}{P_{\text{wave}}} = 8\pi N^2 B^2 L^2 \int (\dot{Z} - \dot{X})^2 \frac{dt}{\rho g^2 a^2 T^2 DR}. \tag{4.53}$$

From this efficiency equation, it can be observed that the system efficiency is a function of both relative speed of buoy and wired loop and the load resistance. In practice, both fluid viscosity and spring constant are small. They exert dumping force way less than that of the induction to the heavy buoy. However, for a lighter wired loop, these effects are almost not negligible. Taking these facts into account, motion equations can be simplified to the following state-space description:

$$\begin{bmatrix} \dot{x}_1 \\ \dot{x}_2 \\ \dot{x}_3 \\ \dot{x}_4 \end{bmatrix} = \begin{bmatrix} 0 & 1 & 0 & 0 \\ \dfrac{-\rho g A}{M + M_a} & \dfrac{-(B^2 N^2 L^2/R)}{M + M_a} & 0 & \dfrac{B^2 N^2 L^2/R}{M + M_a} \\ 0 & 0 & 0 & 1 \\ \dfrac{K}{M_m} & \dfrac{B^2 N^2 L^2/R}{M_m} & \dfrac{-K}{M_m} & \dfrac{-B^2 N^2 L^2/R}{M_m} \end{bmatrix} \begin{bmatrix} x_1 \\ x_2 \\ x_3 \\ x_4 \end{bmatrix} + \begin{bmatrix} 0 \\ 1 \\ 0 \\ 0 \end{bmatrix} \dfrac{\rho g K_p a_a}{M + M_a} \cos(\omega t), \tag{4.54}$$

where the following substitution can be applied:

$$x_1 = z, \quad x_2 = \dot{z}, \quad x_3 = x, \quad x_4 = \dot{x}. \tag{4.55}$$

The general solutions to these linear equations in frequency domain are given by

$$x(j\omega) = (j\omega - A)^{-1} x(0) + (j\omega - A)^{-1} BU(j\omega), \tag{4.56}$$

where A is the state matrix, B is the input, and $x(0)$ represents the initial conditions. The first term describes the transient response of the system and it is not effective in the steady-state response. The second term describes system behavior under particular perturbation. In this case, it takes the form of sinusoidal wave as the assumption that ocean wave moves in this pattern. Consequently, the solution would consist of a pure sinusoidal wave at the same frequency as the input sine wave. The amplitude and phase determined by the system's frequency response at that frequency is given by [38]

$$G(t) = |G_s(j\omega)| \cos(\omega t + \angle \varphi_s). \tag{4.57}$$

In order to model and simulate this system, the parameters of the PM linear generator and the other respective physical parameters can be selected as follows [38]: $M = 1344$ kg, $M_a = 119.5$ kg, $L = 2.64$ m, $K = 1$, $B = 1.4$ T, $A = 4.86$ m^2, $N = 30$, and $D = 2.5$, $K_p = 0.9$, $a_a = 1.5$ m, and $\omega = 2\pi$ rad/s.

By implementation of these parameters, the state-transition matrix can be written as

$$A = \begin{bmatrix} 0 & 1 & 0 & 0 \\ -32.5 & -8.4/R & 0 & 8.4/R \\ 0 & 0 & 0 & 1 \\ 0.005 & 61.47/R & -0.005 & -61.47/R \end{bmatrix}. \tag{4.58}$$

According to the above analysis, the roots of the motion equations can be calculated as follows:

$$x_i = |G_i(j2\pi)| \, 9\cos(2\pi t + \angle G_i(j2\pi)), \tag{4.59}$$

$$G_1(2\pi j) = \left. \frac{s^2 R + 61.47s + 0.005R}{s^4 R + 69.87s^3 + 32.505s^2 R + 1997.7s + 0.1625R} \right|_{s=2\pi j},$$

$$G_2(2\pi j) = \left. \frac{s(s^2 R + 61.47s + 0.005R)}{s^4 R + 69.87s^3 + 32.505s^2 R + 1997.7s + 0.1625R} \right|_{s=2\pi j},$$

$$G_3(2\pi j) = \left. \frac{61.47s + 0.005R}{s^4 R + 69.87s^3 + 32.505s^2 R + 1997.7s + 0.1625R} \right|_{s=2\pi j}, \tag{4.60}$$

$$G_4(2\pi j) = \left. \frac{s(61.47s + 0.005R)}{s^4 R + 69.87s^3 + 32.505s^2 R + 1997.7s + 0.1625R} \right|_{s=2\pi j},$$

These solutions can substituted within the system efficiency equation given in Equation 4.53; then it is observed that system efficiency is a complex function of resistance R. By controlling the value of the equivalent load resistance of R, the system efficiency can be maximized.

The effect of load resistance on the efficiency and power can be observed by using variable load resistance values and using them in Equations 4.61 through 4.63. In the simulation environment, a variable resistance is used to present the effect of the load resistance on extracted power and efficiency. A sinusoidal source is used to represent the ocean waveform. The resistance–average power and resistance–efficiency curves are shown in Figure 4.34 by sharing the same x-axis.

According to the results shown in Figure 4.34, the maximum power and maximum efficiency values can be achieved for a particular value of the load resistance. Ideally, the system should operate at this point. In order to achieve this goal, a current regulation should be provided for the generator's output. The rectifier and a large capacitor as shown

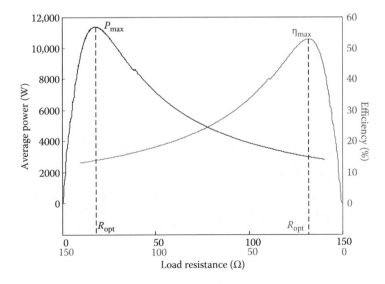

FIGURE 4.34 (See color insert following page 80.) Average power and efficiency variation versus load resistance.

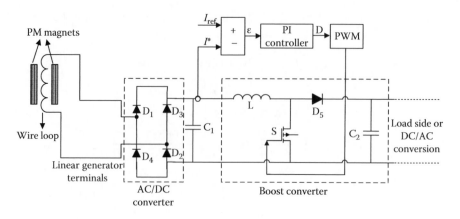

FIGURE 4.35 System-level configuration of the linear generator and power converters.

in Figure 4.35 provide a fixed DC voltage for the input of the boost converter. Assuming a lossless converter, the input current can be controlled to vary the Thevenin equivalent resistance of the system after the rectification.

The generator power and generator voltage are both in sinusoidal form as shown in Figures 4.36 and 4.37, respectively. The generator power has a biased sinusoidal waveform.

In order to regulate the generator's output power and voltage, the output of the linear generator terminals is first converted to the DC voltage and then filtered. Considering the large oscillations in the output, a large capacitor is required to achieve satisfactory voltage regulation performance. A boost DC–DC converter operating in the current control mode is employed, since the input voltage to boost the converter is the output voltage of the rectifier and that is fixed. Current control mode [38] helps optimum loading of the linear generator

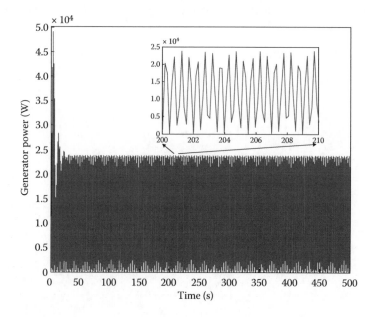

FIGURE 4.36 Generator power.

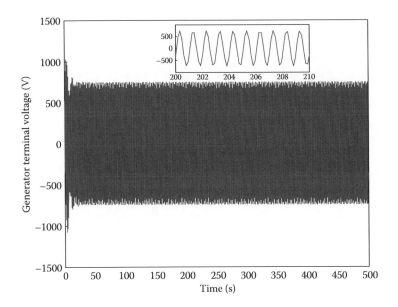

FIGURE 4.37 Generator voltage.

since the equivalent Thevenin's impedance is determined by the converter input voltage and current.

Therefore, regardless of what is connected at the output, by controlling the input current of the boost converter under fixed input voltage, one can control the equivalent resistance. The fixed input voltage for the converter is provided by the rectifier and capacitor connected at the generator terminals. Figure 4.38 shows the current tracking performance of the proposed system. As the current reaches the reference value in order to satisfy the optimal load resistance, the power also increases and reaches its maximum value.

4.3.5.2 Linear, Synchronous, Longitudinal-Flux PM Generators

Linear, synchronous, longitudinal-flux PM generators are one of the methods to convert the wave energy into electrical energy through direct drive. In [39,40], Muller presented some directly driven linear generators as wave energy converters. For WEC, the pistons of linear longitudinal-flux PM generators are generally driven by a buoy. Variable amplitude and variable frequency electromagnetic force is induced in the stator winding by the vertical motion of the piston [41]. The power fluctuation problems caused by the variable amplitude and frequency can be reduced by connecting several units as arrays [42].

A systematic approach for the design and modeling of an longitudinal-flux permanent-magnet machine (LFM) with diode rectifier where the piston is driven by a buoy is described by Wolfbrandt in [41].

The average power of an ocean wave over a wave period of T can be derived from the total energy of a wave as given in Equation 4.61. When a buoy is deployed as a wave rider, the power of that buoy can be approximated as

$$P_{\text{buoy}} = k \frac{TH^2}{2} \min\{d, \omega_a\}, \tag{4.61}$$

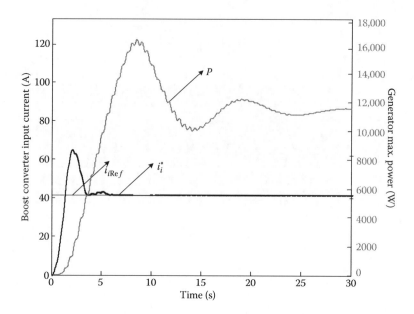

FIGURE 4.38   (See color insert following page 80.) Reference current tracking and power output for the linear PM generator.

where k is the absorption coefficient, d is the buoy diameter, and ω_a is the width of the buoy that can absorb maximum energy.

If the buoy is taken as a point absorber, the wave is assumed to be sinusoidal, and the displacement of the buoy is in one direction, then, for example, for $k = 0.8$, 50% power absorption is possible [43]. The width of the absorption is defined as [44–46]

$$\omega_a = \frac{L_{\text{wave}}}{2\pi},\tag{4.62}$$

where the L_{wave} is the wavelength and is defined as

$$L_{\text{wave}} = \frac{gT^2}{2\pi}.\tag{4.63}$$

This equation is valid for deepwater, that is, the depth is greater than half the wavelength. In deepwater, the kinetic energy of a wave is equal to its potential energy. Wolfbrandt [41] considers a three-phase LFM rectified by a diode bridge rectifier with surface-mounted PMs that are directly driven. Figure 4.39 shows the x–y plane of the cross-sectional view of one pole of the generator.

In this method, the stator of the linear generator is supported by a buoy, which is fixed at the ocean bottom. Upcoming ocean waves drive the piston and this causes the up and down motion of the PM. This generates an electromagnetic field and a voltage is induced on the stator winding terminals.

The magnetic flux intensity in the stator is higher in the coils that are closer to the PM. The flux intensity gets weaker in the outer coil windings. The picture of the magnetic flux intensity created by such a PM system is shown in Figure 4.40.

There are other topologies, which also use the same principle of the linear generator system. For instance, the fixed stator windings can be mounted on a yoke that is fixed to the ocean bottom through a support unit. In this method, the PM is connected through a

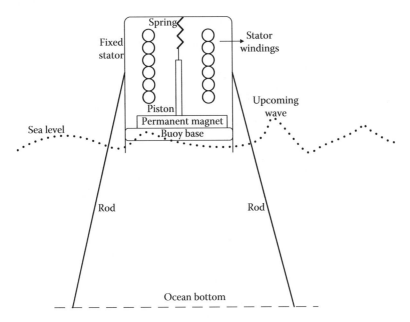

FIGURE 4.39 Schematic of an LFM used for WEC.

rope to a buoy, which is on the ocean level and moves up and down and can be dragged towards different directions. Once a wave results in the motion of the buoy, the buoy pulls the PM piston. The PM moves up and down within the fixed stator windings. This also generates an electromagnetic field generating the electricity. The schematic of this energy conversion device is illustrated in Figure 4.41.

Because of the relatively long wave periods (low frequency in piston motion) displacement current is negligible. Based on Maxwell's equations

$$\sigma\frac{\partial A_z}{\partial t} - \nabla\left(\frac{1}{\mu_0\mu_r}\nabla A_z\right) = -\sigma\nabla V. \tag{4.64}$$

Here, A_z is the z-component of the magnetic vector potential, μ_0 is the permeability of free space, μ_r is the relative permeability, σ is the conductivity, and ∇V is the applied potential.

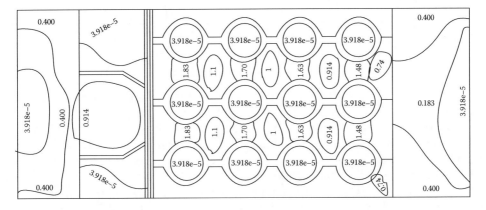

FIGURE 4.40 Approximate magnetic flux intensity (Wb/m^2) created by PMs.

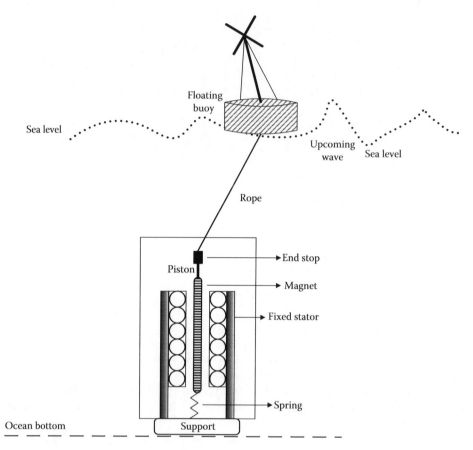

FIGURE 4.41 Another method of using LFMs for WEC.

Current sources can be used for PM modeling. The end effect of the stator windings can be modeled as impedances in the circuit. End effect is an issue that affects the power production performance of a linear generator. Stator winding equations can then be expressed as

$$I_a + I_b + I_c = 0, \tag{4.65}$$

$$U_a + R_s I_a + L_s^{end} \frac{\partial I_a}{\partial t} - U_b - R_s I_b - L_s^{end} \frac{\partial I_b}{\partial t} = V_{ab}, \tag{4.66}$$

$$U_c + R_s I_c + L_s^{end} \frac{\partial I_c}{\partial t} - U_b - R_s I_b - L_s^{end} \frac{\partial I_b}{\partial t} = V_{cb}, \tag{4.67}$$

where I_a, I_b, and I_c are the phase currents, U_a, U_b, and U_c are phase voltages, V_{ab} and V_{cb} are line voltages, R_s is the line resistance, and L_s^{end} is the coil end inductance.

The output phase voltages of the linear generator can be rectified using a three-phase diode rectifier and can be connected to a load. A full bridge six-pulse diode rectifier connected to the load is shown in Figure 4.42. In Figure 4.42, the load components include resistive (R), inductive (L), and back-EMF (E).

The losses in the linear generator can be mainly categorized as copper losses and iron losses. The power flow illustrating these losses is shown in Figure 4.43, where P_{gen} is the ideal generated power and P_{out} is the actual power output of the generator.

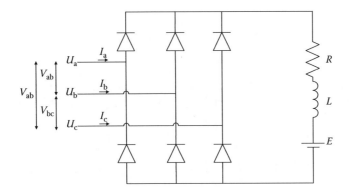

FIGURE 4.42 Three-phase diode rectifier schematic where a, b, and c are the linear generator phases.

The stranded stator winding losses are divided into ohmic losses $\left(P_{Cu}^{ohmic}\right)$ and eddy-current losses $\left(P_{Cu}^{joule}\right)$. These loss equations can be expressed as [41]

$$P_{Cu}^{ohmic} = 3R_s I_s^2, \tag{4.68}$$

$$P_{Cu}^{joule} = \frac{\sigma \omega^2 d^2}{32} \int_S B^2 \, dS, \tag{4.69}$$

where ω is the angular frequency, d is the strand diameter, S is the cross-sectional conductor area, and B is the magnetic field density.

A single-value magnetization curve is used to model the nonlinearity of the laminated iron core. The iron losses can be divided into hysteric losses $\left(P_{Fe}^{hyst}\right)$, joule losses $\left(P_{Fe}^{joule}\right)$, and excess losses $\left(P_{Fe}^{excess}\right)$ [47]. These loss equations are given by

$$P_{Fe}^{hyst} = k_f k_{hyst} f \int_V B_{max}^q \, dV, \tag{4.70}$$

$$P_{Fe}^{joule} = k_f \sigma \frac{d^2}{12} \int_V \frac{1}{T} \int_0^T \left(\frac{\partial B}{\partial t}\right)^2 dt \, dV, \tag{4.71}$$

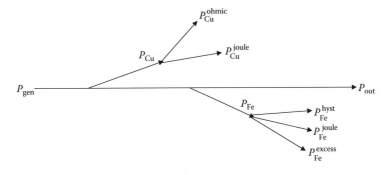

FIGURE 4.43 Power flow and losses of the linear generator.

$$P_{\mathrm{Fe}}^{\mathrm{excess}} = k_{\mathrm{f}} k_{\mathrm{excess}} \int_V \frac{1}{T} \int_0^T \left(\frac{\partial B}{\partial t} \right)^{3/2} \mathrm{d}t \, \mathrm{d}V, \tag{4.72}$$

where $B_{\max}$ is the maximum magnitude of the magnetic field density, f is the frequency, d is the sheet thickness, k_{f} is the stacking factor, T is the time period, and V is the stator core volume. The coefficients k_{hyst} (Ws/T^q/m^3), k_{excess} (W/(T/s)$^{3/2}$/m^3), and the exponent q are determined by fitting of a given loss curve.

Piston losses that are in the back iron, PM, and the aluminum wedge can be calculated using the Poynting vector. Thus, the joule losses $\left(P_{\mathrm{piston}}^{\mathrm{joule}} \right)$ and the hysteresis losses $\left(P_{\mathrm{piston}}^{\mathrm{hyst}} \right)$ can be written as

$$P_{\mathrm{piston}}^{\mathrm{joule}} = \int_V \frac{1}{T} \int_0^T \sigma \left(\frac{\mathrm{d}A_z}{\mathrm{d}t} \right)^2 \mathrm{d}t \, \mathrm{d}V, \tag{4.73}$$

$$P_{\mathrm{piston}}^{\mathrm{hyst}} = \int_V \frac{1}{T} \int_0^T \left(H_x \frac{\mathrm{d}B_x}{\mathrm{d}t} + H_y \frac{\mathrm{d}B_y}{\mathrm{d}t} \right) \mathrm{d}t \, \mathrm{d}V, \tag{4.74}$$

where T is the time period and V is the piston volume.

The electromagnetic force F_{em} induced by the generator can be expressed as

$$F_{\mathrm{em}} = \frac{\mathrm{d}W_{\mathrm{c}}}{\mathrm{d}x}, \tag{4.75}$$

where W_{c} is the coenergy in the volume of the air gap and x is the piston position. Stator current ripple causes the electromagnetic force ripple. These ripples caused by the phase-belt magnetomotive space harmonics can be eliminated by fractional slot pitch winding.

LFM with different voltage loads results in different current densities. The current density should be limited to $i_{\max}$ due to the physical constraints of the linear generator. Because of this limitation, the output generator power will be limited by

$$P_{\mathrm{out}}(i) < P_{\mathrm{buoy}} - P_{\mathrm{losses}}(i), \tag{4.76}$$

$$i < i_{\max}. \tag{4.77}$$

P_{buoy} is defined in Equation 4.61, and P_{losses} is defined as the sum of losses given in Equations 4.68 through 4.74.

Figure 4.44 illustrates these powers with respect to the maximum current. The solid line is the input power (buoy power), the dotted line is the output power, and the dashed-dotted line is the sum of the output power and losses, which represent the total power.

4.3.5.3 A Three-Phase Synchronous Generator for Ocean Wave Applications

The synchronous generators are generally used in power plants with constant speeds. For wave energy applications, the design and model of a wound rotor brushless synchronous generator is presented in [48].

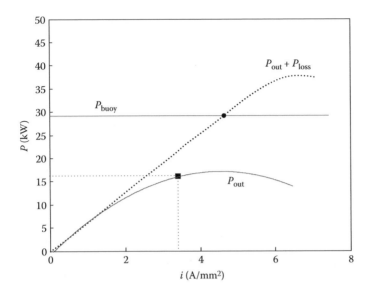

FIGURE 4.44 Buoy power and output power versus current density.

In a synchronous machine, the relation of rotor speed and output power frequency is

$$f = \frac{pn}{120},\tag{4.78}$$

where f is the output frequency, p is the number of salient poles, and n is the rotation speed of the rotor shaft in rpm. Therefore, for a two salient pole machine, the synchronous speed would be 3600 rpm [48], since the output frequency of the system should be 60 Hz for grid connection. The armature excitation generates AC and this current is converted to DC so that armature exciter acts as a DC supply for field winding. The armature exciter has three phase windings in order to get the improved efficiency of three-phase rectification in comparison to single-phase rectification.

In this generator, the salient pole face is taped so that the air gap width reaches its minimum value at the pole center. The magnetic flux is maximum at the minimum air gap. Magnetic flux decreases to zero at the midpoint between the poles and reaches to its negative maximum in the center of the adjacent pole. The flux density in the air gap is sinusoidal as

$$B = B_{max} \cos \beta y \ (T),\tag{4.79}$$

$$\beta y = \frac{p}{2} 2\pi,\tag{4.80}$$

$$y = \pi D,\tag{4.81}$$

where y is the fixed coordinate of the vertical axis with respect to the rotor, β is the rotor flux angle, and D is the air gap diameter.

Therefore, flux density equation can be rewritten as

$$B = B_{max} \cos \frac{py}{D}.\tag{4.82}$$

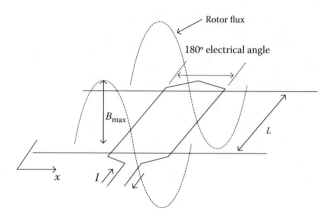

FIGURE 4.45 The traveling flux wave of stator winding.

Figure 4.45 shows the induced EMF in the stator winding to achieve a sinusoidal flux distribution along its periphery. Therefore, the flux in terms of the coordinate x is fixed with respect to the stator [48].

Coordinates and rotor position for stator winding are shown in Figure 4.46.
From Figure 4.46,

$$x = y + st \text{ (m)}, \tag{4.83}$$

where the tangential speed of the rotor, s, is given by

$$s = \frac{n\pi D}{60} \text{ (m/s)}. \tag{4.84}$$

Thus,

$$B = B_{max} \cos\left(\frac{px}{D} - \frac{pn\pi}{60}t\right) \text{ (T)} \tag{4.85}$$

or

$$B = B_{max} \cos(\beta x - \omega t) \text{ (T)}, \tag{4.86}$$

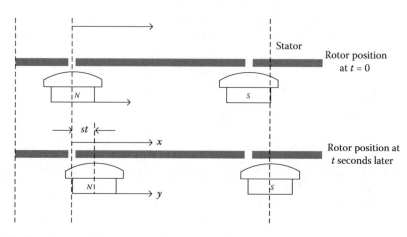

FIGURE 4.46 Coordinates and rotor position for stator winding.

where

$$\omega = \frac{pn\pi}{60} = 2\pi f \,(\text{rad/s}). \tag{4.87}$$

This is the mathematical expression for a traveling wave in the induced EMF stator winding. The differential flux penetrating the differential space window of the width dx is given by

$$d\phi = BL\,dx\,(\text{Wb}), \tag{4.88}$$

where L is the axial rotor length.
   Total flux passing through the coil is given by

$$\phi = \frac{2B_{max}LD\sin\omega t}{p} \,(\text{Wb}). \tag{4.89}$$

Total EMF induced in the equidistant slots can be derived from Equation 4.88 as

$$E = \omega LDNB\cos\omega t\,(\text{V}), \tag{4.90}$$

where N is the number of turns in the stator winding. Generated EMF's effective value can be written as

$$E = \frac{\pi}{60\sqrt{2}}pnNLDB\,(\text{V}). \tag{4.91}$$

A practical design utilizing the total stator surface by placing the stator winding in many slots distributed around the periphery is shown in Figure 4.47.
   The slots are placed by α electrical degree phase angles and totally there are q slots per pole. The total EMF is obtained by connecting the coils in series. The induced EMF depends on the slot α degrees. Thus, the total E is obtained by the summation of complex phasors E_1, E_2, and so on.

$$E = E_1 + E_2 + \cdots + E_q\,(\text{V}). \tag{4.92}$$

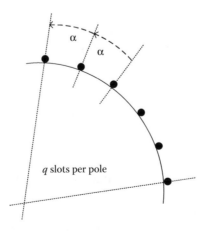

FIGURE 4.47   Total stator surface and q slots around the periphery.

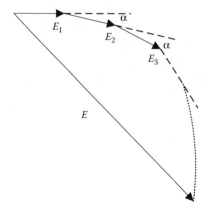

FIGURE 4.48 Induced total EMF phasor.

The distribution effect and the EMF phasor E is shown in Figure 4.48.

Thus, distribution effect can be derived using all induced EMFs for phasor E. Phasor E can be calculated from

$$E = E_1 \frac{1 - e^{-jq\alpha}}{1 - e^{-j\alpha}}. \tag{4.93}$$

The effective value of the complex phasor E is

$$|E| = |E_1| \left| \frac{1 - e^{-jq\alpha}}{1 - e^{-j\alpha}} \right|$$

$$= |E_1| \frac{\sqrt{(1 - \cos q\alpha)^2 + \sin^2 q\alpha}}{\sqrt{(1 - \cos \alpha)^2 + \sin^2 \alpha}} \tag{4.94}$$

$$= |E_1| \frac{\sin(q\alpha/2)}{\sin(\alpha/2)}.$$

Single-phase stator core and winding arrangement is shown in Figure 4.49.

In Figure 4.49, each different couple of poles is associated with the winding inducing EMF components of E_q. The total EMF, E, is the vector summation of all sets of EMF E_q. Figure 4.49 shows that there are two sets of coils and each set of coils contains six coils per pole per phase.

Output power is shared by three phases. When the load is connected to the generator terminals, the voltage drop is given by

$$V = E - IZ, \tag{4.95}$$

where V is the terminal voltage, E is the induced EMF, I is the stator current, and Z is the impedance.

The armature excitation and field excitation are necessary for brushless synchronous generator. To produce more output power and to reduce output frequency more poles can be used for the exciter. Magnetic flux may lose its stability if the frequency ripple is higher.

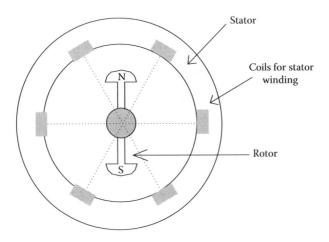

FIGURE 4.49 Stator core design and stator winding arrangement per phase.

In this generator design, six poles for armature exciter and field exciter coil were used for less frequency ripple. The excitation circuit is shown in Figure 4.50.

The resistance and inductance of the generator winding affect the power losses, I^2R, which in turn affects the generator efficiency. Therefore, the number of turns (N) in the stator winding and thickness of the copper wire affect the equivalent resistance and inductance.

The resistance and copper losses can be estimated as

$$R = \frac{0.022}{D_w^2} \ (\Omega/m), \tag{4.96}$$

$$P_w = 0.02 l_w \left(\frac{I}{D_w}\right)^2, \tag{4.97}$$

where D_w is the diameter of the wire, I is the rms current in the winding, and l_w is the winding length. l_w can be calculated by

$$l_w = N \pi D_{av}, \tag{4.98}$$

where D_{av} is the average diameter of the winding.

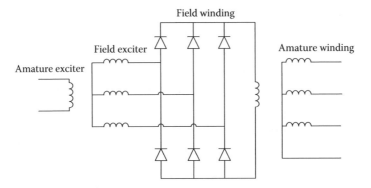

FIGURE 4.50 The schematic of brushless generator wound rotor with armature and field exciter systems.

Another effect that should be considered in the machine design is skin effect. Skin effect is the tendency of the current flow to be closer to the outer surfaces of the conductor. At low frequencies, the current distribution can be assumed as uniform. The copper wire thickness should be determined based on the output frequency.

Eddy current losses and hysteresis losses are the other criteria, which should be considered in machine-laminated core design. Iron-laminated core with higher permeability is preferable to increase the flux density and reduce the size of the generator.

Figure 4.51 shows the flux density, B, versus field strength, H for soft metal. Total output power is increased because high-speed generators (3600 rpm) have less losses in comparison to low-speed generators.

Thinner laminated and soft magnetic materials can be used to develop more efficient machine core design. This can also reduce the mechanical stress and heat losses in the laminated core.

4.3.5.4 Radial Flux PM Synchronous Generator for WEC

A radial magnetic flux electrical generator with a horizontal-axis water current turbine is presented in this subsection. The sketch of a ducted turbine surrounded by an electric generator, presented in Figure 4.52a and 4.52b [49].

Some older technologies are based on the coupling of an axial flow turbine with an electric generator similar to wind turbines. A gearbox is required to shift up the typical rotor speed (10–20 rpm) to the speed of a conventional generator (above 500–1000 rpm). However, direct drive is an attractive solution to eliminate the complexity, failure rate, efficiency, and maintenance cost [49].

In the directly driven turbine topology, the active parts of the machine are surrounded by the periphery of the blades instead of placing them on the axis of the turbine as in the classical methods. This idea is studied in [50–52], which have presented successful results for marine propulsion. Some of those rim-driven propellers are used for autonomous underwater vehicles or vessel propulsion. A similar machine has been tested with a 50-W rating turbine in [49].

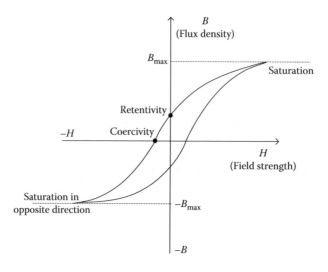

FIGURE 4.51   Flux density B versus field strength H.

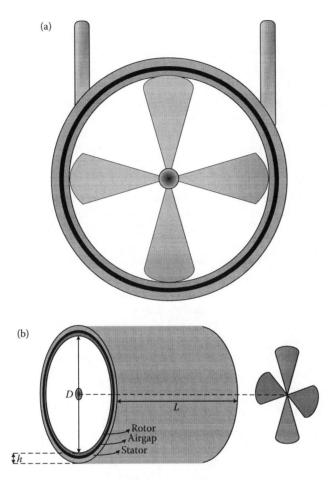

FIGURE 4.52 Radial magnetic flux electric generator that is placed on the periphery of the horizontal axis turbine blades, (a) frontal view and (b) lateral view.

Elimination of the gearbox is one of the benefits of the structure presented in [49]. The gearbox can consist of up to three stages, which brings complexity and is a heavy and expensive component. Directly driven application to a low-speed generator with the blades around results in high rated torque when compared to classical direct drive machines that are usually heavier and less efficient [53]. Also by keeping the generator in a protective duct, the hydrodynamic efficiency of the blades may increase and vibration and cavitation performance may increase [52]. Placing the active part of the generator around the blades results in increased gap diameter of the machine and the machine can have a higher torque. The torque of a machine coupled to a turbine generating the power of P with a rotor speed of Ω would be

$$T = \frac{P}{\omega}. \tag{4.99}$$

The rotor and the stator are separated with a gap in which the electromagnetic interaction occurs. The tangential force density σ_t in the gap depends on the machine technology and cooling system. Using the tangential force density, the electromagnetic torque can be expressed as

$$T = \frac{\sigma_t SD}{2},$$ (4.100)

where D is the gap diameter and S is the gap surface. For a given power and speed performance, increasing the gap diameter D will decrease the required surface S and the size of the machine as well. If the thickness h is independent of a chosen diameter D (this assumption is realistic for a given rotation speed), the required active volume (the surface times thickness, $S \cdot h$) shall decrease by increasing D [49].

Since the generator is a synchronous PM radial flux machine, it has less copper losses within the rotor in comparison to wound rotors. In induction machines a very thin air gap is required in order to limit the magnetizing currents and leakages, so the air gap in these dimensions is not comparable for the propeller diameter several meters long.

The synchronous machine designed in [49] is connected to a pulse width modulation (PWM) voltage converter to control the current wave of the stator and the rotation speed. It can be connected to the electrical grid or load bus through a power electronic interface.

The challenge on design is to achieve the minimum gap. This is due to the fact that the radial induction of the magnets that reaches the stator surface is inversely proportional to the magnetic gap thickness. The other challenge is the immersion of the gap. Letting the water flow through the gap minimizes the sealing problems in comparison to the other hub systems requiring a rotating seal [50–52]. This improves the thermal performance of the machine. On the other hand, active elements need to be covered with corrosion-resistant paint or fiber glass plus epoxy coating, since active elements are corrosive.

The electromagnetic model determines the main physical characteristics of the machine such as size, power rating, and diameters. The most effective parameter is the electromagnetic torque T, which is calculated as given in Equation 4.101 neglecting the mechanical losses. T can also be calculated as

$$T = k_{b1}\sqrt{2}A_L B_1 \frac{\pi D^2 L}{4} \cos \psi,$$ (4.101)

assuming a purely sinusoidal current in the stator. In Equation 4.101, A_L (A/m) is the rms value of the stator current, $B_1(T)$ is the peak value of the stator surface flux density, k_{b1} is the winding factor, and ψ is the angle between the vectors of induced electromagnetic force and stator current. Equation 4.101 is useful to determine the size of the generator active part if A_L, B_1, k_{b1}, and ψ are fixed at classical values for low-speed, high-torque machines [49]. If the angle is set to zero, the size can be minimized. This might increase the size and cost of the converter [54] but for the subwater application, it will be advantageous to reduce the machine size.

The flux density B is affected by the magnet type, gap thickness, and dimensional as well as financial constraints. The magnets have uniform radial magnetization and are surface mounted on the rotor as shown in Figure 4.53.

A simple analytical expression for B_1 as a function of dimensions and magnet characteristics is not easy because of the specific features of the machine. Irregular proportion of leakage flux occurs from one magnet to the next in the case of a large air gap in comparison to short pole pitch. Assuming a purely radial flux density may also give inaccurate results. Using a 2D model developed by Zhu et al. [55] solves the governing equations in the y-axis by separating the polar variables and considering the leakages. It predicts the open-circuit field distribution in the air gap of the surface-mounted PM machine without

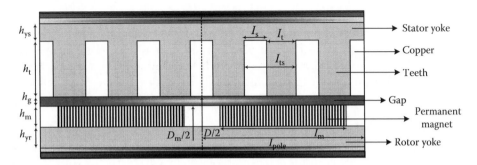

FIGURE 4.53 (See color insert following page 80.) Cross-sectional view of two poles of the generator.

slots. An expression of B_1 can be derived as follows and simplified assuming a large rotor diameter in comparison to the thickness and a high number of poles:

$$B_1 = \frac{2B_r k_\beta R_{sm}^p (1 - x^2 + 2x/p)}{(\mu_r + 1)(R_{sm}^{2p} - x^2) - (\mu_r - 1)(1 - x^2 R_{sm}^{2p})}, \quad (4.102)$$

$$x = \left(1 - \frac{h_m}{0.5D_R - h_g}\right)^p, \quad (4.103)$$

$$k_\beta = \frac{4}{\pi} \sin \frac{\beta \pi}{2}, \quad (4.104)$$

where B_r is the flux density remnant of the magnets, β is the ratio of magnet width to pole width (l_m/l_{pole}), R_{sm} is the ratio of air gap diameter to magnet outer diameter (D/D_m), p is the number of pair poles, μ_r is the magnets relative recoil permeability, h_m and h_g are the magnet and air gap heights, respectively. The slotless machine equivalent can be obtained by Carter factor [55]. Carter factor is a parameter to include the slotting effect on the air gap flux. If B_1, p, and R_{sm} are known h_m can be solved to calculate the required magnet height.

The slot height, h_s, can be calculated as follows for a given loading current A_L [54]:

$$h_s = A_L (J k_f k_s)^{-1}. \quad (4.105)$$

Here, J (A/m) is the density of the rms current, k_f is the slot fill factor, and k_s is the ratio of l_s to l_{t+s} (shown in Figure 4.53). Current density, J, depends on the thermal factors, which are improved by the water surrounding the generator parts in this application. Hence, the overall thickness of the machine can be reduced.

The rotor and stator yoke heights should be chosen to avoid the iron saturation since these heights affect the air gap size. Electric steels generally have a saturation flux density (B_{sat}) around 1.7 T. In order to limit the iron losses, B_{sat} is limited to 1.7 T in [49]. The flux densities produced by magnets and coils can be estimated by classical methods described in [53]. However, due to the large air gap to short pole pitch ratios, 2D finite difference simulations show that these equations miscalculate the dimensions. The mean value of the flux density into yokes is increased by the large proportion of leakage flux between magnets and coils. The yoke heights h_{yr} and h_{ys} can be readjusted by an iterative process [49]. However, the yokes remain thin and mechanical constraints for the integrity of the structure are the main factor for rating. The ratio of tooth width to slot pitch (l_t/l_{t+s}) can also be estimated such that the teeth do not saturate and their shape is also kept realistic. The air gap diameter, D, and the active length, L, can be obtained once the heights of the magnet and rotor yoke are calculated fixing the rotor diameter, D_R, and the gap, h_g.

In order to limit the iron losses the electrical frequency should be limited to a maximum of f_{max}. Thus, the maximum number of poles can be determined. The magnets have to be protected from demagnetization, particularly if low B_1 values are specified. It shows that the magnetic field cannot exceed its coercive field in any point of the magnet [49]. Therefore, the minimum number of poles can be determined.

The electrical part of the analytical model includes the desired voltage and current levels of the generator. Then design parameters such as machine dimensions and coil character-istics are determined. The back EMF in each phase can be derived using ($d\phi/dt$), where ϕ is the magnetic flux of one phase winding and the current can be calculated from the EM torque.

$$T = EI\Omega^{-1}\cos\psi. \tag{4.106}$$

The phase resistance, R, is also calculated by considering the endwindings geometry, taking each endwinding as a half circle with a diameter equal to the pole pitch [56]. The resistance is calculated using the copper conductivity with respect to the nominal temperature [49].

Then the output voltage and the power factor can be calculated as shown in Figure 4.54. The current and voltage levels affect the electronic drive circuit and the electrical network. Moreover, the machine time constant should be determined, which is a function of the converter switching frequency.

The losses and efficiency part of the analytical model can be roughly estimated, using the copper and iron losses of the machine as well as the copper losses. The copper losses are proportional to the coil resistance as

$$P_{Cu} = 3RI^2. \tag{4.107}$$

This equation includes the losses in the active part and endwindings. For a small axial length and large diameter, the endwinding losses may increase and dominantly affect the copper losses [49]. Iron losses can be estimated using loss per unit mass for the stator magnetic circuit as

$$P_{Fe} = P_{Fe_0}\left(f \cdot f_0^{-1}\right)^b \left(B_{max}B_{max_0}^{-1}\right)^c, \tag{4.108}$$

where f is the frequency of the field iron, P_{Fe_0} is the iron losses per mass unit at the given frequency of f_0, B_{max_0} is the flux density, and b and c are 1.5 and 2.2, respectively, for a typical high-quality Fe–Si-laminated steel. The mechanical losses are neglected in this model. Mechanical losses are affected by the choice of the bearing technology to compensate for the axial drag of the turbine and to keep the gap between the stator and rotor fixed.

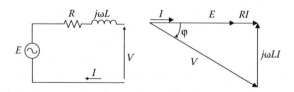

FIGURE 4.54   Electrical phasor diagram for $\psi = 0$.

The thermal model estimates the temperature in different parts of the machine. The thermal model is based on the heat transfer equations under steady-state conditions to derive the thermal resistances. Radial heat transfer represents the overestimation of the temperatures; however, it simplifies the model. In addition, heat transfer in the rotor is negligible due to the fact that water flowing through the gap is a better platform for heat transfer. There are two modes of heat transfer in the system: one is the conductional heat transfer in the solid parts of the machine and the other is the convectional heat transfer, which happens between stator internal and external surfaces and the ocean water.

A unity volume is considered as shown in Figure 4.55 for the conduction heat transfer calculations. The angular width, inner and outer radius, and axial length are α, R_i, R_o, and l, respectively. The temperature difference between inner and outer surfaces can be explained as

$$T_i - T_e = R_1 \phi_i + R_2 P, \qquad (4.109)$$

where T_i is the inner surface temperature, T_e is the outer surface temperature, ϕ_i is the inner flux, P is the volume losses, and R_1, R_2 are the thermal resistances.

R_1 and R_2 can be expressed as

$$R_1 = (R_0 - R_i)(\lambda R_i \alpha l)^{-1}, \qquad (4.110)$$

$$R_1 = (R_0 - R_i)(2\lambda(R_0 + R_i)\alpha l)^{-1}, \qquad (4.111)$$

with a large diameter.

Equation 4.109 corresponds to the electrical representation of the thermal network as shown in Figure 4.55.

The surface resistance is used within the model of the convective heat exchanges as

$$R_{cv} = (hS)^{-1}, \qquad (4.112)$$

where S is the heat exchange surface and h is the convective transfer coefficient.

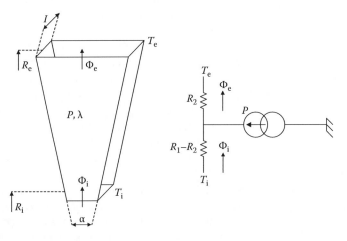

FIGURE 4.55 Unity volume and thermal equivalent circuit.

4.3.5.5 *Induction Machines for Ocean WEC*

In WEC systems, the last conversion is made by generators that convert mechanical into electrical energy. Generally, variable speed turbines are used in WEC systems. The variable speed causes a variable frequency output and the system will require a frequency converter, that is, a rectifier cascaded by an inverter for grid connection, if synchronous machines are used [57]. The drawback of using synchronous machines might be the requirement of a variable speed, constant frequency inverter system to be placed as an interface between the generator and the grid. An automatic synchronizer system is also needed to synchronize the generator output with the grid, during energy supply. In addition, to provide the field current, a variable DC source is necessary. If PMs are not used, brushes are needed. Brushes feeding the slip rings require regular maintenance.

On the other hand, induction generators running at an oversynchronous speed can supply the energy to the grid [57]. Frequency converters, DC field supply, and synchronization mechanisms might be eliminated by using induction generators. In addition, induction machines do not have brushes and they have squirrel cage rotor, which is capable of operating at higher speeds. Furthermore, induction generators require a simple control mechanism and they are more convenient for grid connection. However, they need reactive power from the grid to provide the excitation and magnetizing the generator. In stand-alone applications, large capacitor banks should be connected in parallel to the stator windings in order to provide reactive power to the induction generator.

There are some considerations to ensure the use of induction generators for WEC systems [57].

During a grid fault, the electrical output power will fall to zero but the mechanical power still will be applied to the generator with an accelerating torque. Thus, the turbine's speed will increase and mechanical protection is required. The generator should be protected from the corrosive effects of the seawater and atmosphere. The generator is subjected to be splashed by sea waves; therefore, the internal windings and other parts need to be protected from enclosure. The induction generator has to be designed for higher range of speed variations for coupling to air turbines.

Induction generators are very suitable to be coupled to the Wells-type turbines. Wells turbines can be categorized into two types: self-starting Wells turbines and non-self-starting Wells turbines. Depending on the turbine type, the induction generator coupling and requirements may vary [58]. In self-starting turbine applications, the induction generator does not need to be excited for initial motion start, while non-self-starting turbines need to be cranked by externally energized induction machines or other machines.

The induction machines are one of the most commonly used electric machines in the industry. Prior to advancements in active power electronic devices, fixed capacitors were used to provide reactive power for excitation of induction generators. This is not a flexible solution since orientation is constant all the time regardless of the load or speed deviations [59]. Excitation might be provided from the grid if induction generators are used within a large power system, in which infinite bus is available. Induction generators should be operated with adjustable excitation and isolation with appropriate control of power electronic switching devices.

The structure of an induction generator consists of two electromagnetic components. The first component is the rotating magnetic field constructed using bars with high conductivity and high strength, which are placed in a slotted iron cage. The second component is the stator winding and stator core. Figure 4.56 shows the cross-sectional view of a typical induction generator [59].

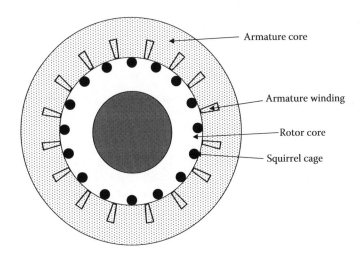

FIGURE 4.56 Cross-sectional view of induction machine structure.

The output of the induction generator is in the form of three-phase AC. The voltage can be controlled to maintain constant output voltage by adjusting the excitation current [59,60]. Thus, the output voltage can be fixed regardless of the load current and speed variations. A controller can be employed to determine the magnitude and frequency of the excitation current. The excitation current should be supplied to the stationary windings from which it is induced into the short circuited rotor windings.

In an induction machine, the synchronous speed and the angular velocity of the rotor can be expressed as

$$n_s = \frac{60}{p} f_n, \tag{4.113}$$

$$\omega_m = \frac{2\pi}{60} n_s, \tag{4.114}$$

where n_s is the synchronous rotations per minute, f_n is the nominal frequency, p is the number of pole pairs, and ω_m is the angular velocity of the rotor.

The mechanical torque produced by the turbines drives the rotor shaft, and this mechanical torque yields the mechanical input to the generator as

$$P_m = T_m \omega_m, \tag{4.115}$$

where P_m is input mechanical power and T_m is the mechanical torque.

The derivative of the angular velocity is

$$\frac{d}{dt}\omega_m = \frac{1}{2H}(T_e - F\omega_m - T_m), \tag{4.116}$$

where H is the combined rotor and inertia constant, T_e is the electromagnetic torque, and F is the combined rotor and turbine friction constant.

It should be noted that this equation is valid for both motor and generator operations of the induction machine. T_m is negative in the generator operating mode.

The angular rotor position is

$$\frac{d}{dt}\theta_m = \omega_m.$$ (4.117)

The relationship between the induced and terminal voltage of the stator winding is given by

$$V_S = E_S - \left(R'_S + j2\pi f L'_S\right) I_S,$$ (4.118)

where E_S is the stator voltage induced per phase, R'_S and L'_S are the combined rotor and stator resistance and inductance, respectively, and I_S is the stator current.

The apparent power output of the induction generator can be written as

$$S_n = 3 V_S I_S^*.$$ (4.119)

The active electrical power output of the machine can be calculated by

$$P_e = 3 \frac{E_S V_S}{\sqrt{R'^2_S + (2\pi f L'_S)}} \sin \delta,$$ (4.120)

where δ is the power angle between V_S and E_S.

4.3.5.6 Switched Reluctance Machines for Ocean WEC

Both linear and rotational switched reluctance (SR) machines are suitable for ocean wave applications, since they are brushless and do not have rotor windings. Therefore, they require less and easier maintenance. PM machines are also very suitable for ocean WEC; however, they are more expensive and have relatively bigger size.

The concept of the SR generator is based on the attraction of opposite poles that are magnetically charged [61]. The numbers of salient poles on the stator and rotor are generally unequal. Laminated electrical grade steel is used in their construction.

Cross-sectional view of a switched reluctance generator (SRG) is shown in Figure 4.57 [59], where there are eight stator poles and six rotor poles. 10/8 and 12/10 stator and rotor poles combinations are also possible.

The rotor does not have a winding in an SR generator. Armature coils are isolated from each other; they are concentric and located on stator poles [59]. The corresponding stator poles are magnetized, when the coils on opposite poles are excited such as rotor poles A–A and stator poles 1–1. Voltage is generated in the stator coil, resulting in power production, if the prime mover drives the rotor in the opposite direction.

The output voltage of the SR generator is a DC voltage with high ripples, which should be filtered and regulated by controlling the duration of excitation current.

Linear SR machines can also be suitable for ocean WEC applications.

Figure 4.58 shows a linear SRG cross-section [62], where the translator (moving part) is placed between two stator sides (fixed parts). The generator is a 4/6 machine, which means there are four stator teeth per six translator teeth. Instead of using 4/6 machine, 6/4 design could also be used but 4/6 configuration allows more space for copper in the stator.

In aligned position, the stator teeth face the translator teeth and in aligned position they do not face each other. The current increases to the desired level in the aligned position and steps down to zero in the unaligned position [62]. The flux density is limited by saturation in the

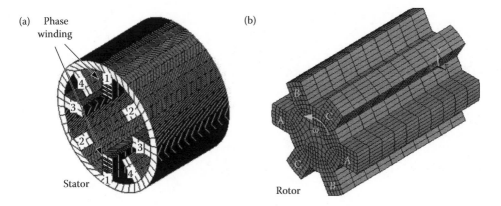

FIGURE 4.57 **(See color insert following page 80.)** 8/6 pole SRG cross-sectional view, (a) stator of the SRG and (b) rotor of the SRG.

aligned position. Flux in unaligned position is lower than the flux of aligned position. The average force can be calculated using the energy conversion area of the flux linkage–current diagram [63] as

$$F = \frac{W_{\text{cycle}}}{x_{\text{cycle}}} = \frac{1}{x_{\text{cycle}}} \int\limits_{\text{cycle}} i \, d\lambda. \tag{4.121}$$

Here, W_{cycle} is the energy converted in a cycle, calculated as the flux linkage–current loop area, x_{cycle} is the displacement in one cycle, and λ is the flux linkage.

4.3.5.7 Ocean Energy Conversion Using Piezoelectric/Electrostictive Materials

Instead of using linear generators or conventional rotating generators, piezoelectric/electrostictive materials can also be used for WEC applications. A new device named Energy Harvesting Eel (Eel) is developed by Taylor et al. [64] to convert mechanical energy of ocean waves into electricity for powering remotely located sensors or devices used for

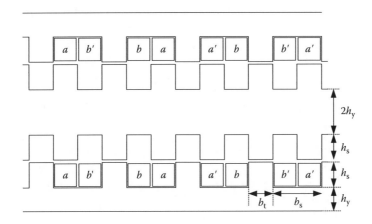

FIGURE 4.58 Linear switched reluctance machine cross-sectional view.

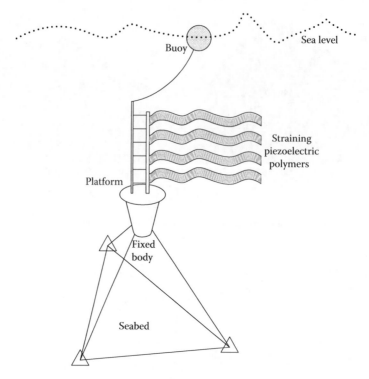

FIGURE 4.59 Eel system structure fixed to the sea bottom.

oceanographic measurements and sampling networks. Long strips of piezoelectric polymer undulates in the water flow are used in this energy harvesting method.

Figure 4.59 illustrates how the piezoelectric polymers are placed at the ocean bottom, using a platform and a fixed body [64].

Figure 4.60 is a schematic showing the forces affecting the Eel.

According to the flow speed and width of the bluff body, bluff body sheds alternating vortices on its either side with different frequencies. The resulting differential pressure caused by the vortices results in the oscillating motion of the Eel. Piezoelectric polymer generates a low-frequency AC voltage along the electrode segment due to the resulting strain on the

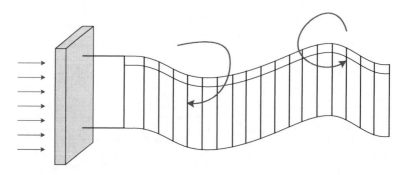

FIGURE 4.60 Eel movement behind the bluff body.

piezoelectric polymer [64]. This voltage can be converted to DC for nearshore transmission or battery charging or supplying power to the oceanographic devices or sensors.

There are multiple layers within the Eel. A central non-native layer (core) and an active layer of the piezoelectric material bonded to each side of the central layer are the three typical layers. The core is a thick and flexible polymer. This flexible structure increases the bending moment around which other layers move. Piezoelectric or electrostrictive materials are used in the generating layer construction. The strain in the active layers should be maximized in order to obtain more power [64].

The Eel undulating in water flow produces the electrical power, P, which is given by

$$P = \frac{\eta_1 \eta_2 \eta_3 A \rho V^3}{2}. \tag{4.122}$$

In this equation, η_1 is the hydrodynamic efficiency and depends on the oscillating frequency of the Eel and the frequency of the vortex shedding behind the bluff body. If these two frequencies match each other, hydrodynamic efficiency increases. In Equation 4.122, η_2 is the conversion efficiency of the strain into electrical energy in the piezoelectric polymer, η_3 is the efficiency of the electrical extraction out of the piezoelectric polymer, which is provided by the resonant circuit, A is the cross-sectional area, ρ is the water density, and V is the water flow velocity.

The flow-driven oscillation of the Eel is similar to a flag behind a pole on a windy day [65]. Strain can be calculated by visually recording the curvature of an Eel in a flow tank and electrically measuring the open-circuit voltage of the segment electrodes [64]. The Eel motion can be described either as a sum of natural frequency modes or as a traveling wave. Changing the Eel system parameters by changing bluff buoy width or Reynolds number in the flow affects the operation of Eel.

To convert the mechanical energy of the water flow into electrical energy, piezoelectric polymers are used. The power obtained from a piezoelectric material per unit volume, P_o, [66] is

$$P_o = \frac{\pi d_{31}^2 s^2 Y^2 f}{2\varepsilon\varepsilon_0}, \tag{4.123}$$

where d_{31} is the mechanical-to-electrical coupling parameter of the piezoelectric material, s is the mechanically strain percentage of the material, Y is the Young's modulus of the material, ε is the material's dielectric constant, and f is the straining frequency of the material. The electromechanical coupling efficiency is given by

$$k_{31}^2 = \frac{d_{31}^2 Y}{\varepsilon\varepsilon_0}. \tag{4.124}$$

Piezoelectric polymer PVDF has been utilized within the Eels. This polymer has advantages such as the following: it is resistant to a wide variety of chemicals and is mechanically strong. In addition, they can be manufactured as continuous reel, which allows them to be fabricated at low costs and different ranges of widths and lengths. The only drawback of the PVDF polymer for the Eel application is the relatively lower piezoelectric constant, d_{31}. This causes the lower conversion efficiency (k_{31}) and, as a result, a lower output power (P_o) produced by Eel [64].

Electrostrictive polymers are an alternative for conventional piezoelectric polymers in Eel for better performance. It is possible to induce a large piezoelectric field by keeping a high DC electrical bias field E_B. The induced piezoelectric effect is a result of Maxwell stress in the dielectric. AC mechanical stress applied to the electrically DC biased polymer causes dimensional differences in the electrostrictive dielectric. Thus, through the electrodes attached to the polymer surface, an alternating electrical charge flows. For an electrostrictive polymer, the effective d_{31} can be defined as

$$d_{31} = \frac{\varepsilon\varepsilon_0}{Y} \frac{V}{\text{th}}. \tag{4.125}$$

So the k_{31} efficiency becomes

$$k_{31}^2 = \frac{\varepsilon\varepsilon_0}{Y} \left(\frac{V}{\text{th}}\right)^2, \tag{4.126}$$

where V is the generated AC voltage and th is the electrostrictive polymer thickness.

Electrostrictive polymers have much higher values of d_{31} and k_{31} and hence output power (P_o) than piezoelectric polymer, PVDF. The electrostrictive polymers can harness more power when they have higher k_{31}^2 values [64].

However, there are some drawbacks of electrostrictive polymers such as weaker mechanical strength, low electrical breakdown strength, high dielectric losses, and lack of commercial sources with desired features. These drawbacks make electrostrictive polymers impractical for use in Eel.

The switched resonant-power conversion technique is used for electrical power extraction from the polymers. Using electrical and mechanical resonant systems, we can overcome the low coupling factor $k_{31}^2 = d_2^{31}(Y/\varepsilon)$ when piezoelectric devices are used as power generators. Eel motion has a very low frequency such as 1–2 Hz, so direct electrical resonance is not practical, since it requires very large inductor values. Therefore, switched resonant power conversion may overcome this limitation. Using this conversion, this technique is capable of high-efficiency operation at the 1–2 Hz range using reasonable inductor values [64].

Figure 4.61 shows the circuit schematic of the switched resonant power converter used for Eel energy harvesting [64].

In this circuit, R_P represents the dielectric loss resistance of the piezoelectric (PVDF), C_P is the Eel capacitance, and V_{OC} shows the open circuit voltage of the Eel with respect to

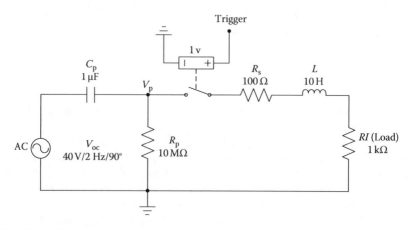

FIGURE 4.61 Switched resonant power converter.

an applied strain [64]. Eel can be modeled using these three fundamental elements. The other circuit components belong to the external power conversion circuit. These are the switch (S), inductor (L), with the series winding resistance (R_S) and R_L represents the load resistance.

It is assumed that the input to the piezoelectric element is a sinusoidal source. The switch S is turned on at the positive and negative peaks of input for exactly one-half cycle of the resonant period of the L–C_P network. The closure time for the switch is

$$T_C \approx \pi\sqrt{LC_P}. \tag{4.127}$$

T_C equals 7.18 ms for the given typical values. The voltage during the closure interval can be calculated as

$$V_C(t) = V_N \exp\left(-\frac{\omega_0 t}{2Q_L}\right)\cos(\omega_0 t), \tag{4.128}$$

with initial voltage, V_N, across the capacitor C_P. Here,

$$\omega_0 = \frac{1}{\sqrt{LC_P}} \tag{4.129}$$

and

$$Q_L = \frac{\omega_0 L}{R_L + R_S} = \frac{\sqrt{L/C_P}}{R_L + R_S} = \frac{R_0}{R_L + R_S}. \tag{4.130}$$

The current of the inductor (L) during this closure period can be calculated as

$$i_L(t) \approx \frac{V_N}{R_0}\sin(\omega_0 t). \tag{4.131}$$

At $t = T_C$ switch S opens and the voltage becomes

$$v_C(T_C) = -V_i \exp\left(-\frac{\pi}{2Q_L}\right) = -aV_i \quad (a \approx 1) \tag{4.132}$$

and

$$i_L(T_C) = 0 \tag{4.133}$$

when the switch opens. The "a" is defined as

$$a \equiv \exp\left(-\frac{\pi}{2Q_L}\right). \tag{4.134}$$

The switch remains open till the next peak (negative) occurs and then recloses. During this open period, V_C keeps charging more negatively because the strained piezo element produces a charge [64]. The variation of $V_C(t)$ during this period is

$$V_C(t) = V_P \cos(\omega_{IN} t) - (aV_N + V_P)\exp\left(\frac{t}{\tau}\right), \tag{4.135}$$

where the time constant of the dielectric loss network of the piezo is

$$\tau = R_P C_P, \tag{4.136}$$

and V_P is the peak value of the open-circuit voltage. S closes for the same period T_C at the next peak and behaves similarly except that $(V_i)+$ is now a higher value. Prior to the next closure period at

$$t = \frac{1}{2 f_{IN}} \omega_{IN} t = \pi, \tag{4.137}$$

the voltage V_C reaches a value V_{N+1}, which can be expressed as

$$V_{N+1} = - (V_P(1 + b) + ab V_N). \tag{4.138}$$

In Equation 4.138,

$$b \equiv \exp\left(\frac{-1}{2 f_{IN} \tau}\right) = \exp\left(\frac{-\pi}{Q_C}\right), \tag{4.139}$$

where

$$Q_C = 2 \pi f_{IN} \tau = \frac{1}{\tan \delta}. \tag{4.140}$$

This process repeats every half cycle till the steady-state condition is reached. It should be noted that the polarity of waveform changes in every half cycle. The steady-state condition occurs when the energy transferred to the load R_L plus losses in the inductor, electrodes, and PVDF dielectric is offset by the energy added by the piezoelectric device.

Equation 4.138 allows us to recursively predict the build-up voltage as a number of half cycles of the input. The build-up of the voltage requires approximately Q_L cycles. The steady-state power conveyed to the load can be determined by the steady-state voltage. The steady-state voltage can be obtained by setting V_{N+1} to V_N in Equation 4.138, which yields

$$(V_C)_{SS} = V_P \frac{1 + b}{1 - ab}. \tag{4.141}$$

A half sine current wave is applied to the load, which is produced by the peak voltage. The width of the current pulse is T_C and peak amplitude can be written as

$$(I_L)_{MAX} = \frac{(V_C)_{SS}}{R_0}. \tag{4.142}$$

Consequently, the average power output can be expressed as

$$P_{OUT} = ((I_L)_{MAX})^2 R_L T_C f_{IN}. \tag{4.143}$$

High-voltage build-up across the reactive elements depends on Q in a switched resonant power converter. The resonant frequency and the input frequency can be different in this switching technique. This is the most important advantage of this circuit. This results in the possibility of reduced component ratings.

If the loss elements can be minimized, the output power can be maximized. The losses are caused by the winding resistance of the inductor, the resistance of the electrode, and the dielectric. As in other resonant systems, the optimum value of the R_L will maximize the power output. Using Equations 4.131 through 4.133, the optimum value of the R_L can be obtained. If we assume that parameters a and b are both 1 for large values of Q_L and $Q_C(>5)$. Therefore, the steady-state value of $(V_C)_{SS}$ can be rewritten as

$$(V_C)_{SS} \approx \frac{2V_P}{\pi \, (1/2Q_L + 1/Q_C)} = \frac{4V_P Q_L Q_C}{\pi(2Q_L + Q_C)}. \tag{4.144}$$

Therefore, the power output becomes

$$P_{OUT} = (V_C)_{SS}^2 T_C f_{IN} \frac{R_L}{R_0^2} \approx \frac{16 C_P V_P^2 Q_L^2 Q_C^2 f_{IN} R_L}{\pi R_0 (2Q_L + Q_C)^2}. \tag{4.145}$$

The mechanical input power is

$$P_{IN} = \frac{C_P V_P^2 f_{IN}}{2k_{31}^2}, \tag{4.146}$$

where the mechanical coupling factor could be recalled as

$$k_{31}^2 = d_{31}^2 \frac{Y}{\varepsilon}. \tag{4.147}$$

Accordingly, the power conversion efficiency can be described as the ratio of P_{OUT}/P_{IN}

$$\eta = \frac{32k_{31}^2}{\pi R_0} \times \frac{Q_L^2 Q_C^2}{(2Q_L + Q_C)^2} R_L. \tag{4.148}$$

Q_L can be substituted with $R_0/(R_L + R_S)$ and maximizing with respect to R_L, the optimum value of R_L can be found as

$$(R_L)_{OPT} = R_S + \frac{2R_0}{Q_C}, \tag{4.149}$$

$$(Q_L)_{OPT} = \frac{R_0}{2\,(R_S + R_0/Q_C)}. \tag{4.150}$$

Integrating Equations 4.149 and 4.150 yields the maximized conversion efficiency as

$$\eta_{MAX} = \frac{8k_{31}^2 R_0}{\pi \, (R_S + 2R_0/Q_C)}. \tag{4.151}$$

Using the typical parameter values given in Figure 4.61, the maximum efficiency can be calculated as 37% for the piezoelectric polymer-based ocean energy harvesting.

4.3.6 Grid Connection Topologies for Different Generators Used in Wave Energy Harvesting Applications

The outputs of the generators generally have fluctuating characteristics due to the variable wave profiles. Thus, the generator outputs need further conditioning prior to grid connection or supply energy to the stand-alone loads. In this subsection, the different grid connection topologies with various generators and power electronic interfaces are discussed.

In this subsection, the grid interfaces for wave energy applications with linear and synchronous generators are described. As given in the general system-level diagram in Figure 4.1, grid or load interface topologies should be used with WEC applications. Figure 4.62 illustrates the system with several wave energy converter devices and an equivalent power electronic interface.

Instead of having a common interface technology for grid connection, individual power electronic interfaces can be employed for each WEC device. Placing several wave energy converters may reduce the output power fluctuations of the generators (represented by G). This may also reduce the need for energy storage devices; however, other methods should be implemented within grid interface technologies for better power smoothing and grid connection issues.

4.3.6.1 Grid Connection Interface for Linear and Synchronous Generator Applications

The wave energy harvesting applications involving generators are generally built in farms structure. Using several power generation units helps reduce the power fluctuations due to different periods and heights of waves [67].

A typical grid connection interface for use with generator applications is presented in Figure 4.63. Although using several units has an effect on obtaining better overall waveforms, further conditioning is necessary for grid connection. Several WEC units have output voltages with different amplitudes, frequencies, and phases. DC bus voltage variations can be lowered if many units are connected together. After the rectification stage, capacitor tanks should be placed for better suppression of the DC bus voltage variations. The capacitor bank acts as a short time energy buffer for the sustainability of energy transfer to the grid, if any of the units does not meet any incident wave for a short time. These capacitors are also used for the determination of the neutral point and the output filter of the inverter. The third stage consists of a six-pulse insulated gate bipolar transistor (IGBT) inverter. This inverter should be controlled to synchronize the inverter output voltage with grid quantities such as voltage amplitude and frequency. The synchronization is generally provided

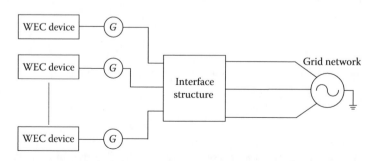

FIGURE 4.62 The interface structure for induction generator-based wave energy converters.

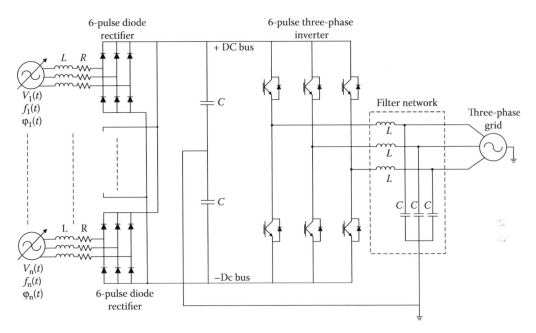

FIGURE 4.63 Grid connection interface for applications using different generators.

using the generator voltage, grid voltage, and the voltage difference between these two. After each phase–phase voltage comparison, the difference is processed for space vector computation for use in the inverter switching controller.

A low-pass LC filtering network is used in the fourth stage. The inductance and the capacitance values for this filter network should be determined to suppress the higher frequency components of the inverter output voltage. Thus, the cut-off frequency of the filter should be a little bit higher than the grid frequency to avoid filtering the fundamental frequency component. After the filtering stage, the output of the network is synchronized, filtered, and qualified for grid connection. This topology is suitable for applications with linear generators and three-phase synchronous generators and squirrel cage induction generators.

4.3.6.2 Grid Connection Interface for Induction Generator Applications

There might be several grid connection interfaces for the systems with induction generators such as an induction generator with a shunt connected static synchronous compensator (STATCOM), an induction generator in series with full converter, and a DFIG [68].

4.3.6.2.1 Induction Generator with Shunt-Connected STATCOM

An induction generator with a shunt-connected STATCOM is shown in Figure 4.64.

STATCOM is capable of injecting and absorbing reactive power to the network to overcome voltage fluctuations. STATCOM is a device with a bidirectional converter and a capacitor connected to the DC side [69]. The AC side of the STATCOM is used to compensate for the voltage fluctuations and low-voltage ride through (LVRT) [70]. A larger energy storage device can be employed at the DC side of the STATCOM for compensation of active power fluctuations originated by the natural behavior of wave power. The magnitude and duration of the voltage drop or swell, generator, and grid parameters are used to determine the rating of the STATCOM.

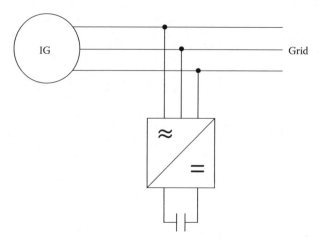

FIGURE 4.64 Induction generator with a shunt connected STATCOM.

The voltage fluctuations magnitude also depends on the strength of the grid and the impedance angle of the connection line. The greater impedance angle and weak grid cause large variations [68].

Direct torque control is not possible on the generator if the STATCOM is used as the interface topology. Hence, this topology can be enhanced with a hydraulic power take-off (PTO) system with an active control and energy buffering system such as a high-pressure accumulator. This is also a similar approach to the Salter Cam method in which a pressurized hydraulic fluid is used to drive the hydraulic motor, which in turn drives the electric generator [4]. This also acts as a mechanical energy buffer [68].

The schematic configuration of the hydraulic PTO system is shown in Figure 4.65. The hydraulic PTO is composed of a hydraulic piston, a high-pressure accumulator, and a hydraulic motor. The high-pressure accumulator smoothes the mechanical power fluctuations caused by the wave power absorbers; thus power input to the induction generator will be smoother but may have more or less fluctuations. The electrical storage may help in smoothing the residual power fluctuations [68]. However, it would be better to handle the power smoothing at the same conversion stage, if the dimensions and cost of the conversion and storage would stay in a reasonable range. STATCOM could also be used as an electrical

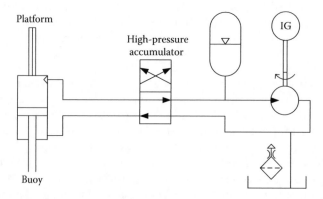

FIGURE 4.65 Energy buffer system using a hydraulic PTO system with induction generator.

power compensation device and provides some residual power smoothing effect due to the energy storage in the DC link [71].

4.3.6.2.2 Induction Generator in Series with Full Converter

Figure 4.66 shows the induction generator connected to a full converter as grid interface technology [68].

There are two cascaded voltage source converters, one of them is an AC/DC converter and the other one is a DC/AC converter. The coupling point of these two converters is associated by a DC link. The generator side converter provides magnetizing current required for generator excitation and to control the electromagnetic torque. The grid side converter compensates for the voltage, power factor, and power flow.

This full converter approach allows the variable speed operation of the generator and active control of the electromagnetic torque. Thus, voltage sags and power unbalance problems can be solved with increased capability of riding through the fault [68].

WEC having full converters cannot significantly contribute to the fault currents if a short circuit occurs in the network, because during transients, they cannot deliver more than a few times of rated current. In terms of LVRT, the optimum technology is a full converter, which has greater margin than the STATCOM [72]. However, the rating of the full converter depends on the flowing current, since they are in series and therefore the components are bigger in size and more expensive than that of the STATCOM.

The electromagnetic torque can be actively controlled by the vector control technique. The generator side converter can provide a latching force in the case of latching. A mechanical brake system could reduce the converter duty in a combined solution.

4.3.6.2.3 Doubly Fed Induction Generator

In the DFIG topology, the rotor windings are not short circuited. Hence, the generator should be wound instead of squirrel caged as seen from Figure 4.67. The outputs of the rotor windings are associated with an AC/DC converter, a DC-link capacitor, and another DC/AC converter for grid and generator stator windings connection points [73].

This converter connection is called back-to-back converter, which is in series with rotor windings and shunt on the grid wires. With the aid of the hydraulic smoothing stage, the generator speed does not change too much, so the converters have lower ratings in comparison to the full converter in series [73]. However, doubly fed topology is not suitable for direct drive. The rotor side converter is used to control the power output of the system and the voltage (or reactive power) measured at grid connection terminals. The grid side converter generates or absorbs the reactive power by regulating the voltage of the DC bus

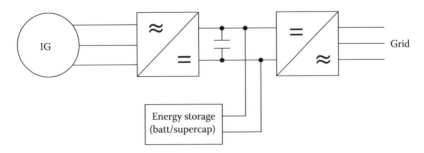

FIGURE 4.66 Induction generator in series with a full converter as grid interface technology.

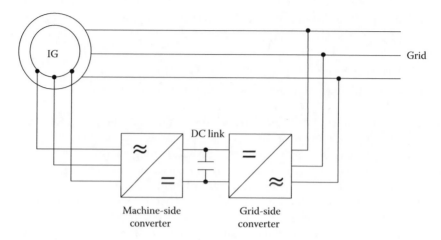

FIGURE 4.67   DFIG topology with back-to-back converters.

capacitor [74]. In other words, the rotor side converter controls magnetizing current and electromagnetic torque, while the grid side converter is controlled similar to STATCOM control [72]. In this topology, the rating of the converter limits the range of the variable speed operation. So it does not have as much capability as the full converter in the active control. The larger control margins can be reached by increasing the converter ratings, which in turn increase the cost. Thus, the full converter provides more flexibility of control.

The system performance can be the same as a fixed speed generator in a STATCOM [73]. This structure is very common in wind farms, and the same technology can be implemented in WEC systems with a hydraulic PTO system.

4.3.6.3   *Grid Connection Interface for SRG Applications*

An overall diagram of the grid connection interface for a SRG is given in Figure 4.68 [75].

The torque is a function of the angular position of the rotor due to the variable reluctance of SRGs. The phase currents of the SR generator should be controlled by the power electronic controller according to the certain positions of the rotor. In order to control the torque and transfer the available power to the grid, the magnitude and the waveform of the phase currents should be regulated by the power electronic interface. The safe operation of the

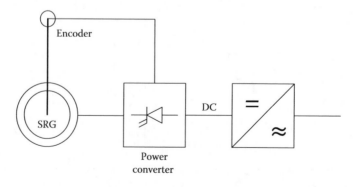

FIGURE 4.68   Overall grid connection diagram of an SRG.

generator is also provided by the power electronic converter. The converter is used to invert the voltages generated by the phases according to the certain positions of the rotor. The converter also provides the phase commutations of the voltages generated by multiphase SRG. The DC-link capacitor connected to the DC bus is used to reduce the oscillations of the output voltage.

The output power over each pulse exceeds the excitation by supplied mechanical power if the losses are omitted [76]. In this condition, the output power of the SRG can be calculated as

$$p(\theta, i_1, i_2, \ldots, i_n) = \left(\frac{1}{2} \sum_{j=1}^{n} \frac{dL_j(\theta)}{d\theta} i_j^2 \right) \omega, \tag{4.152}$$

where p is the instantaneous extracted power, n is the number of phases, j is the phase number, θ is the rotor position, ω is the rotor speed, i_j is the current of the phase j, and $L_j(\theta)$ is the inductance of phase j as a function of the rotor position. This equation is valid if the saturation is neglected and each phase is magnetically independent. The average output power equals the mechanical power without losses. Thus, the average power and the torque of the generator are

$$P = T_m \omega, \tag{4.153}$$

$$T_m = \frac{N_r}{2\pi} \int_0^{2\pi/N_r} \left(\sum_{j=1}^{n} \frac{1}{2} \frac{dL_j}{d\theta} i_j^2 \right) d\theta, \tag{4.154}$$

where P is the average power, T_m is the torque, and N_r is the number of rotor poles.

The grid connection interface of an SRG consisting of power electronic converters and the control circuitry are shown in Figure 4.69 [77]. There are two MOSFET transistors and two diodes per phase in the power circuit, which provide the maximum efficiency and control flexibility with minimum passive elements. The integrated diodes within the MOSFET switches additionally build a full bridge diode rectifier resulting in the shorter charge time of the DC-link capacitor.

The control system of the SRG power electronic interface consists of position sensing elements, synchronization circuitry, commutation logic, and PWM generators [77]. The components of the power electronic interface are shown in Figure 4.69. A magnetic encoder consists of a magnet ring commutator mounted on the rotor and two Hall-sensor ICs used for position detection. The PWM generators are current-mode controllers that are synchronized with digital circuitry and generator voltage and torque references [75]. The block diagram shown in Figure 4.70 presents a current control with torque reference applied to the SRG. The waveforms of the reference currents $(i_1^*, i_2^*, i_3^*, i_4^*)$ are obtained using the trapezoidal model torque $(T_1^*, T_2^*, T_3^*, T_4^*)$ corresponding to each phase.

The converter operates in three different modes, which are charging mode, voltage build-up mode, and generating mode. The DC-link capacitor is charged by the alternating EMF of the PM flux through the diode rectifier in the charging mode. In this mode of operation, the converter operates in a passive mode. The DC-link capacitor is charged up to the peak value of induced EMF of the phase, where the PMs are placed. This voltage varies according to the machine speed and PM flux power.

The capacitor voltage is increased up to the rated voltage during the voltage build-up mode. The MOSFETs are switched actively during this mode. Generally, the excitation

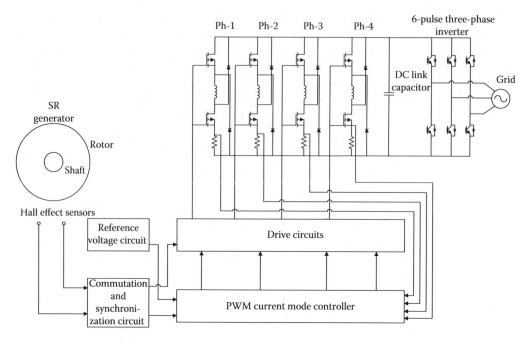

FIGURE 4.69 Power electronic converter and controller for SRG grid connection.

power is first supplied from the capacitor, while transistors are conducting. The capacitor voltage turns at a point slightly before and after the aligned position. So, the bulk of the winding conduction period comes after the alignment. Current begins to rise while the rotor poles approach the stator poles of the next phase, which is going to be excited [77]. The phase current increases with extracting energy from the DC-link capacitor till the switches are turned off. The turn off event happens at the commutation angle, which is determined by the control circuit.

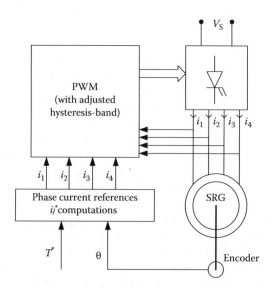

FIGURE 4.70 Block diagram of the operating principle of grid-connected SRG.

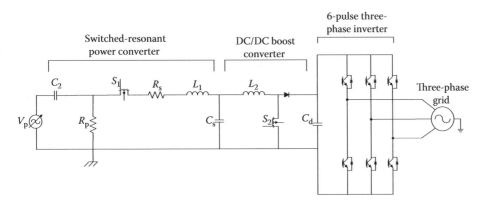

FIGURE 4.71 Power electronic interface schematic for piezoelectric/electrostrictive generators.

The current in the switches is commuted to the diodes, when the switches are turned off till the current reaches zero. The terminal voltage then reverses by the freewheeling diodes. During this defluxing period, the DC-link capacitor receives the returned generated power. The direction of energy flow switches from the machine to the capacitor resulting in the increase of capacitor voltage.

The energy returned during the defluxing period should be greater than the excitation energy supplied when the switches are on for sustainable generating action and to produce a voltage increase during the build-up mode [77]. The prime mover produces the difference between the electrical input energy and the electrical output energy. The energy flow can be regulated by the turning-on times of the switches.

4.3.6.4 Grid Connection Interface for Piezoelectric/Electrostrictive Power Generators

Although the power generated by piezoelectric/electrostrictive generators is relatively low, a power electronic interface can be used for grid connection for the applications composed of a large number of units. The power electronic interface consists of a switched resonant power converter, a DC/DC boost converter, and a DC/AC inverter.

The circuit schematic of this power electronic interface is illustrated in Figure 4.71.

The switched-resonant power converter is used for extraction of maximum energy from the straining materials. The operation principle of this circuit is described in detail in Subsection 4.3.4.5.7. The boost converter is connected to the output of the switched-resonant power converter. This DC/DC boost converter is used to step-up the output voltage of the piezoelectric units, which will be then inverted to the AC voltage. The duty ratio of the boost converter is determined according to the output voltage of the resonant power converter and the grid voltage effective value. The PWM control should be used for the inverter control in order to synchronize the output voltage in terms of modulation index, output voltage magnitude, phase angle, and frequency.

4.4 Wave Energy Applications

There are several types of technologies to convert ocean wave power into electricity. All technologies have advantages and drawbacks in terms of efficiency, maintenance, operation costs, establishment difficulties, and power rating. Some of the mainly used wave energy harvesting technologies are as follows: OWC, Overtopping devices, Pelamis, Wave Dragon, Archimedes wave swing (AWS) device, Wave Star Energy (WSE), and magnetohydrodynamics (MHD) generator.

OWC systems consist of a partially submerged structure, which forms an air chamber, with an underwater opening that allows the seawater to flow into the chamber. The volume of air inside the chamber is compressed as the water rises inside the chamber, driving air through a turbine. As the water level in the chamber subsides, the air is drawn back through the turbine. Bidirectional and self-rectifying air turbines have been developed. The axial-flow Wells turbine is the best-known turbine for this kind of application and has the advantage of not requiring rectifying air valves [78].

Overtopping devices: Overtopping devices guide incoming waves into a reservoir raised slightly above sea level, through a ramp. The water trapped in the reservoir flows back to the sea through a conventional low-head hydroelectric generator.

Float systems: Their common feature is a buoy that sits on the ocean's surface. The motion of this buoy is converted into electricity typically by a hydraulic PTO such as a hydraulic ram. These float systems have various shapes and forms.

Hinged contour devices: Hinged contour devices contain different floating sections, which are hinged together. As the wave passes, the sections move relative to each other and the hinges produce power. The power conversion uses hydraulic elements.

4.4.1 Oscillating Water Column

The OWC is one of the most common and most maturated WEC devices [79]. The conceptual studies on OWC started in the 1970s. Up to date, a number of plants were built in different countries such as Osprey, Scotland, and in Japan, which is called Mighty Whale.

The upcoming waves enter the water column, which is placed in the lower part of the chamber. This wave action on the water column occurs through the submerged entrance located on the device's front side [80].

As the water level increases, the air pressure in the upper part of the chamber increases. In other words, the oscillatory motion of the water causes a difference in air pressure within the chamber. The turbine is located in a channel, which links the air chamber to the outside. The axial airflow direction reverses when the wave level and the air pressure decrease [81]. The turbine should be specially designed in order to rotate in the same direction as air, since the airflow direction is bidirectional. Wells turbines are employed within the most full size OWC applications. These turbines are generally ranging from 500 kW up to 1 MW depending on the diameter sizes of 2 and 3.5 m. When employed together with flywheels, a reasonable kinetic energy can be stored in order to smooth the power fluctuations.

In a typical OWC application, the peak pressures vary in the range of 1.1–1.3 bars. Using relief and throttle valves, the pressure and airflow rate can be controlled and limited not to exceed critical values.

The greatest disadvantage of the OWCs is the large base structure, where its cross-sectional area varies around 100–400 m^2 with the chamber height of 10–20 m. As a result, the cost of a single device is rather high. Integrating several devices in a breakwater structure may increase the overall cost of the power plant. The most important advantage of the OWC is that the moving mechanical parts, that is, the turbine and the generator, are not in direct contact with water. Two large-scale OWCs are in service in Portugal (1999) and in the U.K. (2000), which are called Pico and Limpet power plants, respectively. Other OWCs were built in Japan, Australia, India, and Norway [80,82–85].

4.4.2 Pelamis

Pelamis is designed and applied by an Ocean Power Delivery Company, which is located in Scotland, U.K. The photograph of Pelamis is shown in Figure 4.72. Pelamis is a floating offshore hinged contour device [75].

The Pelamis consists of a long articulated structure with four cylindrical bodies [86,87]. These bodies are linked by hinged joints in series. These joints move under wave action, as the cylinders make the motions of pitch and yaw. The inner hydraulic rams pump high-pressure oil to activate hydraulic motors via smoothing accumulators. The electric generator is driven by these motors [88]. The stiffness of the joints can be adjusted to provide the tuning of the device according to the sea conditions. The reduced cross-sectional area helps the survivability of the device by limiting the drag forces. The longitudinal axis should be fixed parallel to the wave direction by moorings.

The side and plan view of the Pelamis is shown in Figure 4.73 for better understanding of its operation.

In Orkney, U.K., a full scale prototype was built and deployed for testing. In 2004, Pelamis delivered electrical power to the grid for the first time. This prototype is 150 m long and the rated power is 750 kW. The outer diameter of the cylinders is 3.5 m and three of four cylinders hosted inside can independently generate rated power of 250 kW.

The Pelamis wave energy converter is derived from earlier trials such as the Nodding Duck [6] developed by S. Salter in 1974 and Cockerell Raft designed by Sir C. Cockerell [4]. However, Pelamis offers a more cost-effective solution, which is available on market. In Portugal, a small wave power farm of three units with 2.25 MW power rating is under deployment [80].

A single floating Pelamis device can be moored at a water depth of 50–60 m. An umbilical riser cable connects the Pelamis to a junction box on the ocean floor. From this junction box,

FIGURE 4.72 (**See color insert following page 80**.) Pelamis WEC device at sea (Ocean Power Delivery Ltd). (Courtesy of R. Henderson, *Renewable Energy*, 31 (2), 271–283, 2006.)

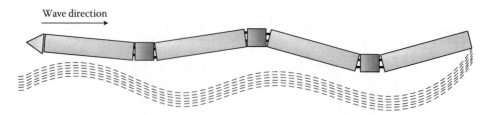

FIGURE 4.73 (**See color insert following page 80**.) General layout of the Pelamis WEC device.

a double-armored three-phase cable is buried into soft sediments along a 3 km route leading to the outfall of the effluent pipe, which is 1 km offshore. The cable is then routed through the 5 km effluent pipe to the International Paper Facility, which is about 4 km inland. An additional cable section connects to the Gardiner substation located next to the property of the International Paper facility.

As shown in Figure 4.74, the commercial system uses a total of four clusters, each one containing 45 Pelamis units (i.e., 180 total Pelamis WEC devices), connected to subsea cables. Each cluster consists of three rows with 15 devices per row. The other state designs are organized in a similar manner with four clusters. The number of devices per cluster varies such that each plant produces an annual energy output of 300,000 MWh/yr. The four subsea cables connect the clusters to shore as shown in Figure 4.75. The electrical interconnection of the devices is accomplished with flexible jumper cables, connecting the units in mid-water. The introduction of four independent subsea cables and the interconnection on the surface provide some redundancy in the wave farm arrangement [2].

Pelamis was used to establish the cost model for a commercial scale (300,000 MWh/yr) wave farm. Levelized cost components are shown in Figure 4.76 [78]. The cost breakdown shows that the impact on the cost of electricity of O&M (operation and maintenance)

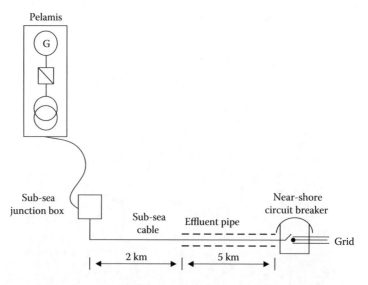

FIGURE 4.74 Electrical interconnection of Demo-Plant—in Oregon. (Redrawn from O. Siddiqui and R. Bedard. "Feasibility assessment of offshore wave and tidal current power production: A collaborative public/private partnership (Paper: 05GM0538)," EPRIsolutions, CA, *Proceedings of the IEEE Power Engineering Society 2005 Meeting Panel Session,* June 2005.)

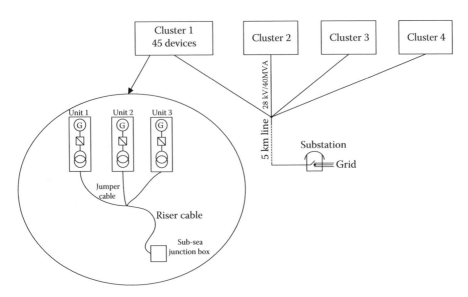

FIGURE 4.75 Electrical interconnection of Demo-Plant—Oregon Example.

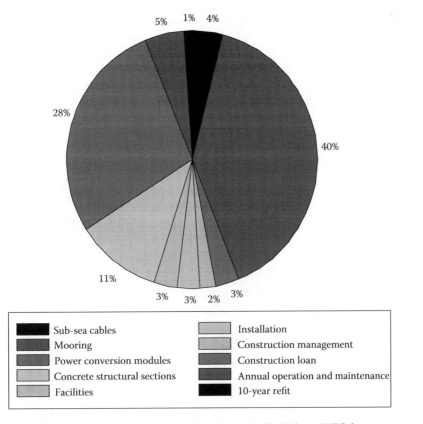

FIGURE 4.76 (**See color insert following page 80.**) Cost pie graph of the Pelamis WEC device.

is significant. The only way to bring the O&M costs down is by building demonstration projects. The second important impact of wave energy establishment is the power conversion modules.

4.4.3 Wave Dragon

The Wave Dragon is an overtopping device, which is developed by Wave Dragon Aps Company in Denmark [89]. Many other partner companies such as Spok ApS, Lowenmark Consulting Engineers FRI, Balslev, ESBI Engineering, and NIRAS AS contributed to the development processes. Wave Dragon is a floating offshore converter without any moving parts except the PTO system. The waves are channeled into a water reservoir with the aid of a wave reflector and a ramp. The reservoir is the main body of the device floating over the mean sea level. The operating principle of the Wave Dragon is shown in Figure 4.77.

The viewpoint of the Wave Dragon from the above is shown in Figure 4.78.

The reflector of the device welcomes the coming ocean waves and water is stored in the reservoir to flow through the turbine outlet channel. The potential energy is stored by taking the water into the reservoir. This potential energy helps in smoothing the generated power by regulating the water flow through the turbines. The reflector has two half-submerged rigid walls. These walls form a short narrowing channel and are wide open towards the sea and they are as high as the filling level of the reservoir. The power conversion stage includes several independent low-head water turbines [89]. These turbines are generally the Kaplan-type turbines discussed in Subsection 4.3.4.4.3 driving synchronous generators. The turbines are actively controlled so that the average amount of water is allowed to flow in order to reduce the power and voltage fluctuations. The front view of the device and the reflectors is shown in Figure 4.79, in which the reservoir is located between the reflectors to collect more amount of water carried by the waves.

In 2003, a prototype was deployed with seven turbines each with 20 kW rated power and driving individual PM generators. In 2004, this Wave Dragon started delivering power to the grid till 2005, which was damaged by a storm due to mooring failure [80].

4.4.4 AWS

Teamwork Technology located in the Netherlands is the developer company of the AWS, which is an offshore submerged device. The device is activated by the oscillations of static pressure caused by the surface waves. Figure 4.80 shows the operating principle of the AWS [80].

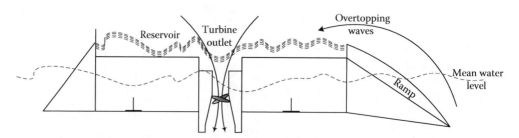

FIGURE 4.77 Operating principle of the Wave Dragon; waves overtop the ramp, water stored in a reservoir above the sea level and hydro turbines rotate as the water discharges.

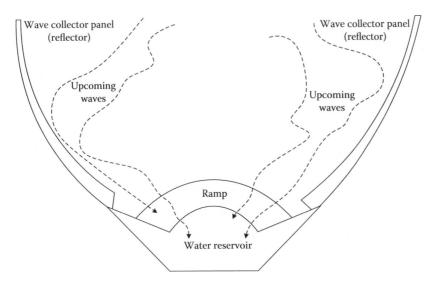

FIGURE 4.78 A bird's eye view of the Wave Dragon.

The AWS is an air-filled cylinder of which the floater is a body heaving up and down to or from the fixed bottom part. The waves create a pressure difference on the top of the device resulting in the force that moves the active part of the device. High water pressure causes the chamber volume to reduce if the wave crest is above the AWS. The floater heaves due to the action of the chamber pressure if the trough is above. Here "trough" represents the concave between the two wave crests. The behavior of the air in the chamber is similar to a spring with a variable stiffness by pumping water in or out of the chamber [80,91]. PM linear synchronous machines are used within the AWS applications as direct drive energy conversion devices. Using an auxiliary energy storage or conditioning, the efficiency of this type of energy conversion can be improved.

FIGURE 4.79 (**See color insert following page 80.**) Front view of the Wave Dragon with wave reflectors in the sides and the reservoir in the middle. (Courtesy of Wave Dragon ApS, available online at http://www.wavedragon.net/)

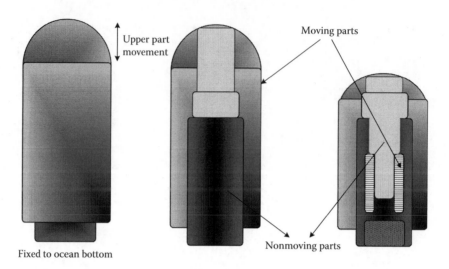

FIGURE 4.80 AWS operation principle.

The water dampers used within the AWS are actuated when the floating part approaches the mechanical end-stops. These dampers reduce the velocity of the floater and avoid the strong collision. Water dampers are also actuated with the electric generator when the floater does not have enough force required for adequate control of the movement of the AWS [91].

A 2 MW full-scaled AWS was built and deployed off the coast of Portuguese in nearly 40 m water depth [91]. The peak force the generator could have 1 MN (Mega-Newton) with 9 m width and 38 m height of the cylinder with a stroke of 7 m and maximum floater speed of 2.2 m/s.

The cylinder was placed within a steel cage on a fixed pontoon. On the fixed pontoon, there exist four ballasts, which have to be filled with water. The structure of the first version of AWS was large and expensive and was built to keep the first prototype safe in a place allowing just the floater to heave. The maintenance reduction via improvements of the generator design and building modular structures is being investigated by researchers [91].

4.4.5 Wave Star Energy

Wave Star is a device developed by Wave Star$^©$ Energy Company. Wave Star is fundamentally different from many other WEC devices. Instead of forming a barrier against the waves, upcoming waves are captured with an optimal wave movement angle. In this way, wave energy can be continuously captured, by the waves passing through the length of the machine. Wave Star$^©$ Energy is currently developing the first series production of a 500 kW wave energy converter [92]. Since July 2006, a small-scale grid-connected system has been in continuous operation. During the winter seasons from 2006 to 2008, there have been 12 major storms without causing any damage to the system. Testing of the first 500 kW device has been started in North Sea in late 2008 and it is expected that grid-connected Wave Star devices will be in operation in 2009.

There are 20 hemisphere-shaped floats that are partially submerged in the water along either side of the device as shown in Figure 4.81 [92]. The WSE is called a multipoint absorber, which means that a number of floating absorbers move upwards and downwards by an

FIGURE 4.81 View of the WSE, a multipoint absorber. (Courtesy of M. Kramer, "The wave energy converter: Wave Star, a multi point absorber system," Technical Report, Aalborg University and Wave Star Energy, Bremerhaven, Denmark.)

upcoming wave. In brief description, these movements activate the pumps, which press the hydraulic fluid to common transmission system, and drive a hydraulic motor.

When a wave rolls in, the floating absorbers in one side of the Wave Star are lifted up. Consequently, the floating absorbers of the other side will be lifted, until the wave subsides. There are individual hydraulic cylinders for each floating absorber. A piston in the cylinder applies pressure to the hydraulic fluid, inside the common transmission system of the device, when an absorber is lifted up. This pressure drives a hydraulic motor. This hydraulic motor drives a generator that produces electric power. Since the device has a length of several wavelengths, the system will operate continuously to harvest energy.

This device is storm-protected. During the storm periods, the absorbers can be lifted up to a safe position, that is, 20 m above the sea surface [92]. In Figure 4.82, the WSE is shown

FIGURE 4.82 (**See color insert following page 80**.) WSE position in storm protection mode. (Courtesy of M. Kramer, "The wave energy converter: Wave Star, a multi point absorber system," Technical Report, Aalborg University and Wave Star Energy, Bremerhaven, Denmark.)

in storm-protected position. Ahead of the device, a sensor on the sea surface measures the waves and ensures the automatic activation of the storm protection system. Moreover, absorbers can be controlled remotely through the Internet.

4.4.6 Magnetohydrodynamics Wave Energy Converter

MHD wave energy generator is a direct drive mechanical electrical energy converter and was developed by Scientific Applications & Research Associates (SARA) Inc. It is a cost-effective solution and promises to cut the costs of WEC systems by a factor of three. SARA Inc. is working on developing a 100 kW MHD unit [93,94]. This device is efficient and has high power density. The main benefit is that it is not composed of any conventional rotating electric generator. Although the output of the device has very high current and low voltage, using power electronic converters the output can be converted to practical and usable current and voltage levels.

The magnetohydrodynamics wave energy converter (MWEC) device couples the up and down motion of the heave caused by ocean waves. The wave motion is transferred by a shaft to the MHD generator that is deep under water. This shaft forces the conducting fluid through strong PMs. Therefore, this MWEC generator works on the principle that flowing seawater can conduct electric current by the aid of a strong magnetic field unlike the other MHD generators. In the operation, the seawater from the over passing waves is taken in. Then, the seawater flows through a hollow tube with flared inlet and outlet sections. These sections boost the water velocity according to the Bernoulli principle. According to this principle, if any fluid moves from a region of high pressure to a region of low pressure its speed increases. Vice versa, the speed of a fluid decreases if the fluid moves from a region of lower pressure to a region of high pressure. Strong electromagnets generate magnetic field perpendicular to the water flow. The electric current is stimulated in the passing seawater by the strong magnetic field. The electrodes placed in the tube are used to collect this electric current.

As claimed by the manufacturer, the MWEC system has several advantages such as high efficiency, capability of operating in shallow sea levels, compactness, rapid and easy deployment, high efficiency, and reliability. There are no gears, turbines, drive belts, bearings, or transmission systems. Therefore, capital needs and maintenance costs are low. MWEC is a quiet and environmental friendly device with zero greenhouse gas emissions [94].

4.5 Wave Energy in Future

Wave energy is a promising renewable energy source for future. Nearshore devices are expected to grow in size. However, due to their higher cost, longer installation period and shortage of available locations, offshore wave energy applications will be more feasible in the long term [95–97].

The wave energy development process consists of three phases. The first phase involves deployment of small-scale prototype devices. In the second phase with more funding from the government and investors, these prototypes will be improved and promising devices will be installed. In the final phase, the production of grid-connected, full scale wave energy converters will be installed in farm-style configurations. Hundreds of prototype devices have been proposed but only 20 of them stepped up to the second phase and some of them are really close to the final stage and commercial deployment [96].

In order to test the durability and efficiency issues, large-scale demonstration projects are required, which have not been available yet. However, systems should be tested in their final state to assess their potential.

Over the next 5 years, the United Kingdom is expected to play a very dominant role with a forecast capacity of 10.6 MW, which is nearly half the market share. Denmark, Australia, and Portugal are the other significant markets with some projected installations. However, they lag behind the U.K. With the reasonable levels of support from the government, a number of advanced wave technologies developed recently in these countries. The U.S. market also shows increased interest in wave energy technologies; however, more research and industry involvement is required [96].

4.6   Summary

This chapter focuses on the ocean wave energy harvesting in which the kinetic and potential energy contained in the natural oscillations of ocean waves is converted into electric power. Nearshore and offshore approaches along with required absorber, turbine, and generator types are discussed. Moreover, power electronic interfaces for grid-connection scenarios for each of the possible topologies are explained. In the final sections of the chapter, commercialized ocean WEC applications and the ocean wave energy development in the future are presented.

References

1. Ocean Energy: Technology Overview, Renewable Development Initiative, available online: http://ebrdrenewables.com/sites/renew/ocean.aspx.
2. O. Siddiqui and R. Bedard. "Feasibility assessment of offshore wave and tidal current power production: A collaborative public/private partnership (Paper: 05GM0538)," EPRIsolutions, CA, *Proceedings of the IEEE Power Engineering Society 2005 Meeting Panel Session*, June 2005.
3. J.N. Newman, "The interaction of stationary vessels with regular waves," *Proceedings of the 11th Conference on Naval Hydrodynamics*, London, 1976.
4. W.J. Jones and M. Ruane, "Alternative electrical energy sources for Maine, Appendix I, wave energy conversion by J. Mays," Report No. MIT-E1 77-010, MIT Energy Laboratory, July 1977.
5. B. Kinsman, *Wind Waves*, Englewood Cliffs, NJ: Prentice-Hall, 1965.
6. S.H. Salter, "Wave power," *Nature*, 249, 720–724, 1974.
7. S.H. Salter, "Characteristics of a rocking wave power device," *Nature*, 254, 504–506, 1975.
8. M. Wolley and J. Platts, "Energy on the crest of a wave," *New Scientist*, 66, 241–243, 1975.
9. H. Kayser, "Energy generation from sea waves," *Proceedings of the Ocean 1974 IEEE Conference*, Halifax, NS, Canada, Vol. 1, pp. 240–243, 1974.
10. J.D. Isaacs, G.L. Wick, and W.R. Schmitt, "Utilization of energy from ocean waves," *Proceedings of the Wave and Salinity Gradient Energy Conversion Workshop*, University of Delaware, pp. F1–F36, May 1976.
11. J.B. Isaacs, D. Castel, and G.L. Wick, "Utilization of the energy in ocean waves," *Ocean Engineering*, 3, 175–187, 1976.
12. M.E. McComick, "A modified linear analysis of a wave energy conversion buoy," *Ocean Engineering*, 3, 133–144, 1976.

13. J.D. Isaacs and R.J. Seymour, "The ocean as a power source," *International Journal of Environmental Studies*, 4, 201–205, 1974.
14. "Ocean Energy," Report of the U.S. Department of Interior Minerals Management Service, available at http://www.mms.gov/mmsKids/PDFs/OceanEnergyMMS.pdf, retrieved in May 2009.
15. S.H. Salter, D.C. Jeffrey, and J.R.M. Taylor, "The architecture of nodding duck wave power generators," *The Naval Architect*, London, 15, 21–24, 1976.
16. R.E. Dingwell, "Predictions of power production by a cam type wave energy converter for various locations," MSc dissertation, MIT, 1977.
17. L.S. Slotta, "Recoverable wave power concepts," ERDA Report No. COO-2946-1, 1976.
18. M.E. McCormik and Y.C. Kim, "Utilization of ocean waves—wave to energy conversion," *Proceedings of the International Symposium: Scripps Institute of Oceanography*, 1986.
19. N.A. Tornqvist, "Theoretical analyses of some simple wave power devices," *International Journal of Energy Research*, 2 (3), 281–294, 2007.
20. A.F. Richards, "Extracting energy from the oceans," *Marine Technical Society Journal*, 10 (2), February–March, 1976.
21. K. Budal, "Theory for absorption of wave power by a system of interacting bodies," *Journal of Ship Research*, 1977.
22. L. Duckers, "Wave power," *IET Engineering Science and Educational Journal*, 9 (3), 113–122, 2000.
23. F. Huang and W C. Beattie, "Modeling and simulation of the Islay wave power conversion system," *Proceedings of the IEEE International Electric Machines and Drives Conference Record*, pp. 2/2.1–2/2.4, May 1997.
24. C.E. Tindall and X. Mingzhou, "Optimising a wells-turbine-type wave energy system," *IEEE Transactions of Energy Conversion*, 11 (3), 631–635, 1996.
25. T.H. Kim, T. Setoguchi, M. Takao, K. Kaneko, and S. Santhaumar, "Study of turbine with self-pitch-controlled blades for wave energy conversion," *International Journal of Thermal Sciences*, 41, 101–107, 2002.
26. T. Miyazaki, "Utilization of coastal seas by floating wave energy device 'Mighty Whale'," *Proceedings of the European Wave Energy Symposium*, Edinburgh, pp. 373–378, 1993.
27. Y. Washio, H. Osawa, Y. Nagata, F. Fujii, H. Furuyama, and T. Fujita, "The offshore floating type wave power device 'Mighty Whale': Open sea tests," *Proceedings of the 10th International Offshore and Polar Engineering Conference*, ISOPE, Seattle, Vol. 1, pp. 373–380, 2000.
28. S. Santhakumar, V. Jayashankar, M.A. Atmanand, A.G. Pathak, M. Ravindran, T. Setoguchi, M. Takao, and K. Kaneko, "Performance of an impulse turbine based wave energy plant," *Proceedings of the Eighth International Offshore and Polar Engineering. Conference*, ISOPE, Vol. 1, pp. 75–80, 1998.
29. A.F. de O. Falcão, T.J.T. Whittaker, and A.W. Lewis, "JOULE II preliminary action: European pilot plant study," *Proceedings of the European Wave Energy Symposium*, Edinburgh, pp. 247–257, 1993.
30. M. Inoue, K. Kaneko, T. Setoguchi, and K. Shimamoto, "Studies on wells turbine for wave power generator (Part 4; Starting and running characteristics in periodically oscillating flow)," *Bulletin of JSME*, 29 (250), 1177–1182, 1986.
31. M. Brezovec, I. Kuzle, and T. Tomisa, "Nonlinear digital simulation model of hydraulic power unit with Kaplan turbine," *IEEE Transactions on Energy Conversion*, 21 (1), 235–241, March 2006.
32. T. Setoguchi, K. Kaneko, H. Hamakawa, and M. Inoue, "Some techniques to improve the performance of biplane Wells turbine for wave power generator," *Proceedings of the 1st Pacific/Asia Offshore Mechanics Symposium*, The International Society of Offshore and Polar Engineers, Seoul, Korea, Vol. 1, pp. 207–212, 1990.
33. T. Setoguchi, K. Kaneko, H. Taniyama, H. Maeda, and M. Inoue, "Impulse turbine with self pitch-controlled guide vanes connected by links," *International Journal of Offshore Polar Engineering*, 6 (1), 76–80, 1996.
34. T. Setoguchi, M. Takao, Y. Kinoue, K. Kaneko, and M. Inoue, "Study on an impulse turbine for wave energy conversion," *Proceedings of the Ninth International Offshore and Polar Engineering*

Conference, The International Society of Offshore and Polar Engineering, Brest, France, Vol. 1, pp. 180–187, 1999.

35. T.H. Kim, M. Takao, T. Setoguchi, K. Kaneko, and M. Inoue, "Performance comparison of turbines for wave power conversion," *International Journal of Thermal Science,* 40 (7), 681–689, 2001.

36. M. Hineno and Y. Yamauchi, "Spectrum of sea wave," *Journal of the Society of Naval Architecture.* Japan, 609, 160–180, 1980.

37. T. Omholt, "A wave activated electric generator," *Oceans,* 10, 585–589, 1978.

38. H. Luan, O. Onar, and A. Khaligh, "Dynamic modeling and optimum load control of a PM linear generator for ocean wave energy harvesting application," *Proceedings of the IEEE 34th Applied Power Electronics Conference and Exposition,* Washington, DC, February 2009.

39. M.A. Muller, "Electric generators for direct drive energy converters," *Proceedings of the IEE General, Transactions and Distribution,* 49 (4), 446–456, 2002.

40. M.A. Muller and N.J. Baker, "A low speed reciprocating permanent magnet generators for direct drive wave energy converters," *Proceedings of the Power Electronics, Machine Drives,* no. 487, pp. 16–18, April 2002.

41. A. Wolfbrandt, "Automated design of a linear generator for wave energy Converters-a simplified model," *IEEE Transactions. on Magnetics,* 42 (7), 1812–1819, 2007.

42. M. Leijon, H. Bernhoff, O. Ågren, J. Isberg, J. Sundberg, M. Berg, K.-E. Karlsson, and A. Wolfbrandt, "Multi-physics simulation of wave energy to electric energy conversion by permanent magnet linear generator," *IEEE Transactions on Energy Conversion,* 20 (1), 219–224, 2005.

43. J. Falnes, *Ocean Waves and Oscillation Systems,* Cambridge, U.K.: Cambridge University Press, p. 221, 2002.

44. K. Budal and J. Falnes, "A resonant point absorber of ocean waves," *Nature,* 256, 478–479, 1975 (with Corrigendium in vol. 257, p. 626).

45. D.V. Evans, "A theory for wave-power absorption by oscillating bodies," *Journal of Fluid Mechanics,* 77, 1–25, 1976.

46. J.N. Newman, "The interaction of stationary vessels with regular waves," *Proceedings of the 11th Symposium on Naval Hydrodynamics,* London, pp. 491–501, 1976.

47. G. Bertotti, A. Boglietti, M. Chiampi, D. Chiarabaglio, F. Fiorillo, and M. Lazzari, "An improved estimation of iron losses in rotating electrical machines," *IEEE Transactions in Magnetics,* 27 (6), 5007–5509, Nov. 1991.

48. C.S. Hoong and S. Taib, "Development of three phase synchronous generator for ocean wave application," *Proceedings of the Power Engineering Conference,* Vol. 15–16, pp. 262–267, December 2003.

49. L. Drouen, J.F. Charpentier, E. Semail, and S. Clenet, "Study of an innovative electrical machine fitted to marine current turbines," *Proceedings of the Oceans 2007 Europa,* 18–21, pp. 1–6, 2007.

50. S.M. Abu-Sharkh, S.H. Lai, and S.R. Turnock "Structurally integrated brushless PM motor for miniature propeller thrusters," *IEE Proceedings of the Electronics Power Applications,* 151 (5), 513–519, 2004.

51. O. Krovel, R. Nilssen, S.E. Skaar, E. Lovli, N. Sandoy, "Design of an integrated 100 kW permanent magnet synchronous machine in a prototype thruster for ship propulsion," *Proceedings of the ICEM'2004,* Cracow, Poland, pp. 117–118, September 2004.

52. M. Lea et al., "Scale model testing of a commercial rim-driven propulsor pod," *Journal of Ship Production,* 19 (2), 121–130, 2003.

53. A. Grauers, "Design of direct-driven permanent-magnet generators for wind turbines," PhD dissertation, Chalmers University of Technology, Goteburg, Sweden, 1996.

54. A. Grauers and P. Kasinathan, "Force density limits in low-speed permanent magnet machines due to temperature and reactance," *IEEE Transactions on Energy Conversion,* 19 (3), 518–525, 2004

55. Z.Q. Zhu, D. Howe, E. Bolte, and B. Ackermann, "Instantaneous magnetic field distribution in brushless permanent magnet dc motors, Parts I to IV," *IEEE Transactions on Magnetics,* 29 (1), 124–158, 1993.

56. H. Polinder van der Pijl, G.J. de Vilder, and P. Tavner, "Comparison of direct-drive and geared generator concepts for wind turbines," *IEEE Transactions on Energy Conversion*, 21 (3), 725–733, 2006.

57. P.V. Indiresan and S.S. Murthy, "Generating electrical power from the wave energy—the Indian experiment," *Proceedings of the 24th Energy Conversion Engineering Conference*, Vol. 5, pp. 2121–2126, August 1989.

58. A.L. Naikodi and G.S. Rao, "Efficient operation of AC voltage controller fed induction machine for wave power generation," *Proceedings of the International Conference on Power Electronics, Drives and Energy Systems for Industrial Growth*, Vol. 1, pp. 265–270, January 1996.

59. J. Vaidya and E. Gregory, "Advanced electric generator & control for high speed micro/mini turbine based power systems," patent pending.

60. P.C. Krause, O. Wasynczuk, and S.D. Sudhoff, *Analysis of Electric Machinery*, New York: Wiley–IEEE Press, 2002.

61. D.A. Torrey, X.M. Niu, and E.J. Unkauf, "Analytical modelling of variable-reluctance machine magnetisation characteristics," *IEE Proceedings—Electric Power Applications*, 142 (1), 14–22, 1995.

62. H. Polinder, B.C. Mecrow, A.G. Jack, P.G. Dickson, and M.A. Muller, "Conventional and TFPM linear generators for direct drive wave energy conversion," *IEEE Transactions on Energy Conversion*, 20 (2), pp. 260–267, 2005.

63. T.J.E. Miller, *Switched Reluctance Motors and Their Control*, Oxford, U.K.: Magna Physics Clarendon, 1993.

64. G.W. Taylor, J.R. Burns, S.M. Kammannm, W.B. Powers, and T.R. Welsh, "The Energy Harvesting Eel: a small subsurface ocean/river power generator," *IEEE Journal of Oceanic Engineering*, 26 (4), 539–547, 2001.

65. J. Allen, *Proceedings of the IUTAM Symposium Bluff Body Wakes and Vortex Induced Vibrations*, Marseille, France, June 13–16, 2000.

66. V. Schmidt, "Theoretical power output per unit volume of PVF and mechanical to electrical conversion efficiencies as function of frequency," *Proceedings of the IEEE Sixth International Symposium on Applications of Ferroelectrics*, Bethlehem, PA, June 8–11, 1986.

67. K. Thorburn and M. Leijon, "Farm size comparison with analytical model of linear generator wave energy converters," *Ocean Engineering*, 34, 908–916, 2006.

68. M. Molinas, O. Skjervheim, P. Anderson, T. Undeland, J. Hals, T. Moan, and B. Sorby, "Power electronics as grid interface for actively controlled wave energy converters," *Proceedings of the International Conference on Clean Electrical Power ICEEP*, pp. 188–195, May 2007.

69. N.G. Hingorani and L. Gyugyi, *Understanding FACTS; Concepts and Technology of Flexible AC Transmission Systems*, IEEE Press, New York, 2000.

70. M. Molinas, J. Kondoh, J.A. Suul, and T. Undeland, "Reactive support for wind and wave farms with a STATCOM for integration into the power system," *Proceedings of the International Conference Renewable Energy 2006*, Japan, October 2006.

71. J. Svensson, P. Jones, and P. Halvarsson, "Improved power system stability and reliability using innovative energy storage devices," *Proceedings of the Eigth IEE International Conference on AC and DC Power Transmission*, ACDC 2006, pp. 220–224, March 2006.

72. B. Naess, M. Molinas, and T. Undeland, "Laboratory tests of ride through for doubly fed induction generators," *Proceedings of the Nordic Wind Power Conference NWPC 2006*, Espoo, Finland, May 2006.

73. F. Blaabjerg, Z. Chen, and S.B. Kjaer, "Power electronics as efficient interface in dispersed power generation systems," *IEEE Transactions on Energy Conversion*, 19 (5), 1184–1194, 2004.

74. R. Pena, J.C. Clare, and G.M. Asher, "Doubly fed induction generator using back-to-back PWM converters and its application to variable-speed wind-energy generation," *IEE Proceedings–Electronic Power Applications*, 143 (3), 1996.

75. P. Lobato, A. Cruz, J. Silva, and A.J. Pires, "The switched reluctance generator for wind power generation," *Proceedings of the IEEE PES Conference Asociación Española para el Desarrollo de la Ingeniería Eléctrica*, 2009.

76. P.J. Lawrenson, J.M. Stephenson, P.T. Blenkinsop, J. Corda, and N.N. Fulton, "Variable-speed switched reluctance motors," *IEE Proceedings*, 127, Part B, (4), 253–265, 1980.

77. V. Nedic and T.A. Lipo, "Experimental verification of induced voltage self-excitation of a switched reluctance generator," *Proceedings of the IEEE Industry Applications Conference*, Vol. 1, pp. 51–56, 2000.

78. M. Presivic, "Wave power technologies (Paper 05GM0542)," Electric Power Research Institute, Palo Alto, CA., *Proceedings of the IEEE Power Engineering Society 2005 Meeting Panel Session*, June 2005.

79. R.S. Tseng, R.H. Wu, and C.C. Huang, "Model study of a shoreline wave-power system," *Ocean Engineering*, 27 (8), 801–821, August 2000.

80. H. Polinder and M. Scuotto, "Wave energy converters and their impact on power systems," *Proceedings of the International Conference on Future Power Systems*, pp. 1–9, November 2005.

81. O. AF Falcao, "The shoreline OWC wave power plant at the Azores," *Proceedings of the Fourth European Wave Energy Conference*, Aalborg, Denmark, pp. 42–48, 2000.

82. J. Falnes, *Ocean Waves and Oscillating Systems*, Cambridge: Cambridge University Press, 2002.

83. M. Vantorre, R. Banasiak, and R. Verhoeven, "Modeling of hydraulic performance and wave energy extraction by a point absorber in heave," *Applied Ocean Research*, 26, 61–72, 2004.

84. C. Josset and A.H. Clement, "A time domain numerical simulator for oscillating water column wave power plants," *Renewable Energy*, 32, 1379–1402, 2007.

85. J.M. Paixao Conde and L.M.C Gato, "Numerical study of the air-flow in an oscillating water column wave energy converter," *Renewable Energy*, 33, 2637–2644, 2008.

86. R. Yemm, R. Henderson, and C. Taylor, "The OPD Pelamis WEC, current status and onward programme," *Proceedings of the Fourth European Wave Energy Conference*, Aalborg Denmark, 2000.

87. Pelamis Wave Power Ltd., available online at http://www.pelamiswave.com

88. R. Henderson, "Design, simulation, and testing of a novel hydraulic power take-off system for the Pelamis wave energy converter," *Renewable Energy*, 31 (2), 271–283, 2006.

89. J.P. Kofoed, P. Frigaard, E.F. Madsen, and H. Chr. Soransen, "Prototype testing of the wave energy converter wave dragon," *Renewable Energy*, 31 (2), 181–189, 2006.

90. Wave Dragon ApS, available online at http://www.wavedragon.net/

91. D. Valerio, P. Beirao, and J.S. da Costa, "Optimization of wave energy extraction with the Archimedes Wave Swing," *Ocean Engineering*, 34 (17–18), 2330–2344, 2007.

92. M. Kramer, "The wave energy converter: Wave Star, a multi point absorber system," Technical Report, Aalborg University and Wave Star Energy, Bremerhaven, Denmark.

93. T.M. Rynne, "Ocean wave energy conversion system," U.S. Patent 5,136,173, to Scientific Applications & Research Associates, Inc., Patent and Trademark Office, Washington, DC, 1991.

94. MWEC, A power take off solution to convert the forces from ocean wave motion to usable electricity, Scientific Applications and Research Associates Incorporated (SARA Inc.), available online: http://www.sara.com/RAE/pdf/MWEC_infosheet.pdf

95. A.T. Jones and A. Westwood, "Recent progress in offshore renewable energy technology development," *Proceedings IEEE Power Engineering Society 2005 Meeting Panel Session*, June 2005.

96. A.T. Jones and A. Westwood, "Economic forecast for renewable ocean energy technologies," *Proceedings of the Energy Ocean 2004*, Palm Beach, Florida, 2004.

97. A.T. Jones and W. Rowley, "Global perspective: Economic forecast for renewable ocean energy technologies," *MTS Journal*, 36 (4), 85–90, Winter 2002.

5

Ocean Thermal Energy Harvesting

The ocean thermal energy conversion (OTEC) system is an energy-generating technology that takes advantage of the temperature difference between the ocean's shallow warm water and cold deeper water. OTEC uses the heat stored in warmer surface water to rotate the steam-driven turbines. Meanwhile, cold, deepwater is pumped to the surface to recondense the steam that drives the turbines [1]. Therefore, the energy is produced from the natural thermal gradient in the seawater layers. This gradient is a self-replenishable source of energy [2].

The density of water is a function of temperature and salinity. For pure water, the water's density reaches its maximum when the temperature is 39.2°F (4°C). Above that degree, the warmer water is lighter. 20°C of temperature difference may occur between the cooler deep level water and the warm surface water in the ocean due to the solar energy collected and stored within the ocean surface water.

The temperature difference between the warm surface water and the cold deepwater has to be more than 20°C (68°F), to have an efficiently operated OTEC system, capable of producing a significant amount of power. It is clear that to extract higher powers, higher temperature difference is required. OTEC has little impact on the surrounding environment. Another attractive feature of OTEC energy systems is that not only the end products but also several other by-products, such as desalinated water, include energy in the form of electricity.

The OTEC system is similar to a heat engine, which operates between the hot and cool water levels of the ocean. The OTEC system uses a fluid that circulates in a cycle, and with the aid of a heat exchanger, the heat of the hot water is absorbed by the working fluid. The fluid gets expanded after taking the heat and drives a turbine, which is mechanically coupled to an electric generator. After rotating the turbine, the working fluid needs to be condensed to continue the cycle.

The block diagram of a general closed cycle OTEC system is depicted in Figure 5.1 [2], which shows the evaporator, condenser, turbine, generator, and the other components used in an OTEC.

The drawback of an OTEC system is the low temperature difference between the hot and cold ends of the engine, which results in a very large volume of water requirement [3]. Therefore, the size and cost of the heat exchanger, evaporator, and condenser become larger. The installation costs of some of the OTEC power plants are really high in cost per kW criteria. The wind turbine costs are around 1/20th of an OTEC system [2]. However, with the technological developments, the OTEC system costs are expected to decrease.

Another drawback of the OTEC systems is the low efficiency, for example, the power required for the auxiliary equipment to pump the working fluid and the ocean water is close to the power produced by the OTEC system [4]. Beyond the "break even point," the net production can be obtained. Net power is the amount of power generated after

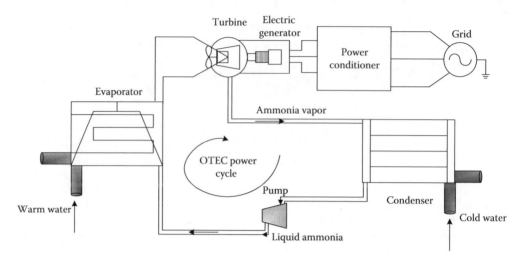

FIGURE 5.1   (See color insert following page 80.) A general overview of an OTEC system.

subtracting the power needed to run the system. The auxiliary power required to pump the cold water from deep sea layers reduces the net electrical power output of an OTEC system. Other parasitic power requirements are the warm water, working fluid pumping, excitation system, and power required by the control system of the OTEC. The cost is also high due to the long and large seawater pipes, pumps, floating platform for offshore applications, special design condensers, and evaporators. In addition, the biofouling, corrosion, and effective storms are the other difficulties of an ocean-based power plant [5]. Nevertheless, energy from the oceans is fuel free, with no emissions, and has low environmental impact.

The oceans cover 70% of the earth's surface, which makes them the world's largest solar collector. Everyday, a tremendous amount of solar energy is absorbed by the oceans. Therefore, OTEC is a vast renewable resource with the potential to produce billions of watts of electricity. However, the extraction of energy is expensive and only has the practical efficiency of about 2% [6]. The theoretical maximum efficiency of OTEC is about 7% [6].

The typical offshore and nearshore OTEC application examples are shown in Figures 5.2 and 5.3.

Usually, cold water intake is provided around 600–1000 m of ocean depth in offshore applications. A floating vessel placed on the sea surface has warm water intake from about 30 m of ocean depth. The floating vessel should be fixed to the seabed. Submarine cable is required for power transmission from offshore to the main land. The system can also be used for lighthouses or buoys. For nearshore applications, a vacuum vessel is required for cold water intake. The cold water intake pipe is relatively longer because of the depth from sea surface in nearshore systems when compared to offshore applications. This is due to the fact that nearshore seabed is much higher. However, it is easier to get the benefit of desalinated water in nearshore systems, which is a by-product of the OTEC systems in nearshore systems.

5.1 History

The concept of OTEC systems was proposed in the 1800s. In 1881, a French physicist, Jacques Arsene d'Arsonval, proposed tapping the thermal energy of the ocean. However,

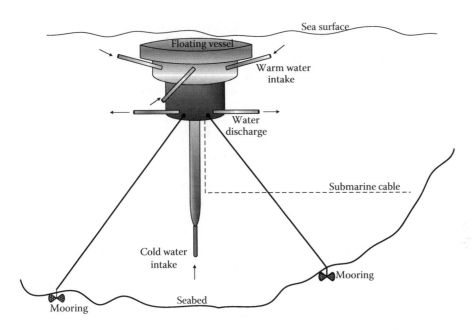

FIGURE 5.2 (**See color insert following page 80.**) An offshore OTEC application.

d'Arsonval's student, Georges Claude, built the first OTEC plant in Cuba, in 1930. The system produced 22 kW electricity with a low-pressure turbine [7].

In 1935, Claude constructed another plant, a 10,000-ton cargo vessel moored off the coast of Brazil. Weather and waves destroyed both the plants before they could generate net power [7].

In 1956, another 3 MW OTEC plant was designed by French scientists for Abidjan, Ivory Coast, West Africa. However, the plant was never completed, due to high cost, and in the meantime, large amounts of cheap oil became available in the 1950s, which made oil-power plants more economical [7].

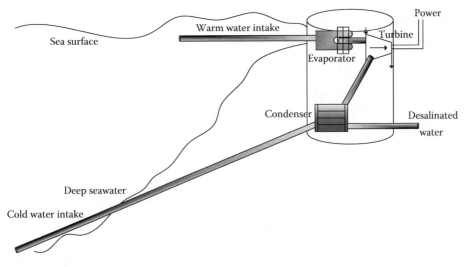

FIGURE 5.3 (**See color insert following page 80.**) A nearshore OTEC application.

In 1972, J. Hilbert Anderson and James H. Anderson, Jr incorporated the Sea Solar Power, Inc., working on the concept of sea thermal power. They focused on developing new, more efficient turbines, pumps, heat transfers (evaporators and condensers), the cold water pipe (CWP), and integrated system designs. In 1967, they patented their new "closed cycle" design [8]. The system was designed to be charged with propylene, a refrigeration fluid, which boils at 67°F under a pressure of 150 psi. Their design was similar to an oil rig offshore. Generated electricity was being transmitted to the shore from the plant-ship via underwater cables.

In the 1970s, the Tokyo Electric Power Company successfully built and deployed a 100 kW closed-cycle OTEC plant on the island of Nauru [9]. The plant started operation in 1981, produced about 120 kW of electricity; 90 kW was used to power the plant itself and used for auxiliary devices. The remaining power was used to power a school and several other places in Nauru [7].

In 1974, the United States started involving itself in OTEC research with the establishment of the Natural Energy Laboratory of Hawaii Authority, at Keahole Point on the Kona coast of Hawaii. Now, it is one of the world's leading test facilities for OTEC technology. Hawaii is often said to be the best location in the United States for OTEC, due to the warm surface water, and excellent access to very deep, very cold water. Moreover, Hawaii has the highest electricity costs in the United States [10], which makes this region attractive for renewable energy generation.

In 1998, another OTEC system project was started by the National Institute of Ocean Technology of India [11]. This 1 MW plant (gross power output) is a closed-cycle system operated with ammonia as the working fluid. It is located southeast of Tuticorin, South India, where the ocean depth of 1200 m is available from 40 km off the main land [11].

In 2002, a qualification test was applied to the plant. The computer analysis, modeling, studies, site surveys, and vendor selection were completed in the early months of 1999. Filling up the ammonia was completed in December 2000. The construction of the intake 1000 m long pipe was also completed in December 2000.

The plant is integrated on a floating barge, moored with an anchor at 1200 m depth. The cold water is taken from about 1000 m depth. The outside diameter of CWP is 1 m [11].

5.2 Classification of OTECs

5.2.1 Closed-Cycle OTEC Systems

5.2.1.1 Structure and Principles of Closed-Cycle Systems

Closed-cycle OTEC process was first proposed in 1881 by French physicist Jacques D'Arsonval [12]. Closed-cycle OTECs use a working fluid with a low boiling point, such as ammonia, to rotate a turbine of heat engine to generate electricity. Warm surface seawater is pumped through a heat exchanger where the low-boiling-point fluid is vaporized. The expanding vapor rotates the turbo-generator. Then, cold, deep seawater—pumped through a second heat exchanger—condenses the vapor back into a liquid, which is then recycled through the system, as shown in Figure 5.4.

In a closed-cycle system, the working fluid is pumped, evaporated, expanded, driven through a low-pressure turbine, and condensed. For the evaporation and condensation, the hot water pipe (HWP) and CWP are used for the respective water intakes. At all times, the

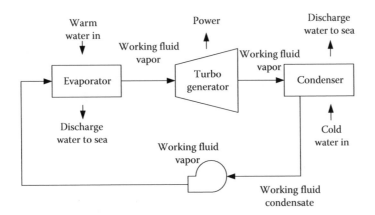

FIGURE 5.4 Operational block diagram of closed-cycle OTEC. (Redrawn from Natural Energy Laboratory of Hawaii Authority, USA, available at: http://www.nelha.org, March 2008.)

working fluid remains in a closed system and is continuously circulated following a Carnot cycle. This technology is essentially similar to the standard refrigeration or air conditioning systems; however, it operates in the reverse direction of the Carnot cycle. Refrigerators and air conditioners consume electricity to create temperature difference, whereas OTEC takes advantage of the temperature difference to generate electricity.

A high level of energy should be extracted per cycle over the temperature limits of 4–26°C due to the thermodynamic specifications of the working fluid used in the closed-cycle configuration [4]. The working fluid in a closed-cycle OTEC system is generally ammonia or propane.

In 1979, the Natural Energy Laboratory and several private-sector partners developed and tested the first 50 kW closed-cycle OTEC demonstration plant called mini OTEC. This OTEC plant is the first successful project which provides net electrical power. The mini OTEC vessel was moored 1.5 miles (2.4 km) off the Hawaiian coast and produced enough net electricity to illuminate the ship's light bulbs, and run its computers and televisions [14].

In 1999, the Natural Energy Laboratory tested another 250 kW pilot closed-cycle plant, the largest closed-cycle OTEC plant ever put into operation. To the best of our knowledge, since then, there have been no tests of OTEC technology in the United States.

5.2.1.2 Thermodynamic Principles of Closed-Cycle Systems

According to the first law of thermodynamics, the energy balance for the working fluid, as shown in Figure 5.5, is

$$W = Q_H - Q_C, \tag{5.1}$$

where Q_H is the amount of thermal energy absorbed from hot reservoir and Q_C is the amount of thermal energy absorbed from cold reservoir, which is negative.

In Figure 5.5, T_H and T_C are the absolute temperatures of the warm and cold sources.

According to the second law of thermodynamics, the system's efficiency is limited by the efficiency of the Carnot cycle. As shown in Figure 5.6, the typical Carnot cycle acting as a heat engine consists of four steps:

1. Isothermal expansion of the gas at temperature T_H (from A to B). During this step, Q_H quantity of heat is absorbed from the high-temperature reservoir.

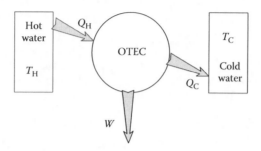

FIGURE 5.5   Schematic diagram of energy transfer.

2. Isentropic expansion of the gas at entropy S_B (from B to C).
3. Isothermal compression of the gas at temperature T_C (from C to D). During this step, Q_L quantity of heat flows out of the gas to the low-temperature reservoir.
4. Isentropic compression of the gas at entropy S_A (from D to A).

Here, S is the entropy. Based on Rudolf Clausius, the entropy is defined as

$$dS = \frac{dQ}{T}.$$ (5.2)

The amount of thermal energy transferred between the hot reservoir and the system is

$$Q_H = T_H(S_B - S_A).$$ (5.3)

The amount of thermal energy transferred between the cold reservoir and the system is

$$Q_C = T_C(S_B - S_A).$$ (5.4)

The maximum possible efficiency would be

$$\eta = \frac{W}{Q_H}.$$ (5.5)

Therefore,

$$\eta = \frac{Q_H - Q_C}{Q_H}$$ (5.6)

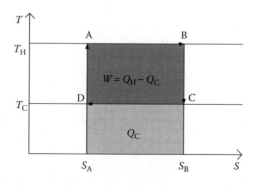

FIGURE 5.6   Temperature–entropy diagram for an ideal Carnot cycle.

or

$$\eta = \frac{T_H - T_C}{T_H} = 1 - \frac{T_C}{T_H}. \tag{5.7}$$

Equation 5.6 defines the principles of the Carnot efficiency. Based on this equation, it can be concluded that ideally the maximum thermal efficiency (or the thermal efficiency of a reversible heat engine) depends only on the temperature of the two thermal-energy reservoirs involved. Therefore, seasonal sea temperature variations may influence the overall OTEC power generation rates.

For warm seawater at 77°F (298.15°K) and cold seawater at 45°F (280.37°K), the Carnot efficiency is $\eta = 0.060$.

A generalized thermodynamic cycle is shown in Figure 5.7.

Based on this cycle

$$Q_H = \int_H T_H \, dS \tag{5.8}$$

and

$$Q_C = \int_C T_C \, dS. \tag{5.9}$$

Therefore, the net thermodynamic cycle work becomes

$$W = \int_H T_H \, dS - \int_C T_C \, dS \tag{5.10}$$

or

$$W = \oint T \, ds. \tag{5.11}$$

Consequently, the efficiency of a nonideal OTEC system would be

$$\eta = \frac{Q_H - Q_C}{Q_H} < 1 - \frac{T_C}{T_H}, \tag{5.12}$$

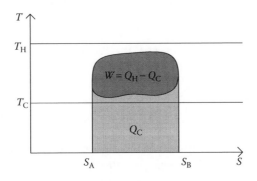

FIGURE 5.7 Temperature–entropy diagram for a generalized thermodynamic cycle.

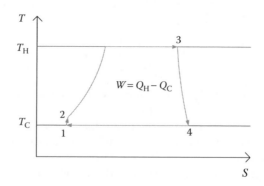

FIGURE 5.8 Temperature–entropy diagram for a Rankine cycle.

where η usually is less than 6%. Another reason for this low efficiency is the operation of pumps in transfer processes. They consume about 20–30% of the power generated by the turbine generator [11]. For other various practical reasons, 3% is the typical efficiency that can be reached.

For an OTEC system, the typical cycle used is the Rankine cycle, as shown in Figure 5.8. In a Rankine cycle, there are four processes:

1. The working fluid is pumped from low to high pressure. Since the working fluid is liquid at this process, the pump requires a small amount of input energy (from 1 to 2).

2. The high-pressure liquid enters a boiler, where it is heated at constant pressure by the warm surface seawater to become dry saturated vapor (from 2 to 3).

3. The dry saturated vapor expands through a turbine, generating power. This process decreases the temperature and the pressure of the vapor. Approximately, it is an isentropic expansion (from 3 to 4).

4. The vapor then passes through a condenser, where it is condensed at a constant pressure by cold water to turn into liquid form (from 4 to 1).

5.3 Technical Obstacles of Closed-Cycle OTEC Systems

5.3.1 Working Fluids and Its Potential Leakage

The working fluid should have a low boiling point, high density, and high pressure. Ammonia and fluorinated carbons are popular choices [15]. Ammonia has superior transportation properties, easy availability, and low cost; however, it is toxic and flammable. On the other hand, fluorinated carbons are potential threats to the ozone layer. Hydrocarbons like ethane, propane, or butane are other appropriate options, but they are also flammable.

A mixture of working fluid, composed of 1% ethane, 98% propane, and 1% normal butane, on a mole percentage basis is proposed as an alternative working fluid [16]. The mixtures can vaporize and condense at varying temperatures in constant-pressure processes, while pure fluids vaporize and condense at constant temperature in constant-pressure processes,

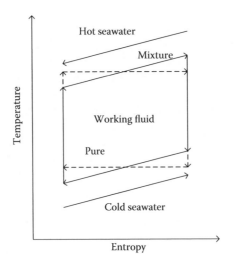

FIGURE 5.9 Comparison of mixture and pure fluid.

as shown in Figure 5.9. The dashed lines represent pure working fluid, whereas the solid lines represent mixture working fluid.

With the assumption of 10°F inlet temperature difference and 3°F exit temperature difference for both the evaporator and condenser, the log mean temperature differences (LMTDs) of the pure fluid cycle is 90% of that of the mixture cycle [16].

Another major advantage of mixtures over pure fluids is that under progressive fouling, the mixture cycle possibly can be maintained operative for a longer time when compared to pure fluids, as shown in Figure 5.10, since the driving force for heat transfer is uniform throughout the heat exchanger for the mixture cycle [16].

In Figure 5.10, ΔT is the temperature difference across the water-side fouling. It increases during the fouling progress.

The evaporator, turbine, and condenser operate in partial vacuum ranging from 1% to 3% atmospheric pressure. Therefore, the system must be carefully sealed to prevent the potential leakage of working fluid and potential in-leakage of atmospheric air.

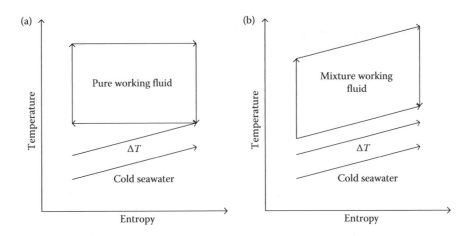

FIGURE 5.10 Comparison of mixture and pure fluid with biofouling.

5.3.1.1 Degradation of Heat Exchanger Performance by Microbial Fouling

Degradation of surfaces by biological entities and deposition of living matter on intake pipes is collectively termed as biofouling. In a closed-cycle OTEC system, since seawater must pass through the heat exchanger, the microbial fouling could degrade its thermal conductivity and affect its performance. It depends on several factors such as water temperature, construction material of the heat exchanger, and water nutrient level [17].

In a study in 1977 [18], some mock heat exchangers were exposed to seawater for 10 months and researchers discovered that although the level of microbial fouling was low, the thermal conductivity of the system was significantly impaired. It is concluded that the apparent discrepancy between the level of fouling and the heat transfer impairment is the result of a thin layer of water trapped by the microbial growth on the surface of the heat exchanger.

Another study, conducted in 1985 at Hawaii, concluded that microbial fouling layers as thin as 25–50 μm can degrade heat exchanger performance by as much as 40–50% [19]. The conclusion is that although the physical cleaning, even simple brushing or passing sponge rubber balls through the cubes, can decrease the rate at which fouling occurs, it is not enough to completely halt microbial growth [19]. Furthermore, it is found that the microbes began to grow more quickly after cleaning under the selection pressure [20].

Nickel or zinc plating can be used for biofouling reduction. However, nickel-based designs cause thicker channel walls, which may reduce the efficiency of the heat exchanger [21]. In addition, instead of taking the actual sea surface water, the HWP intake could be placed at some point below the actual sea surface, for example, 30 m depth, to solve the biofouling problem, because there are less microorganisms there than on the sea surface. However, deeper sea layers have cooler water resulting in OTEC power reduction, due to the lower temperature difference.

Chlorination is another alternative. Berger's study examined this approach and concluded that chlorination levels of 0.1 mg/L treated for 1 h per day slowed microbial growth appreciably and may show effective results in the long-term operation of a plant [19]. However, there is another consideration associated with the environmental impact of discharging chlorination into the ocean. Fortunately, experiments conducted at the Natural Energy Laboratory of Hawaii have demonstrated that very small, environmentally benign, levels of chlorine can successfully control the microfouling [22].

5.3.2 Thermal Energy Conversion for OTEC Systems

The main basis for OTEC is the mechanical power P_m available from the process, which is expressed as [4]

$$P_m = P_{\underset{\text{evaporator}}{\text{from}}} - P_{\underset{\text{CWP}}{\text{from}}}. \tag{5.13}$$

The temperature relationships of the intake temperature of the HWP and CWP can be explained as

$$T_{ai} = T_\omega - dT_\omega - dT_a, \tag{5.14}$$

$$T_{af} = T_c - dT_c - dT_a, \tag{5.15}$$

where T_ω is the intake temperature of the HWP, T_c is the intake temperature of the CWP, T_{ai} and T_{af} are the initial and final temperatures of the working fluid, respectively. In

Equations 5.14 through 5.16, d is the temperature loss factor associated with the HWP and CWP intake temperatures and varies from 0 to 1.

The difference between the final and initial temperatures of the working fluid can be calculated as

$$T_{af} - T_{ai} = (T_\omega - T_c)(1 - d). \tag{5.16}$$

The entropy intake and discharge rates, S_i and S_f, can be expressed as

$$S_i = \frac{P_{\text{from evaporator}}}{T_{ai}} \tag{5.17}$$

and

$$S_f = \frac{P_{\text{from CWP}}}{T_{af}}. \tag{5.18}$$

Using a lumped loss coefficient L, the actual power that can be extracted can be expressed as

$$P_{\text{output}} = LS_i (T_{ai} - T_{af}). \tag{5.19}$$

Generally, L varies around 0.75–0.85. The electrical conversion efficiency is usually in the range of 90% of P_{output}. In order to achieve the cited cold seawater temperature, 600–1000 m depth ranges are required for typical CWPs.

5.3.3   Open-Cycle OTEC Systems

5.3.3.1   *Structure and Principles of Open-Cycle Systems*

In an open-cycle OTEC process, which is shown in Figure 5.11, seawater functions as the working fluid. The boiling temperature of water is a function of pressure [13]. It drops as the pressure decreases. The first step is boiling the warm shallow water by placing it in a low-pressure container of about 2% atmospheric pressure at sea level. Then the expanding steam drives a low-pressure turbine coupled to an electrical generator. The

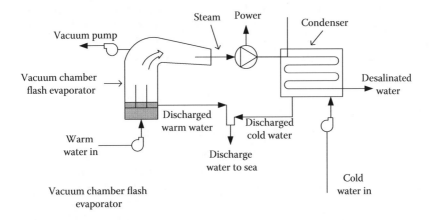

FIGURE 5.11   Block diagram of open cycle.

steam, which leaves its salt and contaminants behind in the low-pressure container, is desalinized. Afterwards, it is chilled and condensed back into a liquid by exposure to cold temperatures from deep ocean water. The by-product, desalinized fresh water, which is suitable for human consumption or irrigation, is valuable especially in local communities, where natural freshwater supplies are limited.

In contrary to the closed-cycle OTEC systems, seawater is used as the effective working fluid in the open-cycle OTEC systems. Some of the warm seawater intake is flashed and boiled by bringing it to a low-pressure chamber (vacuum). The resultant expansion of the steam drives a very low-pressure turbine. The cold seawater intake from the ocean deep is used for condensation. The vacuum vessel is coupled to the low-pressure steam turbine within the low-pressure environment. The steam produced in the vacuum vessel is desalinated; therefore, the condensed discharge water is also desalinated.

The heat exchangers are partially eliminated in the open-cycle design, which results in eliminating the losses associated with the heat exchanger, which is the main advantage of the open-cycle method. However, direct use of the seawater as working fluid and the need for a special design for the vacuum vessel are the disadvantages of the open-cycle system [4].

In 1993, the largest open-cycle OTEC plant was designed by Pacific International Center for High Technology Research. The plant was constructed and operated at Keahole Point, Hawaii, which is a 210 kW plant [23]. Considering the seawater pumps and vacuum systems' electricity consumption of about 170 W, the nominal net output of this experimental plant was about 40 kW. Following the successful completion of experiments, this open-cycle OTEC plant was shut down in January 1999 [24], since it was established for these experiments only.

5.3.3.2 Technical Difficulties of Open-Cycle OTEC Systems

Most of the drawbacks of open-cycle OTEC systems are due to the great turbine sizes. Since the turbine operates at a very low-pressure condition, ranging from 1% to 3% of atmospheric pressure, open-cycle systems require very large turbines to capture relatively small amounts of energy. Georges Claude, the inventor of the open-cycle process, calculated that a 6 MW turbine would need to be about 10 m in diameter. Recent re-evaluation of Claude's work [25] indicates that modern technology cannot improve his design, significantly. Therefore, it seems that the open-cycle turbines are limited to 6 MW, unless some new specialized turbines are developed, which may utilize fiber-reinforced plastic blades in rotors with diameters bigger than 100 m. With current technology, increasing the gross power-generating capacities of a Claude cycle plant above 2.5 MW will incur significant increase in its complexity and cost, and reduce its efficiency [15].

5.3.4 Hybrid Cycle OTEC Systems

5.3.4.1 Structure and Principles of Hybrid OTEC Systems

Another option is to combine the two processes together into an open-cycle/closed-cycle hybrid, which combines the features of both the systems, as shown in Figure 5.12. In a hybrid OTEC design, both seawater and other fluids such as ammonia are used as working fluids [26,27].

In a hybrid OTEC system, warm seawater first enters a vacuum chamber, where it is partly flash-evaporated into steam, similar to the open-cycle evaporation process. Then the steam

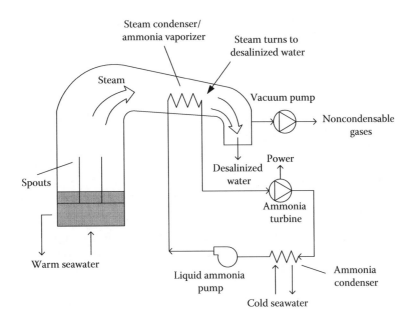

FIGURE 5.12 General structure of the hybrid OTEC process. (Redrawn from Natural Energy Laboratory of Hawaii Authority, USA, available at: http://www.nelha.org, March 2008.)

goes to another vaporizer to release its heat to vaporize the working fluids, like ammonia, in a method similar to the closed-cycle evaporation process. An effervescent two-phase, two-substance mixture is obtained by physically mixing the second fluid with the warm seawater. The evaporated second working fluid is separated from the steam/seawater, which is recondensed similar to the closed cycle. The low-pressure turbine is easily driven by the phase change of the seawater/ammonia mixture. The vaporized fluid then drives a turbine to produce electricity, and the steam condenses within the heat exchanger and turns into desalinated water.

The close coupling of seawater and second working fluid is the main advantage of the hybrid OTEC design. A condenser heat exchanger is needed, even though an evaporator heat exchanger is not required. The separation of two fluids also needs some special design [28,29]. The ammonia vapor/liquid mixture needs to be rendered completely to the liquid phase using a compressor.

Compared to the closed-cycle system, the hybrid system avoids the problem of the degradation of the heat exchanger, caused by microbial fouling, since the cold water heat exchanger has little or even no microbial fouling [19]. In the second vaporizer of this system, desalinized water can be produced as the by-product. When compared to the open-cycle system, power-generating capacity can be improved.

5.4 Components of an OTEC System

Heat exchangers (evaporators and condensers) are critical components utilized in OTEC systems. In this section, the basic structures of these components and vacuum flash evaporators are described and analyzed.

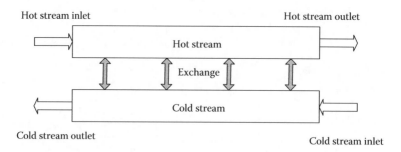

FIGURE 5.13 Countercurrent exchange.

5.4.1 Heat Exchanger

A heat exchanger is used within the closed-cycle OTEC systems to evaporate the working fluid. Heat exchanger can work at two mechanisms: countercurrent exchange and parallel exchange. In the countercurrent exchange, hot and cold streams flow in opposite directions, as shown in Figure 5.13.

In the parallel exchange, hot and cold streams flow in the same direction, as shown in Figure 5.14.

To determine the temperature driving force for heat transfer in a heat exchanger, the LMTD is used. It is the logarithmic average of the temperature difference between the hot and cold streams at each end of the exchanger. For countercurrent flow, it is expressed as

$$\text{LMTD} = \frac{(T_1 - t_2) - (T_2 - t_1)}{\ln((T_1 - t_2)/(T_2 - t_1))}. \tag{5.20}$$

For parallel flow, it is expressed as

$$\text{LMTD} = \frac{(T_1 - t_1) - (T_2 - t_2)}{\ln((T_1 - t_1)/(T_2 - t_2))}, \tag{5.21}$$

where T_1 is the hot stream inlet temperature, T_2 is the hot stream outlet temperature, t_1 is the cold stream inlet temperature, and t_2 is the cold stream outlet temperature.

5.4.2 Evaporator

An evaporator is a heat exchanger, which vaporizes a substance from its liquid state to its gaseous state, as shown in Figure 5.15. The liquid working fluid (low temperature) is fed

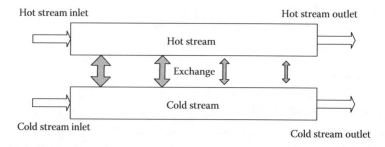

FIGURE 5.14 Parallel heat exchanger operation.

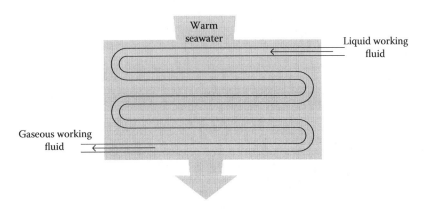

FIGURE 5.15 Schematic diagram of an evaporator.

into the U-type tubes, absorbing heat from the outside heat source (passing warm seawater), which will turn it to gas.

In an evaporator, the main heat transfer passage is *conduction*, not the *radiation* or *convection*. The conduction heat transfer can be defined as the transfer of thermal energy through direct molecular interaction within a medium or between mediums in direct physical contact without the flow of the medium material.

The rate of heat transfer can be expressed by Newton's cooling law [30,31]:

$$\frac{dQ}{dt} = hA(T_0 - T_{env}), \tag{5.22}$$

where Q is the transferred thermal energy, t is the time, h is the heat transfer coefficient, A is the heat transfer surface area, T_0 is the temperature of the object's surface, and T_{env} is the temperature of the environment.

It can be stated that the rate of heat loss of a body is not only proportional to the temperature difference between the body and its surroundings environment, but also proportional to the heat exchanging surface area. The tubes usually are designed in U shape to increase the heat exchanging surface area.

If a plate-fin exchanger is used, the heat transfer coefficient can be calculated as [32]

$$h = \left[\left(\frac{A_p}{A} \left[1 - \frac{A_f}{A_p} \left(1 - \frac{\tan hl(2h_p/k\delta)}{l\sqrt{(2h_p/k\delta)}} \right) \right] h_p \right)^{-1} + \frac{aA}{kA_W} + \frac{1}{h_{sw}} \right]^{-1}, \tag{5.23}$$

where a is the wall thickness (m), A is the total outside area of the plate-fin panel (m^2), A_f is the total fin area of the contact with propane (m^2), A_p is the total area of the contact surface with propane, A_w is the average wall area (m^2), δ is the fin thickness (m), h_p is the film conductance of propane, and h_{sw} is the film conductance of seawater. The film conductance is a parameter that describes the heat transfer capability of fluids. h_p and h_{sw} are functions of the physical and chemical fluid properties such as temperature and salinity. The conductance also depends on the flow rates of the fluids. In Equation 5.23, the other parameters such as k is the thermal conductivity of evaporator material [J/(ms,°K)] and l is one-half of the fin length (m).

A schematic diagram of an evaporator is presented in Figure 5.16, and a dynamic model and its controller are presented in Figure 5.17 [33]. In this model, the temperature of the

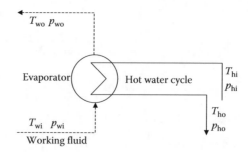

FIGURE 5.16 Evaporator schematic diagram.

hot water is given, and the manipulated variable is the hot water mass flow rate [33]. By changing the rate of the hot water mass flow through the hot water pump, the heat quantity of the outlet working fluid can be controlled.

In Figure 5.16, T_{hi} is the temperature of inlet hot water to the evaporator, T_{ho} is the temperature of outlet hot water from the evaporator, p_{ho} is the pressure of outlet hot water, p_{hi} is the pressure of inlet hot water, where $p_{hi} = p_{ho} = p_h$, and m_h is the mass flow rate of hot water. Q_{hi} is the heat quantity of inlet hot water, Q_{ho} is the heat quantity of outlet hot water, T_{wi} is the temperature of inlet working fluid, T_{wo} is the temperature of outlet working fluid, p_{wi} is the pressure of inlet working fluid, p_{wo} is the pressure of outlet working fluid, and m_w is the mass flow rate of working fluid.

This model is based on the relationship between the heat exchange value of hot water in the evaporator, ΔQ_h, and heat exchange value of the working fluid, ΔQ_w. The heat quantity can be selected as the state variable [34].

In the evaporator, the heat exchange value of the hot water is

$$\Delta Q_h = (h_{ho} - h_{hi})\, m_h \tag{5.24}$$

and the heat exchange value for the working fluid is

$$\Delta Q_w = (h_{wo} - h_{wi})\, m_w, \tag{5.25}$$

where h denotes the water enthalpy. In Equations 5.24 and 5.25, the evaporator inlet enthalpy of hot water (h_{hi}) is a function of the evaporator temperature of inlet hot water (T_{hi}) and the pressure of hot water (p_h):

$$h_{hi} = h(T_{hi}, p_h). \tag{5.26}$$

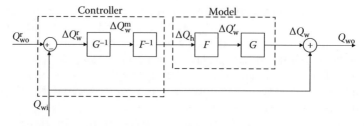

FIGURE 5.17 System model and its controller.

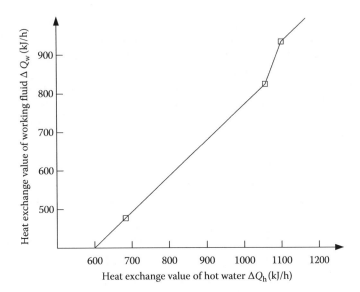

FIGURE 5.18 Nonlinear statics for heat exchange values of the water versus working fluid.

Similarly, the evaporator outlet enthalpy of hot water (h_{ho}), the evaporator outlet enthalpy of working fluid (h_{wo}), and the evaporator inlet enthalpy of working fluid (h_{wi}) can be calculated.

The heat exchange value of the water is not the same as the heat exchange value of the working fluid, due to the losses and efficiencies of heat exchange components. These heat exchange values of the water and working fluid have a nonlinear relation, which is called "nonlinear statics." In steady state, ΔQ_h and ΔQ_w can be measured, and nonlinear statics can be derived by plotting these measured data, as shown in Figure 5.18 [34].

By this figure, the heat exchange value of working fluid (ΔQ_w) can be expressed as a function of the heat exchange value of the hot water (ΔQ_h). The plotted data can be interpolated by a linear function as

$$\Delta Q'_w = F(\Delta Q_h). \tag{5.27}$$

The linear dynamics can be approximated by a first-order transfer function as

$$G(s) = \frac{1}{1 + \tau s}, \tag{5.28}$$

where time constant, τ, can be determined by minimizing the square errors between the actual output and the model output [33].

The outlet heat quantity of working fluid from the evaporator would be

$$Q_{wo} = \Delta Q_w + Q_{wi}. \tag{5.29}$$

As shown in Figure 5.17, the controller can be designed as the mirror image of the evaporator's model [33]. The system dynamics are expressed by F and G functions and the controller dynamics consist of F^{-1} and G^{-1} functions, that is, the inverse dynamics. The mirror image describes that these inverse functions eliminate the system functions and the output accurately tracks the input reference. The outlet heat quantity of working fluid (Q_{wo}) is the controlled variable. The evaporator inlet heat quantity of working fluid (Q_{wi}) is an environmental variable, and the heat exchange value of hot water (ΔQ_h) is the manipulated variable. Q_{wi} is an environmental variable, since its value (heat quantity) depends on the environmental conditions of the fluid that is taken as an inlet to the evaporator.

The reference of the heat exchange value of working fluid in the evaporator (ΔQ_w^r) can be calculated using Equation 5.30, based on the reference of the outlet heat quantity of working fluid (ΔQ_{wo}^r) [33].

$$\Delta Q_w^r = Q_{wo}^r - Q_{wi}. \tag{5.30}$$

The heat exchange value of hot water in the evaporator (ΔQ_h) is the output variable of the controller. It can be generated by the inverse nonlinear statics as Equation 5.31:

$$\Delta Q_h = F^{-1}\left(\Delta Q_w^m\right) \tag{5.31}$$

The dynamic compensated value (ΔQ_w^m) can be calculated as

$$\Delta Q_w^m = G^{-1}(s) \times \Delta Q_w^r, \tag{5.32}$$

where $G^{-1}(s)$ is the inverse dynamics of the $G(s)$ function. $G^{-1}(s)$ is expressed as

$$G^{-1}(s) = 1 + \tau s. \tag{5.33}$$

The ΔQ_w^m is called as the dynamic compensated value because it is multiplied by the inverse dynamics function, that is, $G^{-1}(s)$.

The heat exchange value of hot water, ΔQ_h, is used to calculate the mass flow rate of hot water, m_h, by Equation 5.34 as

$$m_h = \frac{\Delta Q_h}{h_{ho} - h_{hi}}. \tag{5.34}$$

5.4.3 Condenser

A condenser is a heat exchanger, which condenses a substance from its gaseous to its liquid state, as shown in Figure 5.19. The gaseous working fluid is fed into the tubes, discharging

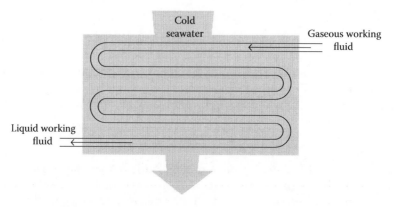

FIGURE 5.19 Schematic diagram of a condenser.

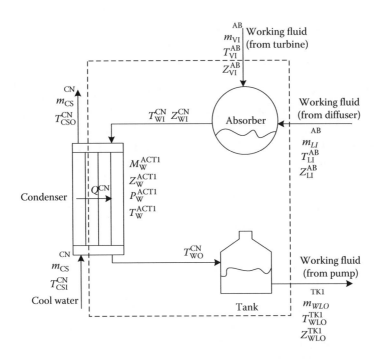

FIGURE 5.20 Schematic of condenser model.

heat to the outside cold source (passing cold seawater), thus it becomes liquid [33]. The operation of the condenser is reverse of the operation of an evaporator.

A condenser model is shown in Figure 5.20. In this model, there is a tank and an absorber. The absorber has two inlets: one from the turbine and the other from the diffuser. The working fluid is a mixture of ammonia and water. In this simple model of a condenser, the transient performance is a function of temperature, pressure, composition, and mass flow rate of the working fluid [35,36].

The enthalpy h (J/kg) and the specific volume v (m^3/kg) are functions of temperature T (°C), pressure P (Pa), and the composition Z. The composition Z is the mass fraction of the ammonia.

$$\begin{cases} h = f_1(T,P,Z) \\ v = f_2(T,P,Z). \end{cases} \tag{5.35}$$

Compositions of the vapor and liquid phase (Z_v, Z_L), the enthalpies of these two phases (h_v, h_L), and the specific volumes (v_v, v_L) of these phases are functions of pressure and temperature, which are given by

$$\begin{cases} (Z_v, Z_L) = f_3(T,P) \\ (h_v, h_L) = f_4(T,P) \\ (v_v, v_L) = f_5(T,P). \end{cases} \tag{5.36}$$

The condensation process can be explained by "mass conservation law" and the "first law of thermodynamics."

There are two mass conservation equations for two working fluids: ammonia and water [36]. The condenser has one input through the absorber and one output to the working fluid tank. It has also cool water inlet and outlet. Thus, the average amount of fluid mass inside the condenser is the summation of input and output masses, given by

$$M_W^{ACT1} = m_{VI}^{AB} + m_{LI}^{AB} - m_{WLO}^{TKI}, \qquad (5.37)$$

where M_W^{ACT1} (kg) is the total mass of the working fluid in the condenser, m_{VI}^{AB} (kg) is the mass of the fluid from the turbine to the absorber (input to the condenser), m_{LI}^{AB} (kg) is the mass of the fluid from the diffuser to the absorber (input to the condenser), and m_{WLO}^{TKI} (kg) is the mass of the fluid from the tank to the pump (the output of the condenser). It should be noted that superscript AB represents the quantities to the absorber, whereas superscript TKI represents the quantity from the tank. The subscripts VI, LI, and WLO represent the working fluid from the diffuser, working fluid from the turbine, and working fluid to the pump, respectively.

Based on the total mass conservation equation, the flow rate of the fluid mass can be written as

$$\frac{d}{dt}(M_W^{ACT1}) = \frac{d}{dt}m_{VI}^{AB} + \frac{d}{dt}m_{LI}^{AB} - \frac{d}{dt}m_{WLO}^{TKI}, \qquad (5.38)$$

where $(d/dt)M_W^{ACT1}$ (kg) is the total mass flow rate of the working fluid in the condenser, $(d/dt)m_{VI}^{AB}$ (kg/s) denotes the mass flow rate of the first inlet, $(d/dt)m_{LI}^{AB}$ (kg/s) is the mass flow of the second inlet, and $(d/dt)m_{WLO}^{TKI}$ (kg/s) is the outlet mass flow rate.

Ammonium mass conservation can be expressed as

$$\frac{d}{dt}(M_W^{ACT1}Z_W^{ACT1}) = \frac{d}{dt}m_{VI}^{AB}Z_{VI}^{AB} + \frac{d}{dt}m_{LI}^{AB}Z_{LI}^{AB} - \frac{d}{dt}m_{WLO}^{TKI}Z_{WLO}^{TKI}, \qquad (5.39)$$

where M_W^{ACT1} (kg) is the total mass of the working fluid in the condensation process, Z_W^{ACT1} is the mean composition of the working fluid in the subsystem, $(d/dt)m_{VI}^{AB}$, $(d/dt)m_{LI}^{AB}$, $(d/dt)m_{WLO}^{TKI}$ are the mass flow rates (kg/s) of the working fluid at the two inlets (from diffuser and turbine) and the outlet tank, respectively. Z_{VI}^{AB}, Z_{LI}^{AB}, and Z_{WLO}^{TKI} are the compositions of the working fluid at related inlets and outlets, respectively.

On the other hand, the volume of the condenser is constant; thus, the volume of the fluid in the condenser V_W^{ACT1} (m^3) is constant. Hence, the input mass flow should be equal to the output mass. Therefore,

$$\frac{d}{dt}\left(M_W^{ACT1}v_W^{ACT1}\right) = 0. \qquad (5.40)$$

The first law of the thermodynamics expresses the energy conservation as

$$\frac{d(M_W^{ACT1}h_W^{ACT1})}{dt} - V^{ACT1}\frac{dP_W^{ACT1}}{dt} = \frac{d(m_{VI}^{AB}h_{VI}^{AB})}{dt} + \frac{d(m_{LI}^{AB}h_{LI}^{AB})}{dt} - Q^{CN} - \frac{d}{dt}m_{WLO}^{TKI}h_{WLO}^{TKI}, \qquad (5.41)$$

where h_W^{ACT1} denotes the mean specific enthalpy of the working fluid, P_W^{ACT1} is the pressure, and h_{VI}^{AB}, h_{LI}^{AB}, and h_{WLO}^{TKI} are the specific enthalpies of the working fluid at the two inlets and the outlet, respectively. Q^{CN} (W) is the heat transfer rate in the condenser, which can be described as

$$Q^{CN} = U^{CN}A^{CN}\Delta T_m^{CN}. \qquad (5.42)$$

In Equation 5.42, U^{CN} [W/(m^2 °C)] is the heat transfer coefficient, A^{CN} (m^2) is the contact area of the heat exchanger, and ΔT_m^{CN} (°C) is the logarithmic mean temperature difference.

The heat transfer coefficient U^{CN} can be approximated as

$$U^{CN} = U_s^{CN1/2} \frac{(d/dt)m_{CS}^{CN}}{(d/dt)m_s^{CN}},$$ (5.43)

where $(d/dt)m_{CS}^{CN}$ (kg/s) is the mass flow rate of the cold water, and U_s^{CN}[W/(m^2 °C)] and $(d/dt)m_s^{CN}$ (kg/s) are the standard heat transfer coefficient and the standard mass flow rate, respectively. These coefficients can be determined experimentally [36].

Logarithmic mean temperature difference can be calculated by

$$\Delta T_m^{CN} = \frac{\left(T_{WI}^{CN} - T_{CSO}^{CN}\right) - \left(T_{WO}^{CN} - T_{CSI}^{CN}\right)}{\ln\left(T_{WI}^{CN} - T_{CSO}^{CN}\right)/\left(T_{WO}^{CN} - T_{CSI}^{CN}\right)}.$$ (5.44)

For given values of h_W^{ACK1}, P_W^{ACK1}, and Z_W^{ACK1}; T_W^{ACK1} and $T_{WO}^{CN} = T_W^{ACK1}$ can be calculated by Equation 5.44. T_{WI}^{CN} can be obtained by using h_W^{ACK1}, P_W^{ACK1}, and Z_W^{ACK1} where h_{WI}^{CN} and Z_{WI}^{CN} are

$$h_{WI}^{CN} = \frac{(d/dt)m_{VI}^{AB}h_{VI}^{AB} + (d/dt)m_{LI}^{AB}h_{LI}^{AB}}{(d/dt)m_{VI}^{AB} + (d/dt)m_{LI}^{AB}},$$ (5.45)

$$Z_{WI}^{CN} = \frac{(d/dt)m_{VI}^{AB}Z_{VI}^{AB} + (d/dt)m_{LI}^{AB}Z_{LI}^{AB}}{(d/dt)m_{VI}^{AB} + (d/dt)m_{LI}^{AB}}.$$ (5.46)

5.4.4 Vacuum Flash Evaporator

Water evaporation rate depends on three factors: (1) The temperature of the water at the air–water surface, (2) the area of the air–water surface, and (3) the H$_2$O molecular concentration of the air, or humidity.

Humidity is the amount of water mass in a particular volume of air, which can be expressed as

$$H = \frac{m}{V},$$ (5.47)

where m is the mass of water vapor and V is the volume of the vessel.

From the ideal gas law,

$$PV = \frac{m}{M}RT,$$ (5.48)

where P is the absolute pressure, M is the molar mass of gas, R is the universal gas constant, and T is the absolute temperature.

For a given volume and temperature, the humidity is proportional to the air pressure, which means that the rate of evaporation can be increased by decreasing the air pressure in a vessel. This is the operational principle of a vacuum flash evaporator.

As shown in Figure 5.21, the vacuum pump is used to keep the air pressure in the vessel in a partial vacuum level. Warm seawater is spouted into the vessel and partially turns to steam.

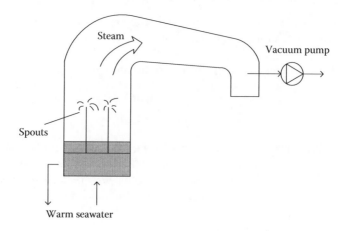

FIGURE 5.21   A vacuum flash evaporator.

5.5   Control of an OTEC Power Plant

In this section, a simple constructed controller model is described for an OTEC power plant by taking the heat quantity as a state variable instead of temperature and pressure.

The controlled variables of an OTEC system are mass flow rate of warm seawater, mass flow rate of cold seawater, and mass flow rate of working fluid, which will be represented as m_h, m_c, and m_w, respectively. Actuation of these mass flow rates is done through three pumps: warm seawater pump, cold seawater pump, and the working fluid pump. Figure 5.22 [37] shows the mass flow intakes and devices with their controllers. In Figure 5.22, MWD represents the megawatt demand, Q is the heat quantity, M is the mass flow rate, superscript ref is for the reference, superscript pv stands for the process variable, C is the nonlinear separation controller, and PI stands for the proportional–integral controller.

According to the hierarchical control strategy shown in Figure 5.20, the upper system controls the heat quantities of the warm seawater, cold seawater, and the working fluid while the lower system controls the amount of adjusted variables. The respective adjusted variables are only taken into account by the lower system and lower system does not consider the heat balance of the OTEC system. Meanwhile, the total balance of the OTEC system is managed by the upper control system.

The OTEC system is a nonlinear system and can be modeled as the cascade connection of nonlinear statics and linear dynamics functions. Nonlinear statics were previously presented in Figure 5.17 and the cascade connection block diagram was shown in Figure 5.18. The nonlinearity can be eliminated by inverting the nonlinear statics. The inverse of the linear dynamics can also compensate the residual characteristics of the linear dynamics.

In this model, warm and cold seawater temperatures are inputs to the system, and warm and cold seawater mass flow rates are the manipulated input variables. Evaporator outlet heat quantity of the working fluid is a function of the warm seawater mass flow rate controlled by the warm water pump. On the other hand, the cold water pump controls the mass flow rate of the cold seawater, which affects the condenser outlet heat quantity of the working fluid.

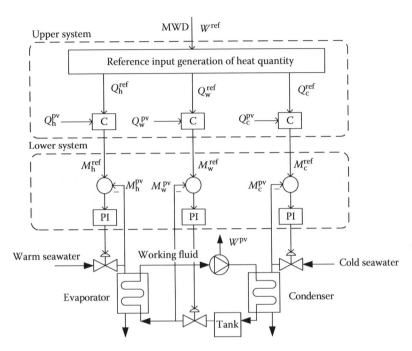

FIGURE 5.22 Block diagram of the OTEC system controller.

Heat quantity is used as the state variable for the upper control system of the OTEC system since it is more robust to the pressure and flow rate variations [37]. If the temperature is chosen as the state variable, the overall system becomes more complicated, since the temperature is very sensitive to the pressure and mass flow rate. Moreover, using the heat quantity as the state variable, highly nonlinear relationship between enthalpy and temperature/pressure can be eliminated. The power generation of the OTEC system is a function of the heat quantity difference between turbine inlet and turbine outlet working fluids. Therefore, the power generation can be controlled by controlling the heat quantity instead of temperature, because electric power can be calculated directly using the heat quantity.

The nonlinear separation controller can easily be implemented in the OTEC systems and it is simpler in compare to other controller strategies based on the physical models [37–40]. The controller requires obtaining the nonlinear statics F using the data plot of the control inputs m_h and m_c, and steady state of the output which is the electric power W^{pv}. In the following process, the linear dynamic transfer function, $G(s)$, should be obtained using the inverse of nonlinear static transfer function, F^{-1}, to get the nonlinear compensated relationship between the reference input W^{ref} and the output W^{pv}. In Figure 5.23, the controller and the model of OTEC pilot plant are demonstrated [37].

The mass flow rate of the warm seawater (m_h), warm seawater temperature (T_h), and the specific heat capacity (C_p) are used to calculate the heat quantity of warm seawater (Q_h). The heat quantity Q_h is given by

$$Q_h = m_h T_h C_p. \tag{5.49}$$

Similarly for cold water heat quantity is

$$Q_c = m_c T_c C_p, \tag{5.50}$$

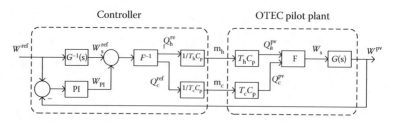

FIGURE 5.23 OTEC pilot model and controller structure.

where T_c represents the cold seawater temperature. The heat quantities of the warm (Q_h) and cold seawater (Q_c) are used as input in Equation 5.51 to calculate the steady-state power output of the OTEC plant, which is W_s.

$$W_s = F(Q_h, Q_c). \tag{5.51}$$

Here, F is the model of the nonlinear statics of the OTEC system. In the nonlinear separation model, a second-order transfer function with a zero can be used as the linear dynamics, which is given by [37–39]

$$G(s) = \frac{b_1 s + b_2}{s^2 + a_1 s + a_2}. \tag{5.52}$$

The a and b parameters of Equation 5.52 can be estimated using the least-squares method. The nonlinear statics F and linear dynamics $G(s)$ can be identified as shown in Figure 5.24. The output power (W) of the OTEC system would be

$$W = G(s) W_s. \tag{5.53}$$

As mentioned earlier, the nonlinear separation controller is designed using the mirror image of the nonlinear separation model. The nonlinearity can be eliminated by the inverse of the nonlinear statics as well as the residual characteristics of the linear dynamics compensated by the inverse dynamics. PI controllers are used to compensate the modeling errors and disturbances and they are used in parallel for the dynamic compensation.

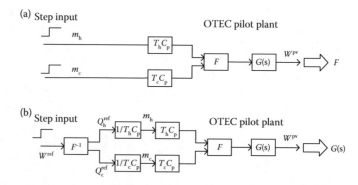

FIGURE 5.24 Construction procedure of the nonlinear separation model: (a) identification of nonlinear statics function, F and (b) identification of linear dynamics function, $G(s)$.

Inverse dynamics of the $G(s)$ can be derived as follows:

$$G^{-1}(s) = \frac{s^2 + a_1 s + a_2}{b_1 s + b_2}.$$ (5.54)

Hence, steady-state power output reference W_s^{ref} can be obtained as

$$W_s^{ref} = G^{-1}(s) W_s.$$ (5.55)

Q_h^{ref} and Q_c^{ref} are the warm seawater heat quantity and the cold seawater heat quantity reference values, respectively. Based on the Q_c^{ref} value, the value of Q_h^{ref} can be calculated. Warm seawater reference heat quantity is

$$Q_h^{ref} = F^{-1} \left(W_s^{ref} + W_{PI} \left| Q_c^{ref} \right. \right),$$ (5.56)

where W_{PI} is the PI controller output. Eventually, the control input of the warm seawater mass flow rate (m_h^{ref}) can be found as

$$m_h^{ref} = Q_h^{ref} / T_h C_p.$$ (5.57)

Similarly, the cold seawater reference mass flow rate can be derived as

$$m_c^{ref} = Q_c^{ref} / T_c C_p.$$ (5.58)

5.6 Control of a Steam Turbine

Similar to the hydraulic turbine control, the main duty of the governor is to control the gate opening. A controller consists of a speed governor, a speed relay, a servomotor, and governor-controlled valves can be used to control the steam turbine [41]. The speed control system of a steam turbine is shown in Figure 5.25.

Figure 5.26 presents an approximate mathematical model of this system. In this model, nonlinearities, except for valve position limits and its rate limits, are neglected.

Here, the gain K_G is the reciprocal of regulation or droop. The control signal, SR, represents a composite load and speed reference. An integrator with time constant T_{SR} and direct

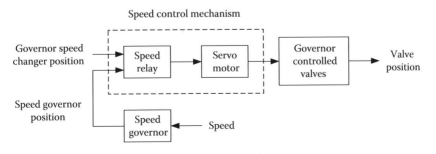

FIGURE 5.25 Steam turbine speed governing system.

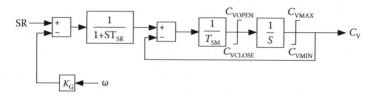

FIGURE 5.26   Approximate mathematical representation of speed governing system.

feedback is used to represent the speed relay. The integrator with time constant T_{SM} and direct feedback represent the servomotor. The servomotor moves the valves and is physically large, particularly on large units. Rate limiting and position limits of the servomotor are shown at the integrator, which represents the servomotor.

This steam turbine control model is simulated in MATLAB/Simulink. A general block diagram of the steam turbine and governor system, excitation system, synchronous machine, transmission lines, and grid is shown in Figure 5.27.

In this model, the reference power generation of the turbine is set as 0.7 (p.u.) and the reference speed is selected as 1 (p.u.). It can be seen from Figure 5.27 that the input of the generator is the mechanical power produced by the steam turbine and governor system and the excitation voltage produced by the excitation system. The output of the synchronous generator is connected to the grid through transmission line, represented by an RL circuit.

The synchronous generator in this model is a salient-pole machine with 200 MW nominal power and 13.8 kV phase-to-phase nominal voltage. Nominal frequency is 60 Hz and the number of pole pairs is 60. Thus, the nominal revolutions per minute is

$$n_S = \frac{60 \times 60}{60} = 60 \, \text{rpm}. \tag{5.59}$$

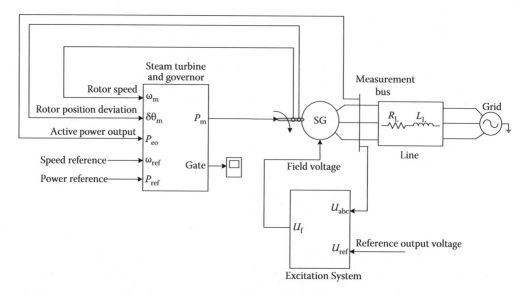

FIGURE 5.27   Steam turbine and governor coupled to a synchronous generator.

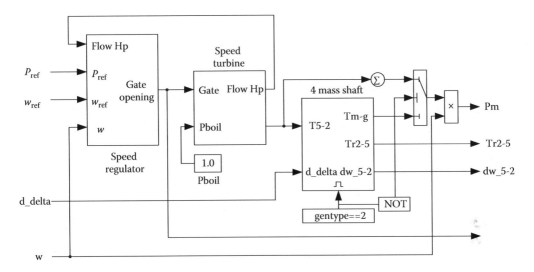

FIGURE 5.28 Complete governor, steam turbine, and shaft model.

The model of steam turbine and governor system includes a speed governing system, a four-stage steam turbine, and a shaft with up to four masses. Their structure is shown in Figure 5.28.

The speed governing system consists of a proportional regulator, a speed relay, and a servomotor, which controls the gate opening to the turbine. It is shown in Figure 5.29.

In the simulation, the governor regulator gain K_p is set to 1, permanent droop $R_p = 0.05$, dead zone $D_z = 0$, speed relay time constant $T_{sr} = 0.001$, and servomotor time constant $T_{sm} = 0.15$. Minimum and maximum gate opening values are set to 0.01 and 0.99, respectively, and gate opening and closing rate values are set to be between -0.1 and 0.1.

Figure 5.30 illustrates a four-stage steam turbine, which are modeled by four first-order transfer functions. The steam chest is represented by the first stage, and reheater or crossover piping is represented by the following three stages.

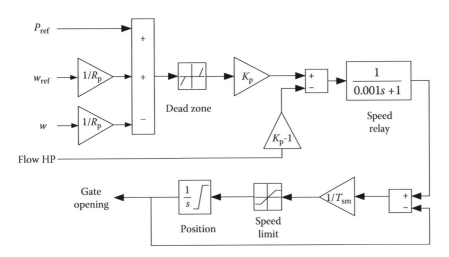

FIGURE 5.29 Speed governing system model.

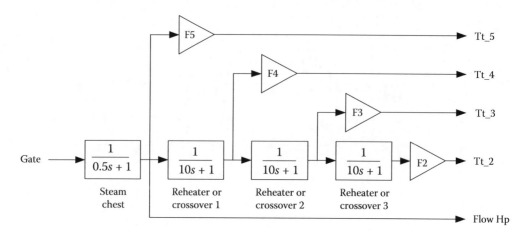

FIGURE 5.30   Steam turbine model.

The excitation voltage control system in Figure 5.27 determines the appropriate excitation voltage level by using the qd-axis components of the generator stator voltage and the reference output voltage by a PI controller. The dq-axis components of the voltage are obtained using Park Transformations from ABC phase voltage quantities.

The simulation time is set to 70 s. Figures 5.31 and 5.32 show the rotor speed and rotor speed deviation. It can be seen that the rotor speed settles at 1 p.u. while the rotor speed deviation damps to zero after some oscillations in less than 10 s.

Mechanical power supplied to the generator's shaft is shown in Figure 5.33, which increases to the reference value of 0.7, which is commanded by the gate opening, as shown in Figure 5.34.

The generator active power output yields up to 0.7 (p.u.) after some oscillations, as shown in Figure 5.35.

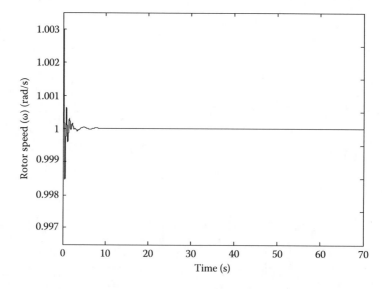

FIGURE 5.31   Rotor speed versus time.

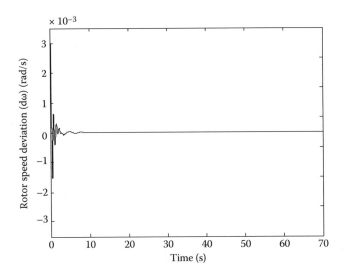

FIGURE 5.32   Rotor speed deviation.

Figure 5.36 shows the output phase-to-phase voltage of the generator. It can be seen that the rated voltage of the generator is 13.8 kV, while the peak-to-peak voltage is 19.52 kV, with a frequency of 60 Hz.

5.7   Potential Resources

Approximately, 70% of the earth is covered by water. In equatorial regions, this number even goes up to 90%. The source of the oceans' thermal energy is the sun, which causes the

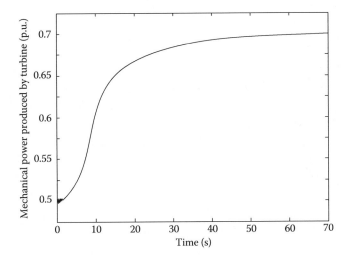

FIGURE 5.33   Turbine mechanical power.

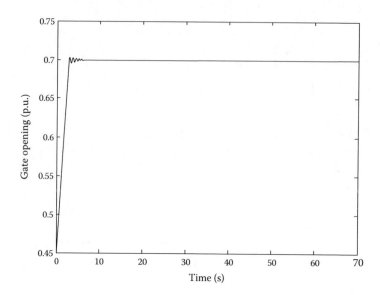

FIGURE 5.34   Gate opening.

temperature difference in ocean water [42]. Oceans collect and store energy from the sun, which is the greatest energy source.

For efficient system operation, commercial OTEC plants must be located in an environment that is thermally stable enough. The temperature of the warm surface seawater must differ about 20°C (36°F) from that of the cold deepwater that is no more than about 1000 m (3280 ft) below the surface. The map in Ref. [43] shows the global distribution of sea surface temperature.

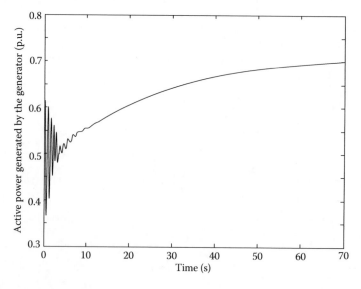

FIGURE 5.35   Generator active power.

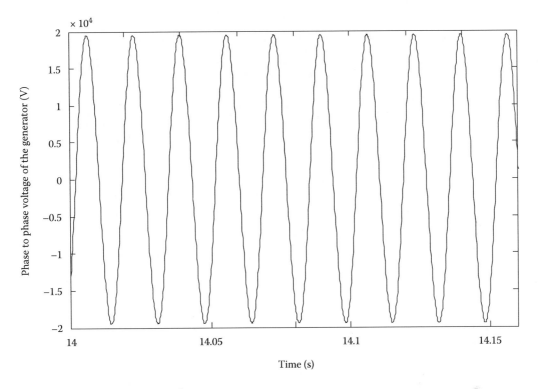

FIGURE 5.36 Generator output voltage.

The available OTEC thermal resources throughout the world can be summarized as follows:

1. Equatorial waters (between 10°N and 10°S), except for the west coast of South America. The deepwater temperature along the east coast of Africa is about 2°C warmer [15].

2. Tropical waters (extending from the equatorial region boundary to, respectively, 20°N and 20°S), except for the west coasts of South America and of Southern Africa [15].

In addition to the United States and Australia, there are 66 developing nations located in this zone. Table 5.1 provides the temperature difference in water between 0 and 1000 m for these developing nations with adequate ocean thermal resources [13].

In addition to temperature difference, many other factors, from logistics to socioeconomic and political factors, must be considered to locate potential OTEC sites. In some cases, floating OTEC plants can avoid the political and socioeconomic issues associated with land-based OTEC systems [15]. Tropical islands with growing power requirements, which are dependent on expensive imported oil, are the best sites for OTEC systems.

TABLE 5.1

Developing Nations with Adequate Ocean-Thermal Resources 25 km or
Less from the Shore

Country/Area	Temperature Difference (°C) of Water between 0 and 1000 m	Distance from Resource to Shore (km)
Africa		
Benin	22–24	25
Gabon	20–22	15
Ghana	22–24	25
Kenya	20–21	25
Mozambique	18–21	25
São Tomé and Príncipe	22	1–10
Somalia	18–20	25
Tanzania	20–22	25
Latin America and the Caribbean		
Bahamas, The	20–22	15
Barbados	22	1–10
Cuba	22–24	1
Dominica	22	1–10
Dominican Republic	21–24	1
Grenada	27	1–10
Haiti	21–24	1
Jamaica	22	1–10
Saint Lucia	22	1–10
Saint Vincent and the Grenadines	22	1–10
Trinidad and Tobago	22–24	10
U.S. Virgin Islands	21–24	1
Indian and Pacific Oceans		
Comoros	20–25	1–10
Cook Islands	21–22	1–10
Fiji	22–23	1–10
Guam	24	1
Kiribati	23–24	1–10
Maldives	22	1–10
Mauritius	20–21	1–10
New Caledonia	20–21	1–10
Pacific Islands Trust Territory	22–24	1
Philippines	22–24	1
Samoa	22–23	1–10
Seychelles	21–22	1
Solomon Islands	23–24	1–10
Vanuatu	22–23	1–10

5.8 Multipurpose Utilization of OTEC Systems

In addition to power production, in order to reduce the cost of OTEC systems, they can be designed for multiple purposes. As shown in Figure 5.37, OTEC systems can be used for desalination of water, aquaculture needs, air-conditioning, and mineral extraction purposes.

5.8.1 Desalination

One advantage of open- or hybrid-cycle plants is that desalinated water can be produced from condensers. In a condenser, the steam is condensed by indirect contact with the cold seawater. For some small islands, pure water is relatively limited and expensive. The water condensed by OTEC is relatively free of impurities and can be collected and dispensed to local communities where supplies of natural freshwater for agriculture or drinking are limited. Block and Lalenzuela's analysis indicates that a 2 MW (net) OTEC plant could produce about 4300 m^3 of desalinated water per day [44].

5.8.2 Aquaculture

Deep seawater discharged from an OTEC plant contains high concentrations of nutrients that are depleted in surface waters due to biological consumption, and it is relatively free of pathogens. This is an excellent medium for growing phytoplanktons (microalgae), such as *Spirulina*, a health food supplement, and a variety of commercially valuable fish and shellfishes, such as salmon and lobster, which thrive in the nutrient-rich, cold seawater [45]. Suitable mixing of the warm and cold water discharges in various ratios can provide

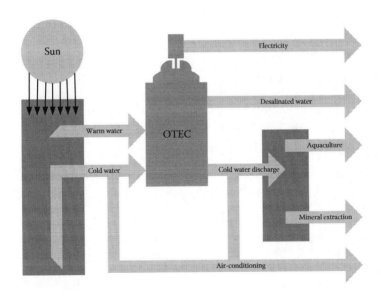

FIGURE 5.37 Multipurpose applications of OTEC systems.

seawater at a specific temperature between surface and deep seawater to maintain an optimal environment for aquaculture.

5.8.3 Air-Conditioning

The cold seawater discharged by an OTEC plant can also be used as a chill fluid in air-conditioning systems to provide air-conditioning for buildings close to the plant. It is estimated that only $1\,m^3/s$ of $7°C$ deep ocean water is required to produce 5800 tons (roughly equivalent to 5800 rooms) of air conditioning. This will typically require a pipeline of about 1 m in diameter and the pumping power required will be about 360 kW, compared to 5000 kW for a conventional air-conditioning system [46].

In 1999, Cornell University installed a "Lake Cooling" system that uses cold deepwater in Cayuga Lake for its campus cooling system and also provides cooling to the Ithaca City School District. This system saves over 20 million kWh annually, even though the air conditioning is needed only in the summer time. The savings would be even more in the tropics where OTEC systems are viable.

5.8.4 Mineral Extraction

In the past, most economic analyses showed that mining the ocean for a variety of elements dissolved in solution would be unprofitable because a lot of energy is needed to pump the large volume of required water and it is very expensive to separate the minerals from seawater. However, since OTEC plants will already be pumping the water, the remaining problem is the cost of the extraction process. Many research groups all over the world are analyzing the feasibility of combining the extraction of uranium dissolved in seawater with ocean energy production [47].

5.9 Impact on Environment

OTEC power plants have some negative impacts on the natural environment, but overall they are relatively clean and environmentally benign power production sources of electricity when compared to conventional options such as fossil fuels or nuclear power. Moreover, the hazardous substances are limited to the working fluid (e.g., ammonia), and no noxious by-products are generated.

In an open-cycle OTEC plant, cold water, released at the ocean's surface, will release trapped carbon dioxide, a greenhouse gas, but its emission is less than 1% of those released from a thermoelectric fossil fuel power plant [15]. The value is even lower in the case of a closed-cycle OTEC plant.

Sustained discharging of the nutrient-rich and bacteria-free deep ocean cold water at the oceans' surface could change local concentrations of nutrients and dissolved gases. However, this impact could be minimized by discharging the cold water at depths greater than 50 m.

In closed-cycle OTEC plants, chlorine might be used to protect the heat exchangers from microbial fouling. According to the US Environmental Protection Agency, a maximum chlorine discharge of 0.5 mg/L and an average of 0.1 mg/L are allowed. The amount of chlorine in a closed-cycle OTEC plant is within this limit [15].

5.10   Summary

In this chapter, OTEC, which is an energy-generating technology that takes advantage of the temperature difference between the ocean's shallow warm water and cold deeper water, is investigated. The chapter consists of closed-, open-, and hybrid-cycle OTEC systems as well as their required components. Thermodynamic principles as well as the dynamic operating principles of OTEC systems are discussed. In addition, potential resources and multipurpose OTEC systems are presented in this chapter.

References

1. R. Pelc and R.M. Fujita, "Renewable energy from the ocean," *Journal of Marine Policy*, 26, 471–479, 2002.
2. S. Rahman, "Alternate sources of energy," *IEEE Potentials*, 7, 22–25, 1988.
3. T. Kajikawa and T. Agawa, "System characteristics of low temperature difference power generation," *Transactions on IEEE Journal*, 98-B, 773–780, 1978.
4. G.T. Heydt, "An assessment of ocean thermal energy conversion as an advanced electric generation methodology," *Proceedings of the IEEE*, 81, 409–418, 1993.
5. C.B. Panchal, J. Larsen-Basse, L. Berger, J. Berger, B. Little, H.C. Stevens, J.B. Darby, L.E. Genens, and D.L. Hillis, "OTEC biofouling-control and corrosion-protection study at the seacoast test facility: 1981–1983," Argonne National Laboratory Report ANLIOTEC-TM-5, 1985.
6. M.A. Girgis and J.M. Siegel, "Open-cycle ocean thermal energy conversion," Florida Solar Energy Center, Technical Report, No: FSEC-FS-28–83, December 1983.
7. M. Mac Takahashi, translated by: K. Kitazawa and P. Snowden, *Deep Ocean Water as Our Next Natural Resource*, Tokyo, Japan: Terra Scientific Publishing Company, 1991.
8. J.H. Anderson and J.H. Anderson Jr., "Sea water power plant," US Patent 3312054, September 1966.
9. V.L. Bruch, "An assessment of research and development leadership in ocean energy technologies," Sandia National Laboratories: Energy Policy and Planning Department, USA, Techical Report SAND93-3946, April 1994.
10. Energy Information Administration Staff, "Average retail price of electricity to ultimate customers by end-use sector, by state, December 2007 and 2006," *Energy Information Administration*, 2008.
11. M. Ravindran and Raju Abraham, "The Indian 1 MW demonstration OTEC plant and the development activities," *Proceedings of the Ocean '02 MTS/IEEE*, Vol. 3, pp. 1622–1628, October 2002.
12. D'Arsonval, "Utilisation de forces naturelles: Avenir de l'electricite," *Revue Scientifique*, 17, 370, 1881.
13. Natural Energy Laboratory of Hawaii Authority, USA, available at: http://www.nelha.org, March 2008.
14. L.C. Trimble and W.L. Owens, "Review of mini-OTEC performance," *Proceedings of the 15th Intersociety Energy Conversion Engineering Conference*, Vol. 2, pp. 1331–1338, 1980.
15. L.A. Vega, "Ocean thermal energy conversion," in A. Bisio and S. Boots (Eds), *Encyclopedia of Energy Technology and the Environment*, Vol. 3, New York: Wiley, pp. 2104–2119, 1995.
16. K.Z. Iqbal and K.E. Starling, "Use of mixtures as working fluids in ocean thermal energy conversion cycles," *Proceedings of the Oklahoma Academy of Science*, Vol. 56, pp. 114–120, 1976.
17. M.G. Trulear and W.G. Characklis, "Dynamics of biofilm processes," *Journal of the Water Pollution Control Federation*, 54, 1288–1301, 1992.

18. R.P. Aftring and B.F. Taylor, "Assessment of microbial fouling in an ocean thermal energy conversion experiment," *Applied and Environmental Microbiology*, 38 (4), 734–739, October 1979.
19. L.R. Berger and J.A. Berger, "Countermeasures to microbiofouling in simulated ocean thermal energy conversion heat exchangers with surface and deep ocean waters in Hawaii," *Applied and Environmental Microbiology*, 51 (6), 1186–1198, 1986.
20. J.S. Nickels, R.J. Bobbie, D.F. Lott, R.F. Martz, P.H. Benson, and D.C. White, "Effect of manual brush cleaning on biomass and community structure of microfouling film formed on aluminum and titanium surfaces exposed to rapidly flowing seawater," *Applied and Environmental Microbiology*, 41 (6), 1442–1453, 1981.
21. F.L. La Que and A.H. Tuthill, "Economic considerations in the selection of materials for marine applications," *Transactions of Society on Naval Architects and Marine Engineering*, 69, 1–18, 1962.
22. J. Larsen-Basse, and T.H. Daniel, "OTEC heat transfer experiments at Keahole point, Hawaii, 1982–83," *Proceedings of the Oceans '83*, San Francisco, CA, pp. 741–745, August 1983.
23. L. Vega and D.E. Evans, "Operation of small open cycle OTEC experimental facility," *Proceedings of the Oceanology International 1994 Conference*, Brighton, UK, Vol. 5, 1994.
24. T.H. Daniel, "A brief history of OTEC research at NELHA," Natural Energy Laboratory of Hawaii Authority, 1999.
25. B.K. Parson, D. Bharathan, and J.A. Althof, "Thermodynamic systems analysis of open-cycle ocean thermal energy conversion (OTEC)," Solar Energy Research Institute, Technical Report, SERI TR-252-2234, 1985.
26. P.Y. Burke, "Compressor intercoolers and aftercoolers: Predicting off-performance," *Journal of Chemical Engineering*, September, 107–109, 1982.
27. C.B. Panchal and K.J. Bell, "Simultaneous production of desalinated water and power using a hybrid-cycle OTEC plant," *Journal of Solar Energy Engineering*, 109, 156–160, 1987.
28. T.J. Rabas, C.B. Panchal, and H.C. Stevens, "Integration and optimization of the gas removal system for hybrid-cycle OTEC power plants," *Journal of Solar Energy Engineering*, 112 (1), 19–29, 1990.
29. A. Thomas and D.L. Hillis, "First production of potable water by OTEC and its potential applications," *Proceedings of the Oceans '88 Conference*, Baltimore, MD, 1988.
30. M.C. Chapman, "OTEC power plant transient analysis: an approach to digital computer simulation of the evaporator," Perdue University Technical Report, PCTR-84-79, TR-EE-79-34, September 1979.
31. M. Chapman and G. Heydt, "A detailed transient model of an OTEC evaporator," *Journal of Engineering Power*, 103, 539–544, 1981.
32. G.T. Heydt and M.C. Chapman, "The transient modeling of an ocean thermal energy converter boiler," *IEEE Transactions on Power Apparatus and Systems*, 100, 4765–4773, 1981.
33. S. Goto, M. Nakamura, M. Kaijwara, Y. Ikegami, and H. Uehara, "Design of nonlinear separation controller based on heat relation for evaporator in STEC plant," *IEEE Power Engineering Society Winter Meeting*, Vol. 1, pp. 378–382, January 2002.
34. M. Nakamura, N. Egashira, and H. Uehara, "Digital control of working fluid flow rate for an OTEC plant," *ASME Journal of Solar Engineering*, 108, 111–116, 1986.
35. P.A. Mangerella, "An analysis of the fluid motion into the condenser intake of a 400 MW (E) ocean thermal difference power plant," Report: NSF/RANN/SE/GI-34979/TR/75/3, Massacushets, 1975.
36. Y. Zhang, Y. Ikegami, and M. Nakamura, "Dynamic model of condensation subsystem in STEC plant with working fluid of binary mixtures," *Proceedings of the SICE 2003 Annual Conference*, Vol. 1, pp. 512–515, 2003.
37. S. Goto, S. Kondoh, Y. Ikegami, T. Sugi, and M. Nakamura, "Controller design for OTEC experimental pilot plant based on nonlinear separation control," *Proceedings of the SICE 2004 Annual Conference*, Vol. 1, pp. 4–6, August 2004.
38. M. Nakamura, N. Egashira, and H. Uehera, "Digital control of working fluid flow rate for an OTEC plant," *ASME Journal of Solar Energy Engineering*, 108, 111–116, 1986.

39. M. Nakaruma, S. Goto, and T. Sugi, "A methodology for designing controllers for industrial systems based on nonlinear separation model and control," *Journal of IFAC, Control Engineering Practice*, 7, 347–356, 1999.

40. Z. Zhong, I. Yasayuki, and N. masatoshi, "Diffuser multi-objective control for STEC plant," *Proceedings of IEEE International Conference on Systems, Man, and Cybernetics*, Vol. 1, pp. 148–153, 2003.

41. IEEE committee report, "Dynamic models for steam and hydro turbines in power system studies," *IEEE Transactions on Power Apparatus and Systems*, PAS-92 (6), 1904–1915, 1973.

42. C.E. Rudiger Jr. and L.O. Smith, "OTEC—An emerging program of significance to the marine community," in *Proc., OCEANS*, Vol. 9, pp. 581–587, 1984.

43. Space Science and Engineering Center-University of Wisconsin Madison, available at: http://www.ssec.wisc.edu, March 2008.

44. D. Block and J. Valenzuela, "Thermoeconomic Optimization of OC-OTEC," Golden, CO: Solar Energy Research Institute, Techical Report, SERI/STR-251–2603, 1985.

45. T.H. Daniel, "Aquaculture using cold OTEC water," *Proceedings of the Oceans*, San Diego, CA, Vol. 17, pp. 1284–1289, November 1985.

46. T.K. Leraand and J.C. Van Ryzin, "Air conditioning with deep seawater: A cost-effective alternative for West Beach, Oahu, Hawaii," in *Proc., MTS/IEEE Challenges of Our Global Environment Conference (OCEANS '95)*, Vol. 2, 1100–1109, San Diego, USA.

47. Solar Energy Research Institute Staff, "Ocean thermal energy conversion: An overview," Solar Energy Research Institute, Techical Report SERI/SP-220–3024, November 1989.

Index

A

Ah. *See* Ampere-hour (Ah)
Air pressure ring buoy, 236
Ampere-hour (Ah), 66
Anemometer, 103
   principle, 103
Archimedes wave swing (AWS) device, 290, 294
   operation principle, 296
Asynchronous generator, 202–205
   doubly fed induction generator (DFIG), 199
      circuit connection, 203
      dynamic model, 204
      generator's slip, 202
Automatic voltage regulator (AVR), 127
AVR. *See* Automatic voltage regulator (AVR)
AWS device. *See* Archimedes wave swing
      (AWS) device

B

Balance of system (BOS) components, 2
Barrage, 167, 169, 170
Bernoulli principle, 298
Betz's law, 107
Biplane wells turbine with guide vanes
      (BWGVs), 243, 244
BLDC. *See* Brushless DC (BLDC)
BLDC machines, 116, 117
   small scale model, 122
   speed, 123
BOS components. *See* Balance of system (BOS)
      components
Brushless DC (BLDC), 113
BWGVs. *See* Biplane wells turbine with guide
      vanes (BWGVs)

C

Cadmium telluride (CdTe), 4
Caisson. *See* Tidal fence
Carnot cycle, application, 309
Carter's coefficient, 124
   factor, 269

CdTe. *See* Cadmium telluride (CdTe)
CIS. *See* Copper indium diselenide (CIS)
Closed-cycle OTEC, 308–309
   block diagram, 309
   energy transfer diagram, 310
   obstacles
      leakage, 312
   thermodynamic principles in, 309–310
   working fluid in, 309
      advantages of mixture, 313
      mixture and pure, 312–313
      mixture versus pure, 313
Cold water pipe (CWP), 308
Condenser, 322
   principle, 323–324
      flow rate of fluid mass, 324
      schematic diagram, 322, 323
Converter
   multilevel topology, 54
   topology, 50, 51, 52
Copper indium diselenide (CIS), 4
Coriolis force, 102
C_p. *See* Power coefficient (C_p)
CT. *See* Torque coefficient (CT)
Current–voltage (I–V), 7
CWP. *See* Cold water pipe (CWP)

D

Dam. *See* Barrage
DC transmission link, high voltage, 225
DD energy. *See* Direct-drive (DD) energy
DFIG. *See* Doubly fed induction
      generator (DFIG)
Digital signal processor (DSP), 28
Direct-drive (DD) energy, 124
Doubly fed induction generator (DFIG),
      137, 199, 285
   back-to-back converters, 286
   benefits of, 138
   conventional control, 134, 135
   electrical circuits of, 138
   operation modes of, 143

Doubly fed induction generator
(DFIG) (*continued*)
power transfer in, 141
topology of, 147
torque-speed characters of, 142
total power and power flow in, 143
DSP. *See* Digital signal processor (DSP)

E

Economic viability, 167
Eel. *See* Energy Harvesting Eel
Electric vehicle (EV), 81
Electromotive force (EMF), 116
Electrostrictive polymers, 277. *See also*
Piezoelectric polymers drawbacks, 278
EMF. *See* Electromotive force (EMF)
Energy Harvesting Eel, 275
movement, 276
structure, 276
switched resonant power converter, 278
Entropy, 310
intake and discharge rate, 315
temperature-entropy diagram, 310
EV. *See* Electric vehicle (EV)
Evaporator, 318
heat exchange value, 320–322
system model, 320
principle of, 319
rate of heat transfer in, 319
schematic diagram, 319, 320
vacuum flash, 325, 326
Excitation forces, 250

F

F. *See* Nonlinear statics function (F)
Falnes buoy-turbine, 235
FF. *See* Fill factor (FF)
Field orientation control (FOC), 128
Fill factor (FF), 12
Fixed speed (FS), 113
FOC. *See* Field orientation control (FOC)
FS. *See* Fixed speed (FS)
Fuzzy Rule Base Table, 29

G

GaAs. *See* Gallium arsenide (GaAs)
Gallium arsenide (GaAs)
advantages of, 4–5
GCC topology, modified
inductor voltage in, 59
Gearbox, 205

Generators
asynchronous generator, 202–205
synchronous generator, 199–202
Grid connection interfaces
grid-connected systems, 212–216
inner loop control system, 214
outer loop control system, 214
Grid connection synchronization, 216–217
phasor diagram for, 217
PLL method, 212–216
synchronization conditions, 216
zero crossing method, 216–217
Grid-connected PV Systems
centralized inverter system, 41
multistring inverter system, 42, 43
single-stage multilevel, 54–56
single-stage multimodule, 53–54
string inverter system, 41, 42
topology, 55
two-stage multimodule, 56–61
two-stage single module, 43–54
Grid-connected systems, 212–216
inner loop control system, 214
outer loop control system, 214
Grid side converter (GSC), 127
objective, 203
$G(s)$. *See* Linear dynamics function ($G(s)$)
GSC. *See* Grid side converter (GSC)

H

HAWTs. *See* Horizontal axis wind turbines
(HAWTs)
Heat exchanger, 314
counter current exchange in, 317
parallel heat exchange in, 318
performance degradation in, 314
Heat transfer coefficient U^{CN}, 319, 325
Horizontal axis wind turbines (HAWTs), 114
Hot water pipe (HWP), 308
Humidity, 325
HWP. *See* Hot water pipe (HWP)
Hybrid cycle OTEC, 316–317
advantages of, 317
block diagram of, 317
principle of, 316–317

I

IGBT inverter. *See* Insulated gate bipolar
transistor (IGBT) inverter
Induced total EMF phasor, 264
Induction machines, 130-134
electrical equivalent circuits, 133

self-excitation phenomenon, 131
voltage and frequency control, 135–136
    with load regulation, 134
Insulated gate bipolar transistor (IGBT)
        inverter, 282
Inverter
  dual-stage configuration, 57
  HBDC
    output inductor current for, 56
    output voltage, 55
    output voltage for, 57
  input voltage, 60
  output voltage, 61
  single-stage topology, 53
  utility interactive, 58, 59
I_{SC}. *See* Short-circuit current (I_{SC})
I–V. *See* Current-voltage (I–V)

L

LFM. *See* Longitudinal-flux permanent magnet
        machine (LFM)
Linear dynamics function ($G(s)$), 328, 329
Linear generator, 230
  power flow and losses, 259
  system-level configuration, 254
LMTDs. *See* Log mean temperature differences
        (LMTDs)
Log mean temperature differences (LMTDs),
        313, 325
Longitudinal-flux permanent magnet machine
        (LFM), 257, 258
Low-voltage ride through (LVRT), 283
LVRT. *See* Low-voltage ride through (LVRT)

M

Magnetohydrodynamics (MHD), 290
  generator benefits, 298
Maximum power point (MPP), 6
Maximum power point tracking (MPPT),
        6, 206–210
  control system Block diagram, 209
  controller Block diagram, 208
  current speed calculation method,
        208, 209
  lookup table method, 207
  P&O-based method, 208
  techniques
    current sweep–based, 32–33
    DC link capacitor droop control–based,
        33–34
    fractional open-circuit voltage-based,
        26–27

fractional short-circuit current-based,
        27–28
    fuzzy logic control–based, 28–30
    incremental conductance, 21–23, 77
    linearized I–V characteristics–based
        controller, 24–26
    neural network–based, 30–31
    perturb & observe (P&O), 23–24
    ripple correlation control–based,
        31–32
  use of, 21
Mechanical power (P_m), 143
MHD. *See* Magnetohydrodynamics (MHD)
Mighty Whale. *See* Oscillating water column
        (OWC)
MPP. *See* Maximum power point (MPP)
MPPT. *See* Maximum power point tracking
        (MPPT)

N

Nearshore OTEC diagram, 307
Net power, 305–306
Neural network (NN), 30
NN. *See* Neural network (NN)
Nodding duck. *See* Salter cam
Nominal revolutions per minute (ns), 330
Nonlinear statics function (F), 328
ns. *See* Nominal revolutions per minute
Ocean thermal energy conversion (OTEC), 305
  advantages of, 306
  applications
    air-conditioning, 338
    aquaculture, 337
    desalination, 337
    mineral extraction, 338
  by-products, 305
  classification
    closed-cycle OTEC, 308
    open-cycle OTEC, 315
  components
    condenser, 322
    evaporator, 318
    heat exchanger, 318
    vacuum flash evaporator, 325, 326
  drawbacks of, 305–306
  effect of season on, 311
  efficiency of, 306
  efficiency possibility in, 310
  power plant
    function of power generation in, 327
    impact on environment, 338
    nonlinear separation model, 328

Ocean thermal energy conversion
(OTEC) (*continued*)
pilot model, 327
state variable of, 327
system controller diagram of, 327
variables in, 326
principle of, 305
requirement for, 305
thermal resources for, 335–336

O

Ocean wave
characteristics of, 225
energy harvesting, 224
Offshore OTEC diagram, 307
Open-circuit voltage (V_{OC}), 7, 8
Open-cycle OTEC
block diagram, 315
desalination, 315–316
drawbacks of, 316
principle of, 315
Oscillating water column (OWC), 233, 290
advantage, 290
disadvantage, 290
OTEC. *See* Ocean thermal energy conversion
(OTEC)
OWC. *See* Oscillating water column
(OWC)

P

Pelamis, 291. *See also* Salter cam
at sea, 291
cost pie graph, 293
general layout, 292
electrical interconnection, 292, 293
Permanent magnets (PMs), 116
Permanent magnet synchronous generators
(PMSGs), 116
Permanent magnet synchronous machines,
123, 128
axial flux generators, 128
with back-to-back inverter, 128
with rectifier/inverter, 127
transverse flux generators, 128
Phase-locked-loop (PLL), 212
Photoelectric effect, 1
Photovoltaic (PV) array, 3
Photovoltaic (PV) cell, 1, 3
BOS components, 2
circuit, 12
current-power curves of, 8
current-voltage curves of, 8, 11

dual-diode model, 10
effect of, 17
n-type semiconductor, 2
p–n junction of, 2
principle of, 1–2
p-type semiconductor, 2
semiconductors in, 3–5
polycrystalline thin films, 4
silicon, 4
single-crystalline thin films, 4
series and parallel connection of, 9
shading effects on, 35–39
efficiency due to, 39
fill factor due to, 38–39
I–V characteristics due to, 36
MPP relocation due to, 37, 39
to reduce, 35
single-diode model, 9–12
Photovoltaic (PV) module, 3
ambient irradiation on I–V of, 15
cell temperature on I–V of, 15
current determination in, 14–15
operational conditions in, 13
Piezoelectric polymers, 276
advantages, 277
PLL. *See* Phase-locked-loop (PLL)
P_m. *See* Mechanical power (P_m)
PMs. *See* Permanent magnets (PMs)
PMSGs. *See* Permanent magnet synchronous
generators (PMSGs)
Power coefficient (C_p), 191
Power electronic interfaces, 40
classification, 40
Power take-off (PTO) system, 284. *See also* Salter
cam
PTO. *See* Power take-off (PTO) system
Pulse width modulation (PWM), 127, 268
PV cell. *See* Photovoltaic (PV) cell
PV systems
illumination area in, 17
I–V characteristics of, 7–8, 10
factors affecting, 21
output voltage in, 7
sunlight collection
concentrator method in, 3
flat panel method in, 3
PVDF. *See* Piezoelectric polymers
PWM. *See* Pulse width modulation (PWM)

R

Rankine cycle
process, 311–312

RCC. *See* Ripple correlation control (RCC)
Ripple correlation control (RCC), 31
Rotor side converter (RSC), 137
RSC. *See* Rotor side converter (RSC)

S

Salter cam, 231–232
  fixing of, 231
  forces affecting, 232
  operating principle, 232
SAUV. *See* Solar-powered autonomous
     underwater vehicle (SAUV)
Short-circuit current (I_{SC}), 7
Sinusoidal pulse width modulation (SPWM)
     generator, 155
Sizing PV panel and battery pack, 66
  autonomy in, 67
  load calculation in, 67
  procedure, 70
  PV array sizing
    design verification, 69
    module selection, 68
    parallel modules number
      determination, 69
    series modules number determination,
      68–69
    system losses, 68, 71
    solar radiation in, 68
    sun hour in, 67
SMES. *See* Superconducting magnetic energy
     storage (SMES)
Solar cell. *See* Photovoltaic (PV) cell
Solar energy, 1
  electric vehicle, 81–86
    block diagram, 83, 85
    block diagram of hybrid solar, 84
    Solar Miner II, 84, 86
  naval applications, 86–89
    boat, 88–89
    SAUV, 87
    test-bed block diagram, 87–88
  potential of, 1
  residential use of, 72
    bidirectional inverter/converter, 78–79
    boost converter in, 74, 75, 76
    factors in, 73
    MPPT control, 77
    operation modes in, 73–74
    power conditioner in, 79–80
    PV system in, 72, 73, 75
  space applications, 89–94
    proposed airplane electrical system, 93

    rechargeable aircraft, 90–91
    satellite energy system, 90
    solar airplane, 91–92
    unmanned aerial vehicle (UAV), 92–94
  system, 6
    active, 5
    components of, 5–6
    passive, 5
Solar-powered autonomous underwater
     vehicle (SAUV), 87
Solar tracking systems, 15–20
  algorithm, 19–20
  block diagram, 19
  intelligent control basis, 18–20
    addition of currents as, 19
    cloud interference as, 20
    current variations as, 19
    photodiode illumination, 17–18
    sensor module structure, 17
    structure, 16
SPLL. *See* Synchronous phase-lock-loop
     (SPLL)
SPWM generator. *See* Sinusoidal pulse width
     modulation (SPWM) generator.
Squirrel cage induction. *See* BLDC machines
SRG. *See* Switched reluctance generator
     (SRG)
Stand-alone PV Systems, 61
  type 1 connection, 61–63
  type 2 connection, 63–64
  type 3 connection, 64
  type 4 connection, 65
  type 5 connection, 65–66
STATCOM. *See* Static synchronous
     compensator (STATCOM)
Static synchronous compensator (STATCOM),
     283
Stator
  core design, 265
  surface around periphery, 263
  winding
    arrangement per phase, 265
    coordinates and rotor position, 262
    traveling flux wave, 262
Stator voltage (V_S), 204
Steady-state power output (W_S), 328, 329
Superconducting magnetic energy storage
     (SMES), 144
Switched reluctance generator (SRG), 274
  configuration, 274
  cross-sectional view, 275
Synchronous generator, 151, 260, 199–202
  advantages in wind turbines, 153

Synchronous generator (*continued*)
  circuit, 200
  cross-sectional view of, 269
  d-axis voltage controller, 155
  with DC link regulation, 155
  excitation circuit of brushless, 264–265
  field voltage control, for, 154
  flux density B vs. field strength H, 266
  multipole, 154
  PMSM vector control scheme, 201
  q-axis voltage controller, 155
  of radial flux PM for WEC, 266, 267
  synchronous speed, 200
Synchronous phase-lock-loop (SPLL), 213

T

Temperature-entropy
  Rankine cycle, 311–312
  thermodynamic cycle, generalized, 311
Terminal voltage of PV cell (V_{PV}), 8
Thermodynamics, law of
  carnot cycle efficiency, 309–311
  energy balance, 309
Tidal energy conversion system, 199–211
  gearbox, 205
  generators
    asynchronous generator, 202–205
    synchronous generator, 199–202
  maximum power point tracking, 206–210
    controller block diagram, 208
    control system block diagram, 209
    current speed calculation method, 208, 209
    lookup table method, 207
    P&O-based method, 208
  optimal running principle, 205–206
Tidal energy harvesting
  advantages, 167
  energy Calculation, 189
  environmental Impacts of, 218–219
  grid connection interfaces, 211–216
    block diagram, 212
    grid-connected systems, 212
  kinetic, 188
    tidal current approach, 192
  potential, 173
    tidal barrages approach, 174
    tidal lagoons concept, 176–177
  potential resources of, 218
Tidal energy, kinetic, 188
  tidal current approach, 192
    horizontal axis turbines, 192–197
    linear lift-based device, 198–199
    vertical axis turbines, 197–198

Tidal energy, potential, 173
  tidal barrages approach, 174
    ebb generation, 174, 175
    flood generation, 174, 175
    two-way generation, 174, 176
  tidal lagoons concept, 176–177
Tidal fence, 167, 169
  installation, 167, 168
  pros and cons, 167
Tidal plant, 167, 170
Tide
  ebb tide, 167
  energy conversion system, 199–211
    gearbox, 205
    generators, 205
    maximum power point tracking, 206–210
    optimal running principle, 205–206
  flood tide, 167
  funneling effect of, 172
  neap tide, 171
  semidiurnal, 172
  spring tide, 171
  variation in a week, 172
Tip speed ratio (TSR), 191, 205
Torque coefficient (CT), 191
TSR. *See* Tip speed ratio (TSR)
Turbine, 167, 168
  air-driven turbines, 229, 233–234
  efficiency, 246
  hydraulic, 181, 184, 185, 243
    characteristics, 181
    synchronous generator with, 185
  impulse turbine
    efficiency of ISGV, 248
    with fixed guide vane (IFGV), 243, 245
    with self-pitch-controlled guide vanes
      (ISGVs), 243, 244
  kaplan turbines, 241
    double-regulated turbine model, 242
    hydraulic turbine, nonlinear model, 243
  nearshore applications, 233
  self-pitch-controlled blades turbine, 239–240
    wave energy, overall view, 241
  specifications, 245
  speed-power characteristics, 210
  starting characteristics
    impulse-type, 248
    non-self-starting, 272
    self-starting, 272
    wells-type, 248
  steam
    gate opening in, 334
    generator active power, 334

generator output voltage, 335
mechanical power of, 333
model, 331, 332
rotor speed deviation, 333
rotor speed versus time, 332
speed governing system in, 329, 330, 331
tidal
  bulb turbine, 178
  complete model, 186
  flow rate in, 182
  governor model, 184
  helical turbine, 198
  linear model, 183
  nominal revolutions of, 185
  nonlinear model, 181
  rim turbine, 179
  tidal current turbine, 191–199
  tubular turbine, 179–180
wells turbines, 237–238
  biplane, 243, 244
  non-self-starting, 272
  output torque waveform of, 239
  self-starting, 272
  turbine efficiency vs. flow coefficient, 238
wind, 101, 105
  classification, 114
  concepts comparison, 157
  connection scheme, 115
  electrical machines types, 116
  power, 107
  siting, 105
  specifications, 121, 129
  systems, 112
Turbine and generator's control, 180
  active power vs. time, 189
  excitation voltage control system, 186
  grid-connected current control, 213
  islanding detection circuit, 213
  mechanical power vs. time, 188
  output power vs. rotating speed, 206
  output voltage vs. time, 190
  PI controller, 186
  power curves vs. tidal speeds, 207
  prime movers, 180
  rotor speed vs. time, 187
  servomotor model, 186
  speed control model, 180, 181
    linear model, 184
  voltage vs. time, 188

U

UAV. *See* Unmanned aerial vehicle (UAV)

U^{CN}. *See* Heat transfer coefficient (U^{CN})
Unity volume and thermal circuit, 271
Unmanned aerial vehicle (UAV), 92–94

V

Vacuum flash evaporator, 325, 326
  factors in evaporation, 325
VAr compensators. *See* Volt-Ampere-reactive (VAr) compensators
VAWT. *See* Vertical axis wind turbines (VAWTs)
Vertical axis wind turbines (VAWTs), 114
V_{OC}. *See* open-circuit voltage (V_{OC})
Voltage source inverter (VSI), 128
Volt-Ampere-reactive (VAr) compensators, 132
V_{PV}. *See* Terminal voltage of PV cell (V_{PV})
V_s. *See* Stator voltage (V_s)
VSI. *See* Voltage source inverter (VSI)

W

Wave dragon, 294–295
  operating principle, 294
Wave energy
  absorbers, 234
    air-driven turbines, 234
    wave contouring raft, 235, 236
  converters efficiency comparison, 249
  development process, 298
  future of, 298
  grid connection topologies, 282
    electrostrictive power generators, 289
    induction generator applications, 283
    interface for different generators, 283
    linear and synchronous generator applications, 282
    SRG applications, 286
  harvesting system, 223–224
  harvesting technologies, 226
    float systems, 290
    hinged contour devices (*see* Pelamis)
    overtopping devices (*see* Wave dragon)
    OWC systems, 290
  nearshore technologies, 233
    air-driven turbines, 233–234
    turbine method, 233
  offshore technologies, 227
    air-driven turbines, 229
    fixed bodies, dynamics of, 228
    floating bodies, dynamics of, 228
    linear generator-based buoy-type, 230
    salter cam method, 231
  systems, 223

Wave energy (*continued*)
  wave power generators, 249, 254
    average power vs. load resistance, 253
    buoy generator system, 251
    buoy power vs. current density, 261
    efficiency variation vs. load resistance, 253
    generator voltage, 255
    induction generators, 272
    linear, 256
    longitudinal-flux PM, 255
    output power vs. current density, 261
    switched reluctance generator
        (SRG), 274
    synchronous, 260
    system-level configuration in, 254
    wave-activated linear model, 249, 250
  wave power turbine types, 236
    kaplan turbines, 241
    self-pitch-controlled blades turbine,
        239–240
    wells turbines, 237–238
Wave energy conversion (WEC), 223
  application
    electrostictive materials, 275
    grid connection topologies, 278
    induction machines for, 272
    switched reluctance (SR) machines, 274
    synchronous generators, 260
  devices, 224
  technology, 226
Wave star energy (WSE), 290, 296
  multipoint absorber, 296, 297
  storm protection mode position, 297
WEC. *See* Wave energy conversion (WEC)
Wind, 101
  kinetic energy of, 103
  map, 105
  power classes, 105
  power vs. speed, 103
  rose, 104
  shade, 107
  speed, 103, 107, 108
  theoretical extracted power, 109

  turbines, 101, 105
  turbulence, 105
Wind energy harvesting
  development, 156
    in control systems, 156
    in distribution, 158–159
    in grid-connected topologies, 158–159
    in machine design, 158
  fundamentals of, 105
  history of, 104
  parameters for, 102
  terrain characteristics
    friction coefficients, 105
    roughness classes, 106
Wind power system (WPS), 114
Wind turbine, 101, 105
  classification, 114
  concept comparison of, 157
  connection scheme, 115
  electrical machines
    BLDC machines, 116, 117
    induction machines, 130
    permanent magnet synchronous
        machines, 123, 128
  location of, 105
    offshore, 107
    onshore, 107
  power, 107
    Betz's law, 107
    coefficient, 110, 111
    curve, 109, 110
  specifications, 121, 129
  system
    basic parts, 112
    low-speed rotor, 113
    nacelle, 112
    pitch control, 114
    tower, 112
    yaw mechanism, 112
WPS. *See* Wind power system (WPS)
W_s. *See* Steady-state power output (W_s)
WSE. *See* Wave star energy (WSE)